THE CAMPAIGN

Olof Hallonsten

The Campaign

How a European Big Science facility ended up on the peripheral farmlands of Southern Sweden

LUND · ARKIV ACADEMIC PRESS · 2020

Arkiv Academic Press is an imprint of

Arkiv förlag
Box 1559
SE-221 01 Lund
Sweden

STREET ADDRESS Lilla Gråbrödersgatan 3 c, Lund
PHONE (+46) 046-13 39 20

arkiv@arkiv.nu
www.arkiv.nu

A list of Arkiv Academic Press titles can be found in the last pages of this book. For up-to-date information on distribution and available titles, please visit:

www.arkivacademicpress.com

Cover photo: Institute Laue Langevin (ILL). Used with permission.
Cover design: David Lindberg

First edition by Arkiv förlag 2020
Arkiv Academic Press international edition 2020
For print information, see the back page of this copy
ISBN: 978 91 986454 0 8

To all my great colleagues, with all their exciting ideas,
without whom this book would have been finished
at least two years earlier.

And in memory of Karl-Fredrik Berggren (1937–2020),
without whom there wouldn't have been a book to write
in the first place.

Contents

Notes on sources, references and style

This book uses reference style pragmatically, to maximize clarity. References are made in Harvard style, which means that author and year are noted for literature (e.g. Edqvist 2009; D'Ippolito and Rüling 2019), whereas for reports, investigations, governmental bills, meeting minutes, press releases, and so on, author and year are sometimes substituted for organizational origin and year (e.g. ESS Scandinavia 2002a; Swedish Government 2008c; Berggren 2004; Melin 2016). References to news articles are made with source and year (e.g. *Sydsvenskan* 2008c) but for opinion letters published in news outlets, author and date are shown (e.g. Söderbergh Widding and Karlhede 2014). Interview sources are simply noted with the surname of the interviewee (e.g. Interview: Johansson).

Several organizations and entities with Swedish names appear throughout the book, and for those most commonly appearing, such as the Swedish Government and the Swedish Research Council (*Vetenskapsrådet*), an English translation is consistently used, including in references to their reports and other documentation. For others, occurring more sparsely, original names are retained (e.g. *Folkpartiet*). The same goes for organizations from other countries, with foreign names (e.g. *Wissenschaftsrat*).

1. Introduction

The Campaign

The *European Spallation Source (ESS)* is a state-of-the-art accelerator-based neutron source and a scientific research facility for neutron scattering experiments which is currently under construction outside Lund in Southern Sweden. It is a collaborative project, run by thirteen European countries who are jointly funding the construction (estimated at 1.843 billion Euro at 2013 prices, or approx. USD 2.5 billion) and will fund its operation (estimated at 140 million Euro annually, at 2013 prices, or approx. USD 196 million) through their memberships in the European Spallation Source ERIC (European Research Infrastructure Consortium, a new type of legal entity invented by the European Commission for the purpose, see below and chapter 11). Neutron scattering is the common name for a number of experimental techniques that use bursts of neutrons (a heavy elementary particle) to study the properties of matter on atomic and molecular level, with a wide range of applications in physics, chemistry, biology, and engineering, and cross-disciplinary fields including materials science, life science, and environmental science. The ESS has been under planning since 1990, and was given its name in 1993 when a number of research centers in ten European countries formed the ESS Council to work towards its realization as a way to ensure that Europe's international lead in neutron-based materials science would be maintained into the 21st century (Kaiserfeld 2013: 29; Berggren and Hallonsten 2012: 23).

This book chronicles and analyzes the *campaign* to build the ESS in Lund, a small town in Southern Sweden (with some 90,000 inhabitants), home to Sweden's largest university and some knowledge-intensive industry, but generally on the periphery of the knowledge-based economy of Europe and the world, and far away from both the European and global hotspots of neutron scattering technology and the utilization of neutrons in scientific work, and the political and organizational centers of gravity on the European continent. The Danish capital, Copenhagen, and its international airport can be reached from Lund by train in less than an hour, and this was a major selling point in the Swedish ESS campaign, but Copenhagen is still small by European standards, even when Malmö and Lund are included in the greater Copenhagen metropolitan area. Previous collaborative European Big Science facilities comparable to ESS have been located in Geneva, Grenoble, Hamburg, Oxford, and Berlin. Those with remote locations outside Europe, or with multiple smaller sites, have their headquarters in Munich and Paris. That a European Big Science facility ended up in peripheral Lund, Sweden, is truly remarkable and calls for an in-depth inquiry: How could this happen?

The short answer is: Because of a successful campaign. The long answer will be laid out in detail in this book.

But a campaign it was. The ESS ended up in Lund as the result of a clever and well-executed political effort that lasted for fourteen years, mobilized and utilized available resources, and enjoyed quite a bit of luck, but most of all was driven by some devoted enthusiasts who managed to be in the right place at the right time, alternating between the manipulation of existing institutional arrangements, in line with the campaign objectives, and overcoming the constraints of the same institutions, to the benefit of the campaign. This group was neither homogeneous nor static, but enrolled people with different competences and in different capacities, as the focus of the campaign shifted from local, via European, to national, regional, and back again to European. Their ultimate success did not amount to establishing Lund or Sweden as a global leader in the use of neutrons for scientific experimentation, or anything like that, since this would require a wholly different effort (see the discussion in the concluding chapter), but to *bringing the ESS to Lund*, a remarkable achievement in its own right, and the result of a highly complex process that the campaign evidently mastered. Once the campaign was over, another effort had to commence, namely shepherding a gargantuan construction project to finalization roughly within budgetary and temporal limits, and meeting the expectations of all its various stakeholders.

This book is not primarily about that, but about *the campaign*, meaning that the main story begins with the forming of the *ESS Scandinavia Initiative* in October 2000 and ends with the start of construction of the ESS in September 2014. The story, nonetheless, overlaps with other processes that stretch out over other time periods, not least the European effort to have the ESS built at all, regardless of site, which began decades ago, and the efforts to actually build the ESS in Lund and make it a resource for science and innovation, which are still ongoing. Therefore, there are no time stamps in the title of this book – these are saved for the chapters.

In a sense, *the campaign* is also slightly misleading as a title. First of all, the word must be understood broadly, not as referring to a political campaign with, say, election victory as the goal, but in a figurative sense, as *an organized attempt aiming at a definite result*, mainly using political means and operating in the political arena. It should also not be misread as asserting that the effort was coordinated and coherent, from start to end. In reality, it was the opposite. The Swedish ESS project, and the campaign itself, changed considerably in the fourteen-year period from 2000 to 2014. It took very different forms in different time periods, and in its later stages it even had multiple simultaneous forms. An attempted periodization based on the campaign's major character and focus yields three main phases: A small-scale lobbying effort by enthusiastic scientists and some local policy and decision makers (2000–2007; chapter 6), a campaign actively backed by the Swedish Government and aimed at building support among other European governments and scientific communities (2007–2009; chapter 7), and an effort to revise the technical design, build a scientific case, achieve a multilateral funding solution, build the organization, and do the necessary PR work (2009–2014; chapters 8, 9, 10). Chapters 1 to 5 are used for introductory notes, conceptual framework, contextualization, and historical background, pertaining to both previous European collaborative Big Science and the Swedish science policy context. Chapter 11 contains an epilogue and concluding discussions, organized around different themes, including the two-edged sword of displacement effects and strategic priority in research policy and in large projects like the ESS.

The three phases were separated by specific turns of events. The ESS Scandinavia Initiative, which was soon to be organized in an *ESS Scandinavia Consortium*, was launched in October 2000 by a number of scientific organizations in Sweden and Denmark. In its first six years, its operations were truly small-scale and conducted at grassroots level, although it also enjoyed the support of several universities and other important organizations (including financial support from various Swedish research funding bodies). In 2007, after the 2006 Swedish general election brought a change of government, the new four-party center-right government threw its weight behind the ESS Scandinavia proposal to bring the ESS to Lund and began a government-sponsored campaign, which meant that negotiations and lobbying on the European stage were elevated to diplomatic level and officially supported by the Swedish Government. Around the same time, the former director of the European collaborative neutron scattering facility *Institute Laue Langevin* (ILL) in Grenoble was recruited as project director for ESS Scandinavia, which meant an intensification of the campaign in some important aspects. In 2009, the decision was made that if the ESS was to be built, it would be built in Lund, and the campaign metamorphosed into a negotiation game, mainly on European intergovernmental level, with the rules largely invented along the way. By then, however, the local marketing campaign had also accelerated in parallel with the actual effort of designing the facility, recruiting the workforce needed, and building up the organization that could take on the task of making the ESS a reality.

Chapters 6, 7, and 9 do not follow chapter 5 or each other entirely chronologically, but are instead organized according to these and other processes and themes. While chapter 5 does not end until the European ESS project imploded politically around 2002–2003, chapter 6 begins with the launch of the ESS Scandinavia Initiative/Consortium in 2000, and ends with the international recruitment of a new project leader, just before the change of government that proved decisive for the Swedish ESS campaign. Chapter 7 begins with the change of government and chronicles the negotiation efforts up until the site decision in May 2009. Chapter 7 also contains a brief detour to the history of the MAX IV project, another large science facility built next door to the ESS in Lund but with a stronger local and national anchoring and a completely different political and financial situation. The MAX IV facility had its own campaign, that both requires and deserves its own analysis in its own book, but it has some relevance for the ESS story. In chapters 8, 9, and 10, three parallel processes are each chronicled in one chapter – the planning, design, and establishing of the ESS and ESS organization in Lund, the intergovernmental negotiations to reach a funding deal, and the marketing campaign around the ESS – and all three end in 2014, when a deal was finally reached for the funding of the construction of the ESS by thirteen partner countries, and the actual construction began. Chapter 11, entitled "Epilogue and conclusions", accounts briefly for the start of construction and the organizational transition of the ESS into an ERIC, and continues with discussion and conclusions under a number of thematic headlines.

Sweden is a small country, home to some 10 million people, and Swedish science policy is naturally small-scale in international comparison. The campaign to bring the ESS to Lund, and the parallel work in later years of launching an ESS organization in Sweden and preparing for the construction of the ESS facility on Swedish soil, were difficult, even traumatic, for several of the organizations and actors within the Swedish science policy system. The ESS was, and is, a completely unprecedented project for Sweden, and the system as such was essentially unprepared to make the necessary decisions and stake out apt strategies. Moreover, Swedish science and science policy has traditionally been decentralized, and generally lacks central steering or such *aggregation mechanisms* as can give priority to grassroots initiatives that prove themselves scientifically (e.g. Benner 2008). MAX-lab, the predecessor of MAX IV which started user operation in 1986, seems to be a rare exception, but managed to grow into an international scientific user facility in spite of the system rather than thanks to it (Hallonsten 2011; Hallonsten and Christensson 2017a). After World War II, a massive buildup of science and science policy occurred across the West, and Sweden was no exception. Unlike many (or most) countries, however, the Swedish mobilization in science relied almost entirely on the universities, where the professoriate retained control over governance and funding. Although some non-academic research institutes were created in the 1940s and 1950s, these remained small compared to the universities and compared to their equivalents in other countries (Hallonsten 2018). The research councils, founded in the 1940s and later, also remained governed by academic representatives and had no mandate to make system-wide strategic decisions or priorities (Premfors 1986: 13). Government funding for the universities, usually called "block grants" or "first stream funding," was distributed among research areas (according to the faculty organizations of the universities) for the whole of the 20th century, which left university Vice-Chancellors and central management with little room for maneuver in terms of redistribution of money or investments in areas deemed to be of special importance (Benner and Sandström 2000; Melander 2006: 133). The unwillingness of the Swedish Government to take major initiatives in research policy (Premfors 1986: 15–41) lasted until the early 1990s, and while recent reforms have profoundly altered the governance and funding of the publicly funded Swedish R&D system (chapter 4), both the government and its funding agencies seem uneasy and ambivalent toward their partially new roles as agenda setters.

The Swedish science policy context of the ESS project, outlined in the previous paragraph, renders the study some additional relevance beyond its contribution to contemporary history of science, the study of the politics of contemporary Big Science, and the analysis of how grassroots campaigns metamorphose when aggregated to governmental and intergovernmental level. In itself, the ESS is illustrative of Swedish science policy in the time period covered, and thus very suitable as a prism through which the recent and current state of Swedish science policy can be studied and an interesting analysis made. Close attention must, therefore, be paid to context and background, to enable conclusions on the very specific level of the workings of the Swedish ESS campaign, as well as on general levels of the politics of contemporary Big Science and how it evolves, and how Swedish domestic research policy handles issues. It is my hope that those readers who persevere through to the final chapter of this book will gain a significantly better understanding of these topics.

Method, material, and basic viewpoints

The chronicling and analysis in this book rely on five main categories of source material.

First, mainly for context, it draws on a body of literature on Swedish science policy and its transformation in the most recent decades (e.g. Benner 2008; Eklund 2007; Edqvist 2009; Hallonsten 2018; Holmberg and Hallonsten 2015; Widmalm 2013), the Swedish experience of Big Science domestically and abroad (e.g. Hallonsten 2011, 2013; Hallonsten and Christensson 2017a; Widmalm 1993), and similar developments in other countries and on supranational level (Cramer 2017, 2020; Hallonsten 2014; Hallonsten and Heinze 2015, 2016; Krige and Guzzetti 1997; Krige 2003; Westfall 2003, 2008, 2010, 2012). Conceptually, the book makes use of different shades of institutional theory (e.g. Friedland and Alford 1991; North 1990; Streeck and Thelen 2005), the theory of bounded or limited rationality in decision making (e.g. Simon 1957; March and Simon 1958; Lindenberg 1985), the conceptualization of a changed relationship between science and society (Berman 2014; Radder 2010; Ziman 1994), and theory of the power of advertising and its mobilization of expectations in creating support for scientific and technological projects (van Lente 1993, 2000; Alvesson 2013; Hobsbawm and Ranger 1983).

Second, the supply of printed first-hand material on the ESS project, made available through personal communication with some key actors in the process as well as by the Swedish Government and its agencies, who keep extensive records of both decisions and background material for decisions, is impressive. The printed material of this type encompasses several thousand pages of text, most of which have been analyzed in the work on this book. The Swedish legal principle of public access to information ("*Offentlighetsprincipen*") is a great friend of the meticulous historical researcher, as is, of course, the generosity and sense of historical responsibility among people who not only keep documents and files but also share them.

Third, and similarly impressive, is the number of reports, brochures, and press releases issued by the ESS Scandinavia campaign, the ESS organization in Lund, the Swedish Government and its agencies, the regional authorities, and other stakeholders. This material has also been extensively used.

Fourth, the news coverage of the ESS, especially (but not exclusively) by the local daily newspaper in Southern Sweden *Sydsvenska Dagbladet* (henceforth *Sydsvenskan*) enables the easy and credible recreation of timelines of events, and access to quotes from key actors in the process. Therefore, although *Sydsvenskan* is known for its partisan stance, especially in relation to the ESS, and can be criticized for a lack of depth and nuance in its reporting on the issue, it is a formidable source of basic information about the progress of the ESS project, especially in the early years, when the ESS Scandinavia campaign was very much a local or regional operation, and the ESS a local or regional issue. As part of the effort of gathering material for this book, more than five hundred news articles from *Sydsvenskan* (and other newspapers) in the years 2000–2014 were collected and downloaded; some 120 were eventually used as direct sources.

Fifth, to complement the various printed sources, a number of interviews have been held with key actors in the story, with the focus on the explanation and analysis of topics and processes in three phases: in 2010, in 2015–16, and in 2018–19. This type of source material also includes the published personal memoirs of some key actors in the story, which, in perfect timing, have emerged in print just before the completion of this book.

The next chapter, "Conceptual framework", is chiefly theoretical, drawing up the analytical framework for the book by establishing a key conceptual duality which is both intuitively straightforward and analytically demanding, namely the use of a *(rational) actor perspective* alongside, and in contrast to, an *institutional perspective*. The latter in particular has grown immensely popular in the social sciences in recent decades and largely overshadowed the actor-centered approaches that are based on methodological individualism, namely the basic theoretical premise that all processes and events in society have individual actors at their center, who take decisions and actions that accumulate and aggregate to broader developments. This need not stand in conflict or contrast to institutional analysis, which uses the collected rules, habits, norms, and organizational arrangements in society as the main explanatory cause and guiding framework for any action or process. The analytical effort in this book is based on the realization and firm conviction that a combination of these two perspectives is necessary to give a proper account of the relevant aspects of the story being chronicled and analyzed. There is enough evidence in the material used to suggest that individuals, in the right place at the right time, and with the right convictions, ambitions, and competences, were absolutely instrumental in the campaign. But institutions also played important roles, sometimes as inhibitors and sometimes as enablers of individual ingenuity and action, and it is an essentially empirical matter to find out when it was one or the other. In addition, in order to do justice to the central theme of an ESS *campaign*, the conceptual framework includes insights from social scientists, including historians, who have studied the use of advertising, imagery, and the *invention of tradition* in political campaigns and in society at large. Such conceptual insights are helpful in understanding the fervent public relations campaign undertaken by the ESS organization and other regional actors from 2007 and on.

As the table of contents shows, considerable effort has been put into explicating the context(s) of the ESS campaign – especially scientifically and politically. Chapter 3 contains rudimentary descriptions of basic technological, scientific, organizational, and political aspects of neutron facilities and European Big Science, including recent policy hypes and new interventions in research policy by the European Union. In chapter 4, the institutional landscape is further outlined in a historical account of the postwar developments of Swedish research policy, as well as the Swedish pre-ESS experience with Big Science. The chapter also contains a summary and basic synthesis of the important transformation of Swedish research policy that has taken place from the turn of the millennium, and which forms a natural and necessary framework for the ESS campaign from a governmental/political point of view. Chapter 5 chronicles the history of the ESS facility on the European stage, including its roots in other collaborations and scientific achievements.

The book has a clear Swedish perspective and campaign perspective. This means, among other things, that the preparatory work of other countries, and their internal deliberations pertaining to the possible hosting of the ESS, investment in the ESS, membership in the ESS, and scientific or technological contributions to the ESS, have been omitted unless they have a direct relevance for the Swedish campaign to bring the ESS to Lund. This includes Denmark, whose work as a presumptive equal partner in the bid to host the ESS, and eventually as formal co-host, is given only limited attention, on par with other (prospective) partner countries. There are, of course, good reasons for a closer look at the Danish perspective, but the topic has been left to other, future studies. The decision to focus on Sweden and on the *campaign* also means that while considerable effort is made to provide the proper historical, political, scientific, and technological context of the whole story, some aspects that would appear to pertain directly to the history, politics, science, and technology of the ESS have also been left out or toned down unless they have a specific relevance for the *campaign* that is the key focus of the book. The reasons for these delimitations are as simple as they are self-evident; of course, the book needs to be fairly coherent in its account, but most of all, it needs to be a manageable read and limited in the number of pages, as well as the number of perspectives and subtopics covered.

The book is not anti-ESS. There is no reason why a historical-sociological work undertaken in an academic context, within the framework of a university post and funded by a grant from a research council, would manifest antipathies of any sort towards its study object. Yet this disclaimer is necessary. As the reader of this book will soon learn, the campaign to locate the ESS in Lund was strongly political, with the result that criticism has been a sensitive matter, and nuanced discussions of the type that are commonplace in academic seminar rooms were misunderstood as attacks and accusations. Although this book builds on a vast collection of empirical source material and years of experience of studying and analyzing the politics, organization, and history of large research facilities abroad that are very similar to the ESS, there are still elements in the analysis and the conclusions that will be interpreted as negative toward the ESS or toward key people in the campaign. This has happened before, and it is in no small part due to the contrast between a thorough scholarly analysis and the strongly biased, embellished, and even hagiographic style of the memoirs of people involved and the official histories presented by, among others, the ESS organization. That the latter are used as reference points for the appraisal of the contents of this book and previous works by the same author signals shocking ignorance regarding the nature and purpose of social science in an academic context, but it is, unfortunately, not unusual among people in power. The reason is as simple as it is obvious: politics is politics. The ESS campaign was, to a great degree, driven by politicians, ex-politicians, and scientists acting like politicians, none of whom necessarily have a natural understanding of, or sympathy for, critical and impartial analysis of their work or the greater purposes of their work. This is especially true with regard to the politically sensitive phases, for instance when undertaking precarious negotiations with powerful interests in Swedish public administration or on the European stage, with foreign governments. Moreover, all three types of politicians have, with some variation, strong interests in controlling the message and the public image of the operations and projects they are involved in, both in the midst of the operations and years later, when they write their memoirs. It should also be borne in mind that the ESS project was launched and took root in Lund, and Sweden, at the height of the influence of very alluring and potentially deceptive theories of *innovation systems*, *learning regions*, *networks*, and *clusters*, and dynamic economic effects of co-location of universities, research institutes, and knowledge-intensive industry (e.g. Florida 2002; Castells 1996; Törnqvist 2002; Maskell et al. 1998). While there is a lot to be learned from these theories, whose influence has certainly also contributed to a revitalization of economic policy, innovation policy, and regional development policy, and hence to the renewal of economies and societies, much of their use, at least in Sweden, remained on a very shallow level, and contributed to creating false impressions of spectacular socio-economic effects of both the policies themselves and of the location of knowledge-intensive and high-tech operations in specific regions. Greatly exaggerated forecasts of economic growth and the creation of new jobs as a result of establishing the ESS in Lund were allowed to strongly influence the discourse that accompanied the ESS campaign as it went from a small-scale operation to a government priority in the years 2004–2007, and in later stages, when local and regional marketing picked up speed (chapter 10). Neither should it be ruled out that the strong importance of certain entrepreneurially-minded, politically talented, and hard-working individual actors in the campaign to bring the ESS to Lund also meant these individuals identified strongly with the project and the campaign, and thus interpreted any criticism of the project (fair or unfair) as attacks on them personally.

All in all, for a long time, this created a situation where any form of criticism, and any form of stringent scrutiny of the claims made as part of the marketing campaign around the ESS, were mistaken for expressions of an anti-ESS sentiment and, at best, ignored. Meanwhile, resentful criticism of the government investment in the ESS was brewing in scientific communities in Sweden, but never surfaced (except on two occasions, see chapters 7 and 9), and it is hard to judge if this was merely an expression of envy and selfishness on the part of representatives of other fields and other projects, or of genuine concern that Sweden was ill-prepared to reap the benefits of the future ESS, scientifically, and should make other priorities. There are good grounds for both interpretations, but it does not fall within the scope of this book to investigate this particular matter further. It should, however, be emphasized that this behind-the-scenes and sometimes faceless, but probably strong, criticism of the ESS project might very well have been perceived within the ESS campaign as a threat. In combination with the sometimes recklessly ignorant local opposition to the project (see chapter 6), the criticism probably contributed to a situation where ESS champions, marinated in regional innovation systems and learning regions discourse, and dazzled by the exaggerated forecasts of socio-economic effects, believing strongly in their own marketing slogans or pretending to do so out of pure political calculation, and under heavy pressure to succeed, mistook fair and responsible scrutiny and critical analysis for hostile attempts to induce scandal.

This book is written with great respect for the people involved in the campaign, and a strong appreciation of their achievements. This involves a deep understanding of the particular and peculiar circumstances of political campaigning and the roles that the people involved must play for such a campaign to succeed. Fortunately, years have now passed since the height of the campaign, and the story can not only be told, but also fairly, critically and (to the extent it is at all possible) unbiasedly analyzed. The aim is not to cater to any special interest or actor group, or to use it as criteria for the selection of angles to explore and episodes to account for anything other than scholarly interest and the availability of empirical material.

This book is a piece of contemporary history, and as such a sociological analysis of a political campaign to establish a hugely complex and expensive technical infrastructure for experimental science, and the necessary organizational arrangements to embed it, and is therefore primarily aimed at an academic audience of historians, sociologists of science, political scientists, and science and innovation policy scholars. The hope is, nonetheless, that practitioners of science policy, and the actors involved in the campaign itself (whether identified in the book or anonymous), will also find some use in reading it.

2. Conceptual framework

The transformed Big Science

The ESS is a recent example of a well-known phenomenon in post-World War II science and science policy, namely *Big Science*. The term itself was coined in the early 1960s by the director of the Oak Ridge National Laboratory in Tennessee, Alvin Weinberg (1961, 1967), who saw a worrying development much similar to what President Eisenhower warned about in his January 1961 speech on the topic of the "Military-Industrial Complex." Weinberg noted how publicly funded research (especially in physics) in the United States had produced an unprecedented growth in the sizes of the equipment used and the teams required to use it, and warned of bureaucratization that would corrupt classic academically-oriented science and strip it of its crucially important and thus far inherent creativity and serendipity (Weinberg 1961: 162). Another giant in the early sociology of science, Derek J. De Solla Price (1986/1963), used the term Big Science to denote the growth of science as human activity in all measurable aspects: money, manpower, and publications. Price called science "big" simply to point at an unprecedented volume of science in society, relative to other enterprises, and while he mentioned giant instruments and unprecedentedly big team sizes, he did not brand these as threats to science itself but concluded that Big Science was a mere historical interlude: The growth was inherently unsustainable and would level off by itself (Price 1986/1963: 28–29).

Since the 1960s, Big Science has become an everyday term and a concept that many scientists, scholars, and laypeople use colloquially to describe scientific activities or technical installations that are considered bigger than the usual, whatever that means. There are several interpretations of what Big Science might mean conceptually (Capshew and Rader 1992; Galison and Hevly 1992; Cramer et al. 2020), but it is clear that the use of the term is very broad, and that unless some kind of precision work is done, it is "analytically unworkable" (Hallonsten 2016a: 1). The list of phenomena that scholars have named Big Science – from 16th century astronomy thru 1960s space programs, to the project to sequence the human genome – is long. Well-executed efforts to define Big Science have ended up none the wiser (Hevly 1992: 355), and commentators have noted that what Big Science means depends more than anything on the agenda behind the use of the term (Westfall 2003: 32).

An exhaustive effort was recently made to sort out conceptually what Big Science might mean, and how well, or badly, alternative meanings work for analytical purposes, given the vast changes to the dynamics of science and science policy on a global level that accompanied the geopolitical transformations of the end of the Cold War, the rise of East Asia on the international markets, and the multifaceted but uncompromising forces of globalization (Hallonsten 2016a). For the present purposes, it makes less sense to dwell on the issue of what Big Science can, or should be, taken to mean, based on what, and for what purposes. The "narrow" interpretation of Big Science, that focuses on "big machines, big organizations, and big politics," and that acknowledges the dramatically changed circumstances for science and Big Science in the past fifty years and changes to the use of large scientific facilities like the ESS, is sufficient (Hallonsten 2016a: 13ff). For while there is probably no doubt that the ESS is Big Science, it is also clear that it is something dramatically different from the Big Science that Alvin Weinberg (1961) issued dire warnings of, and that Galison and Hevly (1992) studied in close detail, which was a Cold War phenomenon. On August 6, 1945, all doubts whether science and technology had the power to directly influence geopolitics effectively ended, when the United States dropped the atom bomb on Hiroshima. In the years after the end of World War II, all remaining doubts as to whether science and technology *would* directly influence geopolitics similarly ended, as the Iron Curtain descended on Europe, and the Soviet Union commissioned a successful test of their own nuclear weapon, and thus ushered in the era of superpower competition in politics, the military, economy, culture, and, of specific interest here, science and technology. This superpower competition was allowed to dominate and define science and technology policy across the Northern Hemisphere for several decades, and within its framework, Big Science was born and raised. This Big Science was above all the use of very large machines – reactors and accelerators, for the most part, but also ground-based telescopes, spacecraft, and early computers (McCray 2006; Smith 1989; Westwick 2003), and the organization of science in very large teams and hierarchically-structured organizations, for example in the National Laboratories that were erected across the North American continent out of the assets of the wartime Manhattan Project, namely the collected efforts to develop the first atom bomb. Due to the bomb's spectacular and horrendous manifestation of the capacity of nuclear physics, the postwar efforts to (re)build society on the basis of technological progress centered on physics, and especially subatomic physics, and as long as some connection to nuclear energy and weaponry was retained (e.g. Hiltzik 2015; Stevens 2003), these fields of scientific discovery, and many auxiliary activities, enjoyed almost limitless resources. Though the military connection was in part lost in the 1950s, as nuclear physics developed into particle physics and the essentially esoteric hunt for elementary particles became its foremost occupation, the huge particle accelerator complexes that continued to be built throughout the 1960s and 1970s became its own branch of superpower competition. Military and civilian exploration and exploitation of the sub-nuclear world were separated in governmental budgets and, at least in Europe, also in organizational and institutional arrangements (Hewlett and Holl 1989), and civilian Big Science became a quest to answer the most fundamental questions of the origins of the universe and the structure of matter (Hoddeson et al. 1997). The science policy doctrine that guided the expansion of Big Science was governance by political and vocational elites, and the *Linear Model of Technological Innovation* that built on John Desmond Bernal's prewar ideas of science as a source of good (Elzinga 2012; Greenberg 1999/1967; Guston 2000).

The end of the Cold War in 1989–91 meant the end of the binary geopolitical world order, and it also brought the growth of costs and sizes of the accelerator complexes built for particle physics to a halt, although the development had already started to slow down in the 1960s, and did not come to an immediate halt after the fall of the Soviet Union. Particle physics had remained a relatively small field throughout the 1950s and 1960s, with university-based teams visiting particle accelerators to do experimental research on a temporary basis. But as both accelerator complexes and the detector equipment necessary for particle physics experimentation grew beyond what was possible for single university departments to design and operate, particle physics entered what has been called "Megascience." This term can be given at least two analytically relevant meanings in this context, leaving aside the purely political use of the word as a (possibly more rhetorically attractive) synonym to Big Science, most prominently in the name of the OECD Megascience Forum, which was a committee under the auspices of the OECD and active in the 1990s, and not without importance for the ESS (chapter 5). The first meaning of "megascience," promoted especially by Hoddeson et al. (2008), pertains to a development in the 1970s in which experiments in particle physics were prolonged for several years and started to engage teams of hundreds of researchers, engineers, and technicians (the growth has continued still further, and today teams number thousands of members). This microsociological development (although the prefix "micro" is, of course, somewhat misleading) also meant that experimental particle physics became further specialized and, indeed, also increasingly disconnected from the rest of science (Hallonsten 2016a: 9–10, 64), with its political and organizational connections. Thus, second, the transformation of particle physics into "megascience" also entailed an increased concentration of resources to a very small number of sites and accelerator laboratories: The costs of remaining at the forefront of subatomic physics grew so high that it was only possible to keep one or two labs per continent in operation. In the United States, the Fermi National Accelerator Laboratory (Fermilab) was the flagship facility; the Brookhaven National Laboratory on Long Island and the Stanford Linear Accelerator Center (SLAC) in Northern California remained in the field for a few more decades. In Europe, the European Organization for Nuclear Research (CERN) almost monopolized funding for particle physics and most (but not all) national accelerator labs were forced to close. A bureaucratization of (Big) science also followed the advent of megascience; old governing bodies were abolished and replaced by new agencies, and regulations were launched to keep excessive budgets under control and impose professional project management on Big Science. Alvin Weinberg lived to see the demise of the Superconducting Super Collider (SSC), the largest scientific infrastructure project ever proposed, and had he lived to see the analysis of what went wrong with the SSC, exemplarily carried out and published by Riordan et al. (2015), he would probably have noted "I told you so" in the margin of that book. There is much evidence to suggest that the SSC failed largely because of "triple diseases – journalitis, moneyitis, administratitis," (Weinberg 1961: 162) although, of course, several other factors also contributed.

First of all, the postwar growth in spending on science had slowed down in the early to mid-1970s, largely due to the economic downturn but also because of new priorities. Superpower détente made superiority in nuclear physics less of a priority both in the US and in Europe, and the 1960s and 1970s also saw the rise of other sciences to prominent and prestigious roles, in particular life science and materials science, both of which had the potential to have a more direct impact on people's lives, and relevance for new challenges to society and the world, such as environmental concerns and sustainable growth. When the SSC project was terminated in 1993, after 8 km of tunnel had been dug in Texas and over USD 2 billion spent, the Cold War was over, and new projects and programs in life science (e.g. the Human Genome Project) and materials science (e.g. nanotechnology initiatives) had become top priorities of western governments (Johnson 2004).

In the context of the historical chronicle relevant in this book, the demise of the SSC was for the most part symbolic. Particle physicists had grown used to a seemingly endless growth, and for them the initial cost estimation of USD 4 billion to construct the SSC (and the eventual increases of these estimations) was merely the next step in a natural progression (Hoddeson and Kolb 2000: 308). For the Washington establishment, the broader scientific community, and to some extent the taxpayers, it certainly was not. When the ax fell, in October 1993, the world of physicists found themselves for the first time without a "next big machine" in sight. CERN would, eventually, embark on building the LHC and find the Nobel Prize-winning Higgs Boson with it, but particle physics would have to play by the same rules as any other field of science, and compete politically for priority in national and international funding programs.

Meanwhile, and certainly not unrelated, other uses of large particle accelerators had grown in importance and promise, and started to compete for funding. In the late 1940s, it had been shown that the electromagnetic radiation (light) that accelerated elementary particles inevitably emit when their trajectory is bent could be extracted and possibly used for various studies of the electronic and molecular structure of materials. The first such studies were carried out in the early 1960s, and with a series of technological developments in the 1960s and 70s, radiation became manageable and safe to extract, manipulate and put to use, and "parasitic" synchrotron radiation programs were launched at several particle physics labs in Europe, North America, and Asia (Hallonsten 2015b; Hallonsten and Heinze 2015). In the decades since, technology has been refined and methods developed; the organizations around the exploitation of synchrotron radiation have matured, and demand in the sciences has grown exponentially; and, most importantly, dedicated (purpose-built) synchrotron radiation facilities have emerged in several countries and established themselves as crucially important resources in a range of sciences, from physics and chemistry to biology, medicine, environmental science, and geoscience – MAX IV in Lund, sited next to the ESS, is a prime example (Hallonsten and Heinze 2015).

The use of neutrons, the purpose of the ESS, had had a similar development. Neutrons, the unavoidable byproduct of nuclear reactor operation, had already been meticulously extracted and studied during World War II, as part of the reactor programs of the US nuclear weapons program. The basic properties of the neutron – a heavy elementary particle, with no charge (neutral, as the name suggests) but some ability to induce radioactivity – had been known since its discovery by James Chadwick in 1932. During the war, the penetration abilities of neutrons were further studied and also exploited in early uses of neutron scattering in studies of the properties of materials. After the war, the great supply of neutrons secured by the launch of nuclear energy and weapons programs across the industrialized world enabled extensive exploration and exploitation of the experimental techniques of neutron scattering. Gradual improvements in reactor technology and reliability greatly enhanced the performance of neutron scattering experiments, and although neutron scattering remained an auxiliary activity at research reactor facilities, by the 1960s it was "a well established branch of international science," with users predominantly in solid-state physics and solid-state chemistry, but with a clear course set towards "crossing of the threshold into biology" (Bacon 1986: 5). The global lead in the area was retained by the US National Labs until the 1970s, when investments in Western Europe secured new dedicated experimental facilities and the consolidation of a European neutron scattering community. The *Institute Laue Langevin* (ILL) in Grenoble became the global powerhouse of neutron scattering and remains in a leadership position to this day, with the United States still lagging behind (Rush 2015; D'Ippolito and Rüling 2020).

The development of neutron scattering as an experimental technique for studies of materials on atomic and molecular level is numerically inferior to the use of x-rays from synchrotron radiation facilities, but part of the same broader development of Big Science leaving the military-strategic realm to become a force for scientific and technological development in the service of the economy and broader society. Thus, while the bona fide example of the *transformed Big Science* is probably the synchrotron radiation and free electron laser facilities popping up across Europe (Hallonsten 2016a), including in Lund (the MAX IV), neutrons are, and have been, just as important for the significant shift in the science-society relationship that the world has witnessed in the past few decades, and the role of Big Science in it.

Essences of decision

It is quite clear that the analysis of complex historical processes, like the ones outlined very briefly in the previous section, is much helped by being framed by adequate theory. But theory should not be misrepresented either as postulations to be verified or nullified, or as all-encompassing belief systems in which empirical material is to be sorted, but instead be used pragmatically, as a *toolbox* from which different tools can be selected to assist the analysis and explanatory task at hand. Different problems require different tools, and the most complex problems require a set of tools used in parallel to analyze different aspects, or the same aspect in a variety of ways. In this book, which is mainly empirical and follows a relatively strict chronology, theory is in the background and the colorful landscape is presented in its own right. But the topic and the historical processes under study are enormously complex, which means that certain viewpoints and conceptual tools are nonetheless necessary to guide interpretations and assist in sense-making, as part of the analysis.

Great inspiration can be found in an unlikely place. A 1971 book by political scientist Graham Allison, revised and expanded together with Philip Zelikow in 1999, analyzes what was probably the most dramatic and simultaneously well-documented geopolitical event of the Cold War Era, the Cuban Missile Crisis, to understand the framework, logic, and procedure of political decision making in the face of serious threat and under heavily institutionalized conditions. Allison and Zelikow (1999/1971) go about this in a tripartite fashion, where the fateful events of October 1962 are recounted and analyzed first from the viewpoint of the rational actor, then from an institutional perspective, and finally with the use of political theory of power and decision making. The ESS campaign was, of course, nothing like the confrontation of two superpowers and their horrendously forceful militaries in the context of a bipolar world order, and, empirically, the book by Allison and Zelikow (1999/1971) and this book differ in most respects and on most levels. But lessons can be learned and inspiration found in the method and structure of their analysis. The combination of institutional and actor perspectives is controversial for some, but powerful and viable, and in this book they are used side by side to account for different perspectives on the same phenomena, events, and processes, and complemented by relevant concepts from the study of expectations, trust, and promises in political campaigning that are selected from a less coherent theory tradition, but that nonetheless have great explanatory value here.

Quite clearly, the history of the ESS is a history of powerful individuals who are in the right place at the right time and manage to turn events in their favor, or in accordance with the agendas they represent. All the key actors that will be identified in the coming chapters acted rationally, or *boundedly rationally* (see below), which in this context means that they purposefully and deliberately made decisions and took actions in order to achieve a desired outcome. Their agenda might have been personal gain, if not monetary then at least reputational, and/or related to career development and power, or altruistic. Their ambitions were possibly different in different time periods or situations, or even dual or multiple at the same time. The latter is not unusual in science, which is about both personal achievement and the advancement of knowledge for the benefit of humankind, or in politics, where dealmaking by mutual backscratching not uncommonly degrades into nepotism or even corruption, while, of course, simultaneously often leading to outcomes that benefit large groups of people.

But no actor, regardless of how powerful, acts in a vacuum. Institutions matter, and while not easy to define (see below), they are durable, inert and not easily changed. They play an important part in determining the logics of events and processes, and they are as present in the history of the ESS campaign as the individual actors. The Swedish research policy system and science system are highly institutionalized systems of rules, regulations, behavioral codes, beliefs, habits, and norms, all of which can typically be summarized as *institutions* and, not least, fruitfully studied with the help of institutional analysis. The same is true for the international scientific community, both in general and in the specific fields that relate to neutron scattering, and thus the ESS. A key concept is *path dependence*, meaning the tendency of past decisions and actions to set constraints and open opportunities for present and future decisions and actions. The frequent clashes between attempts to stage potentially discontinuous events, like the disruptive nature of a campaign to place an unprecedentedly large and advanced research facility in Sweden, and the path-dependence of inert institutional reality, like the procedures of national Swedish political administration and the organizational culture of universities or governmental agencies, is a very fruitful conceptual framing of the history accounted for and analyzed in this book.

But clearly, the history of the ESS and how it ended up in Lund is also a history of politics, and politics has its own logics that means it differs not only from science but also from many other forms of organized human activity. There are few other realms of society where persuasion and compromise are so symbiotically combined as in politics, while simultaneously governed by very strict procedural rules (constitutional law) and operating in so many arenas (policy areas) and on so many levels (local, national, supranational) at the same time. Though ultimately resting on the legitimacy of democratic rule of the people and by the people, most political decisions do not have the active support of the electorate but (at best) its *passive support.* In other words, pertaining to the topic of this book, very few Swedish voters ever pondered the question of whether they wanted to see the ESS built in Lund, and, in fact, nobody ever asked them. This is not to say that the decision making around the siting of the ESS in Lund was undemocratic or in any way faulty or wrong, only that the analysis of politicized decision-making processes and campaigns requires careful attention to be paid to the cynism, opportunism, and haphazardness of political lobbying. These have the power of altering agendas and mobilizing support for decisions with far-reaching consequences without ever being backed by an electoral majority, and often also without the electorate being particularly familiar with the issue at hand.

These three conceptual viewpoints – *rational actors*, *institutions*, and *politics* – are the theoretical foundations for the chronicle and analysis in this book. The remainder of this chapter is devoted to their explication, with the key aim of equipping the reader with tools for understanding the case and its various shadings. But it is also important to acknowledge and establish the interconnectedness between the three conceptual viewpoints. Rational actors are bounded by institutions and enmeshed in the games of politics. Given that institutions are "shared beliefs and norms" (see below), someone quite clearly needs to share these beliefs and norms in order for institutions to exist and make sense, and this someone is an individual actor that acts rationally (within certain boundaries). Likewise, politics is a game that individual actors play. And politics quite clearly has institutionalized practices, just as institutions are changed both by rational actors and by politics. In other words, the perspectives are overlapping and complementary. For the most part, they will be used as a backdrop to the empirical accounts that dominate this book, and returned to in the discussion and analysis in the concluding chapter, where theory will be used selectively, creatively, and variedly with no other underlying ambition than to enhance the explanatory value of the analysis.

Institutions, actors, politics

The omnipresence in society of structures that both restrict and empower the actions of individuals is strikingly obvious. Douglass North (1990) identified *institutions* as the "rules of the game" of economics, that go beyond laws, regulations, norms, and conventions to also include stable and persistent structures that set boundaries for human action and enable it by providing means for it (Giddens 1984: 24; Scott 2001: 48). In a very rudimentary and general meaning, anything can be an institution, but for institutions to make sense as a conceptual tool with analytical usefulness, the concept must be modified or specified. In this book, two uses of the concept of institutions are relevant.

First, *institutional logics*. Different games obviously have different rules, and rule systems can be understood as sets of institutional logics that differ between societal sectors (or society's subsystems) but also covariate with cultural differences and thus differ between countries, and, conceivably, between organizations or even units of the same organization, depending on what purposes and aims these entities have and what they reproduce in terms of guides for behavior. The institutional logics framework was developed as a means of "bringing society back in" to organization theory (Friedland and Alford 1991), for example demonstrating to organizational scholars that there are other realms or subsystems of society than the market of private enterprises, and, not least, other basic logics for organizing in society than market mechanisms. Institutional logics are powerful, as they embody and encapsulate everything that is part of the context of a human action: formal rules and regulations (including law) and its commonplace interpretation and enactment; culture, norms, and belief systems (cf. Bourdieu 1977); and habit bred through social interaction (cf. Giddens 1979). But the merit of the concept lies not in its capability to assist in a detailed mapping of constraints for human action; such mapping is difficult, to say the least. Instead, the concept of institutional logics should be used to highlight *contrast* and analyze why it exists, and why, and how, it gives rise to tensions. The concept and its use in this fashion can be traced back to Durkheim's "social facts" (Durkheim 1982/1895: 50ff) and Weber's "value spheres" (Weber 1920: 536–573) and the functionalist view of society as composed of different subsystems that operate according to their own logics and together make up an integrated society (Merton 1968/1949; Parsons 1977; Luhmann 1995). The most obvious examples of such subsystems or value spheres are capitalism, or market economy, state bureaucracy, and political democracy, complemented by the judicial system, science, and religion. It is no surprise that the contrasting *institutional logics* of these give rise to tension and contradiction in decisions and actions, and, by extension, in the functioning of society.

Organizations are among the most institutionalized features of society. Coleman (1982) and Presthus (1979) both show that perhaps the most distinguishing feature of contemporary society is its reliance on, and trust in, formal organizations. But as shown by early institutionalists (e.g. Selznick 1957), many or most organizations also become institutionalized in themselves and tend to lead lives of their own; while usually formed to fill a predefined need or fulfill a predefined function, organizations evolve, and adopt routines that may run counter to their declared or original purpose, or what others expect to be their purpose (Meyer and Rowan 1977; DiMaggio and Powell 1983). Organizations embody institutions and perpetuate institutions, sometimes to the degree that it can be difficult to analytically separate organizations from institutions, or, at least, an organizational field from an institution. This is especially true in the public sector, where the connection between the supply of and demand for goods or services or a societal function is not as clear as in the private sector. Swedish universities, for example, have a near-monopoly on science and higher education in Sweden. The organizational field of research-funding agencies and foundations in Sweden crosses the boundaries between the public and private sectors, and also displays significant heterogeneity in other respects, but has a core *institutionalized* function which binds the actors together and corresponds to expectations from the scientific communities and other stakeholders, such as those involved in the ESS campaign. In the current study, the *institution of science*, as Merton (1973) exemplarily put it, is contrasted with the institutions of national politics, governmental bureaucracy, supranational (European) politics, the local economy, and the mirage of the globalized knowledge-based economy, all of which exhibit highly institutional logics that differ in some respects, and overlap in others. But they also develop in harmony, or disharmony, with each other, and, most of all, they are not static.

A key point of institutional theory is to account for the inertia and resilience of institutions and show how this shapes events and processes in society. But, as shown by *historical institutionalists*, the inertia and resilience of institutions is also a key factor for long-term *change* – institutions, in fact, provide the basis for gradual, incremental, and cumulative renewal of society in dynamic processes on varying time scales (e.g. Streeck and Thelen 2005; Mahoney and Thelen 2010; Fioretos et al. 2016). There are several instances in the story of the ESS campaign that appear discontinuous, major events that change the dynamics completely or push the development forward in a quantum leap. But even those seemingly dramatic and discontinuous events were part of longer-term developments that unfolded in a much slower, step-by-step fashion. This suggests that proper attention must be paid to *path dependence*, both in the campaign as such, where it is clear that certain developments were set on a particular path early on, and in the institutions with which the campaign interacts. Science systems and policy systems are, generally, strongly path-dependent (Hallonsten 2018; Hallonsten and Heinze 2012, 2016; Heinecke 2016; Rothstein and Steinmo 2002; Steinmo 2010). This is true, specifically, for European and Swedish politics.

How, then, do individuals break away from the institutions that embed their lives, and how do they deviate from the paths that make up societal development, to shape it and alter history? This is essentially an empirical question, and the remaining pages of this book are devoted to telling a story dominated by decisive actions by individuals, who were often in the right place at the right time and acting out of the habit developed through decades-long close interaction with strong institutions, which gives them power and influence to shape history, but who sometimes clearly deviate from the accepted patterns of behavior predicted and mediated by those very same institutions. Why, and how? These are questions that can only be answered in a very specific manner, for each individual and each situation – and this will be done throughout the book, not all-encompassing but for the most critical actions and events.

The proper analytical tool to use alongside institutional theory, in order to contrast individual action with structure and context, is sociological rational choice theory, based ultimately on methodological individualism; in essence, an epistemological point of departure that establishes that individuals shape social phenomena, and not the other way around. *Absolute rationality* is, of course, just as unlikely as a guide for human behavior as absolute altruism. Therefore, the fierce critique of rational choice theory that rests on a presupposition that it somehow imagines or conceptualizes a perfectly rational "homo economicus" is futile. This criticism has, consequently, been refuted with a minimum of effort (Opp 1999), but other traditions in the theorizing of human behavior and the organization of society have also launched workable varieties, most importantly "bounded" or "limited" rationality (Simon 1957; March and Simon 1958; Pfeffer 1982) within organization studies and economics, and the more sociologically relevant "procedural" rational choice theory (Esser 1993; Goldthorpe 1998), which moves the view away from the endless philosophical discussions over what rationality means, and instead uses rationality as a concept to explain why and how individuals make certain decisions and take certain actions depending on the situations they are in. As summarized by Lindenberg (1985: 100): There shall be no doubt that humans, especially when in positions of power, are *resourceful, restricted, expecting, evaluating,* and *maximizing* at the same time.

It should be noted that it is the *combination* of institutional theory and rational choice theory that is advocated here, and applied in this study. The power vested in the positions that individuals hold, be that in politics, academia, enterprise, or the media, must be acknowledged as institutionalized and institutionally reproduced. But it would also be difficult indeed to analyze a historical material where actors take prominent roles, without assuming that these actors made rational, or at least purposeful, decisions.

But one piece of the puzzle remains. As has already been established, the campaign to bring the ESS to Lund was a political campaign. Although politics is a conspicuously shallow game, it also reflects underlying societal developments for which politicians occasionally make themselves the spokespersons or interpreters. The emergence of a globalized knowledge economy and the changed role of science and technology in society became fundamental political assets for the Swedish ESS campaign. The developments as such can be interpreted with some variety; what is clear is that the economic downturn of the 1970s spurred a profound shift in the bearing principles of science policy, leading governments (and electorates) in many countries to question the ruling doctrines and start demanding more demonstrable results from publicly funded science. What used to be robust foundations for science policy, at least in (Western) Europe and North America, namely the "social contract for science" (Guston 2000) and the "linear model of technological innovation" (Stokes 1997), were replaced by a strategic agenda with more direct meddling by politicians and decision makers in the procedures for governance, priority-setting, and performance evaluation in science (Irvine and Martin 1984; Ziman 1994; Greenberg 2001). Current science policy is very much characterized by "commodification," meaning that scientific productivity is measured in monetary terms (Radder 2010), "economization," that scientific research is supported and sustained primarily for the sake of promoting economic growth (Berman 2014), direct steering, and strategic prioritization of areas and problems on a political level (Nowotny et al. 2005; Whitley et al. 2010), all of which are clearly seen in the story of the campaign to establish the ESS in Lund. As noted at the end of the introductory chapter, and as will be returned to in the relevant chapters, the ESS campaign also took place at the height of the influence of theories of regional innovation systems and learning regions (e.g. Florida 2002; Törnqvist 2002; Maskell et al. 1998), which has special relevance when analyzing the marketing campaign around it (chapter 10) but also general relevance as one explanatory factor behind the political campaign itself, and its success.

The impatience, and the demands for science to contribute directly to economic and social progress, and to the solving of society's challenges (including sustainability), are, quite naturally, directed at the public science system – private sector R&D is out of bounds for governmental intervention. Discontent with a perceived low return on investment in science and technology, regardless of whether this is conceptualized as the "innovation problem" (Guston 2000: 113), the "European paradox" (Andreasen 1995), or the "Swedish paradox" (Edquist and McKelvey 1998), will become policy in the shape of steering, performance evaluation, and reform agendas. But universities have institutional logics of their own, and their governance models are often laid down in law (in some countries, academic freedom is constitutionally protected), which make them strongly path-dependent and resilient to change. The whims of impatient science policymakers may be mere perturbations on the surface in old and reputable academic organizations, whose institutionalized work practices and organizational structures seem to withstand most reform attempts, good or bad, and can only be changed slowly and gradually. Expectations, demands, and aggressive reform agendas are therefore more likely to succeed and have a strong impact on younger organizations or organizations in the planning stage. But a square reversal of causalities is also possible: Expectations and demands for positive – yet unrealistic – returns from investment in science and technology may be the primary, or sole, reason for investing in a new R&D organization, or putting political weight behind a proposal to establish such an organization within one's borders.

It is not uncommon for Big Science projects to be marketed with references to unstoppable technological progress (Hallonsten 2016a: 185ff).

Although the results of the scientific work that will be undertaken as a result of these investments can only be expected ten or twenty years ahead, it is common to present them as being already real, or at least realistic. The "strategic turn" in science policy in the past four decades has accentuated this reliance on promises and expectations, because only those investments in science that have some kind of articulated expectation of utility tied to them can be considered strategically important, and thus be subject to priority (van Lente 1993: 10; Irvine and Martin 1984: 3–5). It matters little that science is by nature unpredictable, and that it cannot be planned but must rely on serendipity and inherently uncertain individual creativity. Increasingly harsh prioritization between areas probably strengthens the role of articulated expectations and promises, because it becomes increasingly important to market and advertise projects. Likewise, commodification (Radder 2010), which means that scientific productivity is measured in monetary terms, is also likely to increase the role of expectations by promoting strong beliefs in over-simplified measures of productivity, excellence and relevance (cf. Hessels et al. 2009).

The theoretical basis for the analysis of how promises and expectations are used in political campaigns is eclectic, and not found in a pure social science tradition. A fairly recent effort in social studies of science and technology, under the slightly misleading headline "the sociology of expectations," has contributed strongly to the understanding of how promises of the future impact of investments in science and technology can become a resource in the campaigning for these projects, and it appears that the expectations built around these promises can become the deciding factor for their success in a very competitive policy context (Brown and Michael 2003; Borup et al. 2006). The underlying mechanism, as van Lente (2000) shows, is that promises have an inherent imperative that may make opposition to a particular project, if this project is wrapped in enough promises of future gains, seem irresponsible, very much like the sociological mechanism of "self-fulfilling prophecies" (Merton 1948). When promises and expectations become political campaign assets in this way, their uncertainty is turned into an advantage, because forceful action to reduce the uncertainty, by working towards the project's realization, becomes the only politically viable course of action, and is sometimes even presented as the only *moral* course of action, especially if it can be established with reasonable credibility that important values like the earth's climate or an otherwise sustainable future are at stake (van Lente 1993, 2000; Brown and Michael 2003; Borup et al. 2006).

Political campaigns thus institutionalize expectations, and in campaigns around scientific and technological projects, promises typically have a broad appeal and harmonize with *Zeitgeist* by embodying notions of the "knowledge society" and similar grandiose, but essentially empty, catchphrases that reflect the apparent need for society to place trust in more or less grandiose expectations for the future (Alvesson 2013). Neoinstitutionalists have shown that organizations (most certainly including campaign organizations) adopt all kinds of behaviors and traits in order to appear legitimate in some kind of public's eye, sometimes above and beyond their real purpose or mandate (Meyer and Rowan 1977; DiMaggio and Powell 1983). In campaigns for new projects and investments in science, the glorious past is also readily mobilized as a resource. When current scientific or technological advances are marketed with the help of the general notion of scientific and technological progress, it invokes unmistakable parts of the advancement of civilization into the era of modernity: the wheel, the compass, the steam engine, penicillin, nuclear energy, manned space voyage. Or, if the situation calls for a genius or two to lend their posthumous support: Da Vinci, Descartes, Galilei, Newton, Darwin, Curie, Einstein, Turing. If the purpose is to advocate dramatically increased spending on science within the Swedish Government's budget, Swedish innovations such as safety matches, ball bearings, refrigerators, and the drug Losec, or Swedish scientists and innovators like Carl Linnaeus, Svante Arrhenius, Alfred Nobel, and Eli Heckscher, can be mobilized (see chapter 4). Historians have named this the "invention of tradition" – the construction of a quasi-historical continuity that serves contemporary purposes (Hobsbawm and Ranger 1983; Hanel and Hård 2015). While not using this particular term, historians studying particle physics, manned spaceflight, and the race to map the human DNA have similarly shown how projects have been framed by their proponents in public campaigns for support as belonging to proud traditions and uninterrupted advances in humanity's quest to conquer nature (Riordan et al. 2015; Faherty 2002; Davies 2001).

It is important, in this context, to realize that for virtually all of the story chronicled and analyzed in this book, the ESS did not exist in any other form than as an *idea*. The reader will be well served by noting and bearing in mind the (lack of) substance in the mirage used for marketing the ESS project, and the discrepancy between the content and form of the campaign as documented and analyzed in this book and the real ESS facility, which is briefly dealt with in chapter 8 but is above all already physically present in the fields northeast of Lund at the time of writing.

The proper way to fill any such gap between image and reality is, of course, knowledge, facts, and nuanced analyses of the type patented in the social sciences undertaken in an academic context, with the utmost attention paid to details and their context.

3. Science, technology, politics, and organization

Neutron scattering and neutron facilities

The neutron is one of the three elementary particles that make up atoms. As the name suggests, neutrons are *neutral*, which means they have no electrical charge (unlike protons which are positive, and electrons which are negative) but they are quite heavy and can be used as a probe to study the molecular structure of matter. The neutron was discovered in 1932 by British physicist James Chadwick, who was awarded the 1935 Nobel Prize in physics. Neutrons exist in very large quantities in heavy elements and are essential in *fission*, i.e. the nuclear reaction that drives reactors and atomic weapons. In the latter case, an uncontrolled nuclear chain reaction sets off a blast of enormous power, and in reactor operation, the fission process is controlled by the absorption of neutrons by control rods, although excess neutrons can also be extracted. During World War II, a large number of nuclear reactors were constructed as part of the atomic weapons programs of the United States, Germany, and the United Kingdom, meaning neutrons could be studied closely and their possible use in experiments explored (Hewlett and Anderson 1962: 27ff). Neutron *diffraction* or *scattering* means essentially that neutrons are fired at a sample and bounce off it, after which the patterns of the scattered neutrons are detected and several properties of the sample documented, including not just molecular structure but also magnetism and molecular motion; neutrons enable observation of "where atoms are and what atoms do," in the words of Bertram Brockhouse and Clifford Shull, who were awarded the 1994 Nobel Prize in physics for their pioneering work in neutron scattering (Berggren and Matic 2012: 34). Because neutrons are neutral and heavy, they can penetrate solid objects undisturbed and enter into the nuclei of atoms largely without harming them. This is not the case for, say, x-rays, which can destroy samples, especially the high-intensity x-rays produced in *synchrotron radiation sources*, which are complementary to contemporary neutron sources and play a role in the history of neutron scattering and the ESS story in particular (see below). The drawback of neutrons compared to x-rays is that they are difficult to control (and hence focus) and that they tend to induce radioactivity in some elements. Today, neutron diffraction is used in many areas of science, with some emphasis on physics and materials science, but also with applications in chemistry and biology (see below).

The results of the first neutron diffraction experiments, conducted as part of the wartime weapons programs, were published after the war and formed the basis for organized efforts to use neutrons from reactors for studies of materials in several places in North America and Europe. The physical and organizational assets of the US wartime nuclear weapons project, known as the Manhattan Project, were transformed into a system of US National Laboratories in 1947 (Westwick 2003), and neutron scattering experimentation became an auxiliary but regular activity at the nuclear reactor facilities built and operated at Argonne National Lab in Illinois and Oak Ridge National Lab in Tennessee. In the late 1940s, the atomic structures of several materials were mapped with the help of neutrons, including breakthrough results on the distribution of hydrogen atoms in ice (Hallonsten 2016a: 67–68).

The 1940s and 1950s saw rapid reactor development on both sides of the Iron Curtain, and regardless of whether reactors were built to enable weapons production or energy production, neutron scientists were typically there to exploit the neutron beams emerging from the reactors. The works published by Shull, Brockhouse, and colleagues in this era inspired efforts in many corners of the globe. In 1955, experimental neutron scattering programs were underway in most Western European countries, the Soviet Union, and several National Laboratories in the United States (Bacon 1986). The Atoms for Peace program, launched by US President Eisenhower in 1953, enabled peaceful exploitation of reactor technology in many countries, in part through US support in technology development, and thus helped increase the average available neutron *flux* (flow rate per unit area, or, simply, the number of neutrons in a given time unit). In the early 1960s, the prospects of better neutron scattering experiments were for the first time included in the scientific motivation for new reactor facilities in the US National Labs system, at Brookhaven National Lab and Oak

Ridge National Lab (Rush 2015: 137). By this time, neutron scattering was a small but well-established experimental technique in solid state physics and solid state chemistry, and its use had begun also in biology (Bacon 1986: 5).

Atoms for Peace had enabled smaller countries to embark on ambitious reactor-based R&D programs. In the mid-1960s, scientifically excellent research environments had been established in Europe, most notably in Harwell (UK), Garching outside Munich and Jülich outside Aachen (West Germany), and Saclay (France). The global US lead in neutron scattering, largely the result of early adopter advantages and the ubiquity of nuclear reactors in the well-funded National Laboratory system, was nonetheless intact through the 1960s. But an idea for a pan-European, purpose-built neutron facility started to take form in discussions along the French-German "neutron axis," between prominent scientists Heinz Maier-Leibnitz in Munich and Louis Néel in Grenoble and with some involvement from the UK neutron scientists at Harwell. The emerging political alliance between France and West Germany, formalized in the 1963 *Élysée Treaty*, became the natural framework for these discussions once they entered political level. Largely thanks to these two alliances, the agreement to launch the *Institute Laue Langevin* (ILL) in Grenoble was signed by representatives of France and West Germany in 1967. Five years later, the new European neutron facility started operation, and in 1974, after the restoration of UK relationships with France and the European community, Britain joined the ILL (Pestre 1997; Cramer 2017). Although the United Kingdom had played an important role in the political genesis of the ILL and the design work for the reactor, it did not become a founding member of the facility in 1967, which Atkinson (1997: 145) explains by the ambitions of British scientists to build their own state-of-the-art neutron scattering facility, and the serious strains of the political and diplomatic relations between the UK and mainland Europe, especially between the UK and France. Consequently, after the defrosting of the relationships in the early 1970s, which also led to the UK membership in the European communities, the UK joined ILL as a third equal member (Cramer 2017: 399; Atkinson 1997: 146).

Importantly, ILL was the first reactor facility that was purpose-built for neutron scattering, which gave the laboratory and its pan-European user community a competitive edge. In its first decade of operation, the ILL "grew to be almost a synonym for neutron scattering to a large proportion of the neutron community" (Bacon 1986: 7). US competitors were mere "parasitic" operations on reactor facilities built for other purposes, and, furthermore, did not have the user-orientation of ILL, which stifled healthy competition for experimental time and, by extension, scientific quality (Crease 2001; Rush 2015). In parallel with the expansion of the use of the ILL, significant advances were also made in instrument development. Increases in neutron fluxes and in the performance of other vital components in neutron scattering instrumentation "revolutionised the practical manner of carrying out neutron experiments" (Bacon 1986: 9). In the 1970s, new technologies for the production of neutrons, that would greatly improve experiment performance, also emerged on the drawing board. But these would require entirely new facilities to be built.

Reactors produce neutrons in a continuous flow, and while the development of reactor technology was rather dramatic in the first three to four decades after World War II, there was (and still is) a limit to how far the performance of a reactor can be pushed to increase neutron flux. Another technique, that does not use nuclear fission but instead a process where neutrons are knocked out of a heavy element, was introduced in the 1970s: *spallation*. The most obvious advantage over reactor-based neutron production is that spallation sources can produce shorter pulses of neutron beams with a very high flux in their peaks. The first spallation source was built at Argonne National Lab in the 1970s and opened to scientific use in 1981 (Westfall 2010), and four years later, the British spallation source ISIS (no acronym) started operations in Harwell.

Spallation originally means "fragmentation," which is a good description of the spallation process: protons, accelerated in a linear collider and smashed into a very heavy (and thus very neutron rich) target material, usually mercury or tungsten or an alloy of these and other elements, knock out neutrons that are extracted through *neutron guides* to experimental stations (Berggren and Matic 2012: 31). The neutrons cannot, however, be manipulated like x-rays and other light, or electrically charged particles, and therefore spallation facilities, like reactor-based neutron sources, must rely on the effect of the proton accelerator so as to spallate as many neutrons as possible from the target material, so that the flux is on acceptable levels when it reaches the instrumentation where experiments and measurements are made, usually called *experimental stations* or simply *instruments*.

In both reactor-based neutron sources and spallation sources, the neutrons are led through neutron guides to the experimental stations, where neutrons of different "wavelength" (speed) are used for different experiments. Several specialized experiments are typically conducted in parallel, around the reactor or the target station of the spallation source, with the use of different instruments and operating in independence from each other. At the ILL in Grenoble, some forty instruments are operated in parallel, and are used by different groups doing different experiments and measurements within research projects usually planned and funded elsewhere, in universities and institutes across Europe. The current design of the ESS facility has sixteen planned instruments but room for several more, and these will be similarly used by visiting research groups. Importantly, instruments and experiment areas do not follow typical disciplinary categories, and the full range of applications of neutron scattering typically found at a neutron facility is very broad and amorphous, as discussed in the next section. It is also important to note that neutron facilities like the ESS are very complex technical systems where each component is designed not only to perform at as high a level as possible, but also to function together with other components, which makes the overall design effort an extremely complex task involving hundreds of specialist scientists and technicians from a number of fields and areas working in collaboration. The proton accelerator (usually called the linac, which is an abbreviation for linear accelerator) is a complicated piece of technology in its own right, as are, of course, the target station, the control system, the neutron guides, the various experimental stations with their vacuum chambers and detectors, and the data management and storage systems.

The complexity is very well-illustrated by the volume of the *ESS Technical Design Report (TDR)* published in 2013, which encompasses some 600 pages (excluding summary and reference list) and was coauthored by 460 people from 49 organizations in 16 countries. Its most voluminous chapters are those covering "Neutron Science and Instruments" (137 pages), "Target Station" (116 pages), and "Accelerator" (123 pages), while "Integrated Control Systems," "Emission Control," and "Safety and Security" obviously also each make up a large part of the document (ESS 2013b). The TDR was preceded by only one year by the *Conceptual Design Report (CDR)* which is much more condensed and, as the name suggests, conceptual in nature, encompassing 246 pages (ESS 2012a). As will be examined more closely in chapter 8, this is the conventional way of proceeding with the design of a facility like the ESS, and the preparation of these reports is, of course, only the most visible part of the process, which predominantly takes place at desks and in meeting rooms at the ESS and elsewhere. An important ingredient in this consists of reviews of different types, ranging from the informal consultations between staff inside the organization and with trusted colleagues at other places, via regular meetings of the Technical Advisory Committee (TAC) and Scientific Advisory Committee (SAC), where progress is discussed with the respective group's members, who are renowned specialists in the fields and areas concerned, to the regular reviews of temporarily composed expert groups that scrutinize the whole project and its various parts and make their recommendations to the directors and organization.

The gradual growth in the use of neutrons from reactors for studies of materials has been called the "disarmament" of neutrons, which is a conceptually erroneous inference of the history briefly accounted above. Neutrons themselves were never a key resource in weapons development, but a mere byproduct of reactor operation for the enrichment of uranium in weapons development and other nuclear physics research. The growth of the use of this byproduct into a Big Science in its own right, with several purpose-built reactors for neutron production in operation today, is certainly a "disarmament" of the use of reactor technology, but it is both late and relatively insignificant in a broader perspective, compared to the "disarmament" of reactors within the Atoms for Peace program of the 1950s and on, whereby nuclear reactor technology was proliferated for peaceful purposes and enabled the exploitation of nuclear energy across the globe. But reactors have also still not been completely disarmed, at least not in a political sense, given that they are still a key asset in nuclear weapons development. What is more, nuclear reactors – no matter how peaceful the motivations are for building them – are still shrouded in political controversy and surrounded by thorny national and international regulations. The debate in the 1970s and 1980s over the safety aspects of nuclear energy, driven by the environmentalist movement and fueled by the 1979 Three Mile Island accident and 1986 Chernobyl disaster, is quite evidently not dead and was quickly reawakened – with far-reaching political consequences in, among other places, Germany – in the wake of the 2011 Fukushima nuclear reactor accident. While the emergence of non-reactor neutron sources was predominantly technically and scientifically warranted, since spallation sources are significantly more efficient and high-performing, the political dimension should not be neglected. It is interesting to note that in spite of the fact that spallation sources are nowhere near nuclear reactors in terms of risk to the environment and human safety in cases of meltdown or leaks, the ESS project still invoked anti-nuclear sentiments in local opinion in Lund, among members of the Green Party in the Swedish Parliament, and, not least, among Danish politicians and scientists who seemed to rule out locating the ESS in Denmark right from the start, although the Danish neutron user community had a significantly stronger position in the late 1990s, when the idea for a Scandinavian ESS emerged (chapter 6).

The most interesting historical developments in the history of the use of neutrons for studies of materials are not technical, but political and sociological, and interrelated. First, the global center of gravity for neutron scattering shifted from the United States to Europe in the 1970s and on, mainly due to the launch of the ILL and the deliberate policy to make it into a user facility for the whole of Europe. The US simply did not manage to keep up with this development, setting different priorities in the 1970s and 1980s and putting most of its resources into particle physics and synchrotron radiation, at the expense of neutron scattering (Rush 2015: 141–142). Although the Spallation Neutron Source (SNS) at Oak Ridge National Lab is a direct competitor to the ESS on the global stage, and could lead to a revitalization of US capabilities in the area, it should be noted that it was already on the drawing board in the 1980s but funding proposals were apparently ignored (Rush 2015: 144–147; Crease 2001: 51). The European lead was, in a sense, proven by the fact that SNS was only funded and launched after the first preliminary ESS design had been published in 1997, and built strongly on its technical and scientific cases. But the shift of the center of gravity from the US to Europe was not merely a question of politics and (lack of) investment; as mentioned above, the ILL embodied a new organizational principle, where the development and operation of the facility and its instruments were focused almost entirely on cultivating a strong European user community, which included user-friendly operation and administration, and extensive scientific user support. According to some, this is what ensured the success of the ILL in the long run, and, with it, the success of European neutron use (Maier-Leibnitz 1986: 137–139; Atkinson 1997; Crease 2001). This is an important point, because it constitutes a maturing not only of techniques, but also of organizational practices, and an opening of the experimental technique of neutron scattering to non-expert user communities, much like the transformation and broadening of the design and operation of synchrotron radiation facilities in Europe and the United States, which has been shown to have been crucial for the growth of that technique into a mainstream resource for broad scientific communities (Hallonsten and Heinze 2015).

The fact that so many neutron scattering facilities are still operated as auxiliary activities at reactor facilities with other purposes makes it difficult to assess their total number; sizes and scope vary from the purpose-built and internationally-oriented Spallation Neutron Source (SNS) at Oak Ridge National Laboratory and the ILL in Grenoble, to minor facilities with just one or a few instruments at reactor facilities owned and operated by universities or institutes in smaller countries. At least ten neutron scattering facilities with regular user operation and open calls for proposals are currently in operation in Europe and the United States (Hallonsten 2016a: 257–258) and it is a reasonable estimation that a similar number are in operation in Asia and Australia.

With the ESS, Europe seems to be on the path to reinforcing its global lead in neutron scattering, but it is quite clear that this path is neither straight nor smooth. This book is about the campaign, and so the campaign's claims that the ESS would be the world's leading neutron facility will be discussed in later chapters. Here, it suffices to note that first, the prerequisites to be fulfilled in order to claim that position are many and go far above and beyond mere technical superiority over competitors in other parts of the world; second, it is not entirely certain that the ESS will be able to brandish either the technical superiority its designers set out to achieve, or superiority in any other respect (see next section), given its governance and funding model; and third, it is also not entirely clear how a world-leading position is to be claimed, on the basis of what evaluative criteria, and by whom. These claims will be substantiated by the analysis and discussion in the next section.

The politics and organization of user facilities

The use of neutrons for research on materials is an experimental use of large infrastructure not unlike the use of ordinary-size equipment in any science; what makes it special is, of course, the size of the infrastructure needed, which makes costs soar and leads to politics taking a firm grip on planning, funding, and organization. While Big Science is not at all an unequivocal term, its rhetorical attractiveness as a catchphrase also makes it almost unavoidable for those with deep analytical ambitions. But one thing deserves to be reemphasized and can serve as a fundamental premise for the synthesis of the existing knowledge on the politics and organization of neutron facilities in this section: The use of neutrons for research on materials is radically different from classic Big Science such as the particle physics experiments conducted at, for example, CERN, by teams of hundreds or thousands of scientists and engineers. At the ESS, as with all its predecessors since the 1940s, neutrons are made available to *users* in groups of quite ordinary sizes, who may be employed by the same laboratory organization that operates the neutron facility, as was usually the case at the US National Labs in the 1950s and 1960s, or employed by other organizations such as universities or research institutes, as has been the case with the user base of ILL and ISIS since their inception, and will be the case for the lion's share of the ESS users. At the major particle physics facilities like the LHC at CERN, experiments are undertaken on a large scale and with time horizons of several years, and generally designed for the specific facility. At neutron facilities, scientists from universities and institutes do experiments for a few days at a time, within their ordinary research projects. While neutrons may be an absolutely crucial resource in these experiments, the projects and studies as such are conceived, defined, and not least funded and organized in other settings, predominantly university departments and labs.

This fundamental premise for the use of neutrons for scientific research is the same as for synchrotron radiation, which is a sibling resource that both complements and in some cases overlaps with neutrons. The two – neutrons and synchrotron radiation, occasionally complemented with free electron laser – are techniques that have grown dramatically in importance in the past few decades and expanded their use across the disciplinary spectrum of the natural sciences (Hallonsten 2016a: 62ff). The ubiquity of neutron and synchrotron radiation facilities in Europe, Asia, and North America is an important feature of contemporary global science that emerged and grew into prominence as part of a broader transformation of science and science policy away from the military-industrial complex and toward solving grand challenges and innovating for consumer markets, in new boundary-breaking constellations. This also provides a wider context for the growth of neutrons as an experimental tool, that takes into account (geo)political shifts and the broader development of the role of science, technology, and innovation in society, toward interdisciplinarity and increased collaboration across institutional and national borders. The ESS fits very well into this picture; although sometimes described as a physics facility, the ESS will be a resource for a broad spectrum of scientific disciplines and interdisciplinary fields: physics, chemistry, biology, medicine, and varieties of these; materials science (including nanotechnology), environmental science, geosciences, and engineering sciences. In its own presentation of figures on the distribution of *experimental time* (see below) among different areas of science, the ILL shows that physics (in a wide sense, and including the categories "materials" and "soft condensed matter") accounts for roughly 75% of the total use of the facility, but the remainder is distributed among chemistry (12%), biology (9%), engineering (2%), and other (1%) (ILL 2018: 95).

The vast majority of future users of the ESS – estimated at 2,000 to 3,000 annually when the facility is fully built-out – will be external to the ESS, and come from universities, institutes, and companies from across Europe and the world. In other words, neutron facilities "do not produce science themselves – their users do" (Hallonsten 2016b: 486). Importantly, neutrons are, to an increasing degree, used alongside other techniques within the same projects, and users are generally rather promiscuous, meaning that their loyalty to a specific facility is limited. Users travel the world in search of the most favorable experimental conditions for the project they are currently undertaking, and while the instrumentation does of course set some fundamental preconditions for the studies that can be carried out, users generally use neutrons as a resource to achieve the goals of their research. The ESS will have hundreds of scientists employed, and in charge of developing instrumentation and keeping it in a useable condition together with engineers and technicians, and support users. These scientists will also be active users of the instrumentation, but the ultimate purpose of the ESS is to make instrumentation available to scientists from universities, institutes, and the private sector who visit the facility temporarily. It has been shown that a majority of the publications reporting on results from the ILL are coauthored by non-ILL employees (external users) and ILL-employees (Hallonsten 2016a: 174–175). Qualitative inquiries of the relationships between users and facility staff at the ILL show a variety of arrangements, where some users establish long term relationships with facility staff and co-publish with them for several years, whereas others see facility staff as mere service technicians, with all kinds of varieties in between (D'Ippolito and Rüling 2019). Importantly, though, there is a distinct *functional differentiation* between, on the one hand, the facilities providing the experimental opportunities, and, on the other, the users who do the science and achieve the results; a division of labor or separation of tasks that is the norm for contemporary large scientific facilities, but not always acknowledged in policymaking.

The several thousand neutron users in the world, and the dozens of neutron facilities in operation, match supply and demand on a *market* of sorts. Exceptions from the market logic do exist, but the fundamental principle is that users seek the most favorable experiment opportunities and exploit these, with some natural bias of geographical proximity and established relations with facilities and their staff (Hallonsten 2016a: 161ff), and facilities seek to attract the most prominent users in order to have the most impactful science done with the help of their instruments. The critical resource, of course, is experimental *time*. The tradition, at least in Europe and the United States, is that experimental time is awarded free of charge to those who agree to publish the results, and who apply according to predefined standards and pass the peer review assessment of their application. Routines for application, review, and scheduling of experimental time differ between facilities but there are some similarities; most publish detailed information about their instruments online and issue open calls once or twice a year. Experiment proposals are reviewed by panels of experts, in just the same way as when a research funder has a grant application reviewed, or when a journal reviews a manuscript, and a user office is usually in place to take care of scheduling of experimental time (Hallonsten 2016b). It is also possible for users to pay the full cost of experimental time and skip the ordinary process, and they may then also choose not to disclose their results, which is a viable alternative for proprietary research. Such direct industrial use normally only accounts for a small percentage of the total experimental time; at neutron facilities operated within the United States National Laboratories, less than 2% of the registered users in fiscal year 2016 had an organizational affiliation in the private sector, and this figure included every person who was physically present at a neutron facility to conduct an experiment (Department of Energy 2019). A certain symbiosis is usually established between neutron facilities and their user communities, especially at those facilities where use of neutrons was originally auxiliary (or parasitic) and started off very small scale, and where competence and knowhow resided almost entirely with (prospective) users, who consequently had to involve themselves heavily in instrument design and construction, and the development of experiments. This user involvement was institutionalized on an aggregated or structural level at the ILL, which was built largely with the expertise of the leading neutron use and neutron instrument development environments in Europe (in Grenoble, Harwell, and Munich) and thus quickly managed to establish itself as the world's cutting-edge facility, and drove the pan-European neutron user community's rise to global leadership (Rush 2015). At the ESS, a greenfield project that relies on no preexisting organization or infrastructure, the symbiosis between the facility organization and the expertise of the (prospective) users is, in one sense, even more important: The ESS has been designed, planned, built and will be operated by a community of neutron scientists recruited from Europe and the rest of the world, who have networks stretching out in the user communities, with large numbers of people in various organizations and capacities who are also involved in the design and planning of the facility and its instruments (chapter 8).

The future instruments of the ESS have, hence, been developed in *instrument workshops* involving hundreds, or sometimes thousands, of European scientists. These have been guided by the overarching idea that the future ESS facility should provide a battery of instruments and experiment opportunities that match the (potential) demand of the collected European neutron user community, not today but primarily some years down the road, when the instruments are ready to use. This means that the work needs to be strongly forward-looking and hence imaginative, while also using resources in a responsible way, which usually means going for more conservative technical solutions. Adding to this complexity is the extensive use of what are known as *in-kind contributions* in the ESS. This means that instead of cash, member countries contribute with ready-built components for the facility. The in-kind model has become extensively used in recent European Big Science facilities to allow member countries to retain some of their investment in their local economy (Hallonsten 2014). While in-kind contributions are very complicated and administratively burdensome from an organizational and governance point of view, not to mention the troubles they appear to create in terms of taxation (Hilling et al. 2017), this is a formalized structure for securing the input of top-notch competence and knowledge from specialized R&D units in universities, institutes, and companies internationally, which is absolutely crucial in most large and complex Big Science projects, but especially so in a greenfield project like the ESS, i.e. a project built outside of established structures and in a place and a country with limited (or non-existent) competence and tradition in developing cutting-edge technologies of the kind found in the ESS.

In-kind contributions are only one of several means of correcting the imbalance of benefits from the construction and operation of large scientific facilities. While several European countries contribute financially, it is generally believed that the largest benefits – scientifically, but also in socio-economic terms – stay in the region and country hosting the facility. The notion was a key argument put forward in the 2005 investigation by Allan Larsson on the prospects of establishing the ESS in Sweden, and although the conclusions of that investigation with respect to these benefits were greatly exaggerated (see chapter 6), it is clear that host countries reap the greatest direct benefits of the large investments in Big Science facilities, which is shown not least in the considerable efforts made to correct this imbalance in recent and contemporary Big Science facilities in Europe, including the ESS. In fact, the imbalance in socio-economic impacts, which benefits the host country and region, has been a major hurdle in the process of reaching agreements on European collaborative Big Science facilities since the 1970s (Hallonsten 2014). Among the measures in place to rectify the imbalance are the *host premium*, meaning that the host country is expected to pay significantly more than other countries toward the construction budget (Sweden pays 35% of the construction of the ESS), in-kind contributions, and the use of *fair return* on procurement of goods and services, and in scientific use. While fair return (or *juste retour*) on procurement is well-established and has been used at CERN since its founding in the 1950s, its application to scientific use, namely the distribution of experimental time among applications based on the country of residence of research groups, is quite controversial. Fair return on procurement of goods and services means that procurement contracts (in the operations phase) are awarded to companies in part based on the country of their origin, so that in the long term, the share of the budget of various countries is reflected in the shares of the total value of procurement. This can have its drawbacks, such as inefficiencies, suboptimal solutions, and lock-ins (Hallonsten 2014: 44) but is generally accepted as an apt means of balancing returns for investment. Fair return on scientific use, on the other hand, means the same principle for the allocation of experimental time, namely a correction mechanism that redistributes experimental time slots to research groups based on their country of residence. It is controversial because it goes against the idea of allocation purely on the basis of scientific quality, which is a very strong ideal in the international neutron (and synchrotron radiation) user communities. The fair return correction mechanism for distribution of experimental time is in place at the ILL (and other European science facilities), and it is considered highly likely that it will be used, in some form, at the ESS (Interview: Yeck).

Of course, the risk of scientific fair return compromising the ideals of open competition for access to experimental time is not the only goal conflict in the governance of a research facility like the ESS. It is fair to say that the organizational complexity of the ESS is at least as great as the technical complexity of the infrastructure, and several interests are competing to define what the optimal output and measure of quality standards should be. While a high-performing technical system is, of course, a prerequisite for scientific success, it doesn't guarantee it, since it is ultimately the users that do the science. A myriad of arrangements must be in place to provide the best possible opportunities for users to undertake successful experimental work with the instruments, not all of which are within the powers of the facility organization to guarantee. The most important are a competent scientific support staff, a flexible and service-oriented user office that schedules experimental time, the proper resources for sample preparation and handling, user-friendly and effective IT systems for storage and processing of data, regulations that facilitate import of samples, close proximity to public transportation and other communications, functional and cheap hotels/hostels on site or very close by, and a canteen or restaurant that is open to meet the needs of scientists working in shifts (Hallonsten 2016a: 183). Still, the issue of productivity and quality is highly subjective. From the perspective of any of the (to date) thirteen European countries that fund the ESS, success might very well be defined both as a fine record of outputs in general, for Europe as a whole or even for humankind in a very wide meaning, and a maximized turnout for the particular domestic scientific community. Needless to say, these are not always entirely overlapping. Furthermore, some decision makers or bureaucrats in member countries may look only at the value of procurement contracts vis-à-vis membership fee, with or without fair return on procurement, when evaluating the success of value of investment from the perspective of their particular country or organization. Other countries, or the facility's management, may consider the same balance between procurement and budget contributions a bureaucratic nuisance that locks the facility organization into suboptimal deals with vendors in specific countries.

An example of how goal conflicts of this type accompany facility projects like the ESS almost from first proposal to full-scale operation, and can create ripple effects down the road or in unforeseen areas, is the fact that the huge anticipated tax revenues from the ESS to the Swedish state (e.g. on electricity) were an important selling point in the preparatory investigations that paved the way for the government's decision in 2007 to pursue the project; but some eight years later the ESS became an ERIC and nominally VAT-exempt, and as a result the issues around taxation of the ESS remain very complicated, to this day (Hilling et al. 2017; Yu et al.

2017). Another example concerns the political demands reportedly made by German representatives in the ESS Steering Committee in 2011 for a suite of instruments that would meet the demands of the German neutron user community and, not least, suit German research institutes and high-tech companies and give them a pole position in delivering equipment to the facility; however, that went squarely against the idea of a collection of instruments at the future ESS that would meet the balanced demands of the current and future neutron user community as a whole, which had been a guiding principle for the ESS project right from the start in the 1990s. The issue, and how it was resolved, is discussed in chapter 8.

These are but some examples of how governance of a facility like the ESS is a complex matter, and how the purpose of constructing and operating a state-of-the-art neutron facility can be very differently interpreted and defined by different stakeholders. Politics (national and intergovernmental), legislation (national and supranational), technical capabilities, and scientific ambitions do not always cohere, and different scientific communities can, quite obviously, have different priorities, just as different countries can have different (research) policy agendas, and complex organizations with hundreds of employees from a number of countries and a variety of divisions and departments also have inconsistencies and conflicts of interest built-in, perhaps especially when a significant share of these employees are high-skilled workers with scientific career ambitions alongside loyalty to the employer organization. To say nothing of the fact that neutron facilities like the ESS are hugely complex technical systems with some vital parts designed and constructed at the absolute forefront of technical and scientific development in a variety of fields and areas (see previous section), another aspect which certainly causes challenges of compatibility and harmonization in a very material and technical sense.

The expectations for the socio-economic impacts of neutron facilities and the contribution they can make to innovation are great, and as will be shown in chapters 6–7, and 10, they were decisive in the Swedish ESS campaign. Although a series of "studies" and "investigations" into the future socio-economic impact of the ESS have been made, the general image of the future benefit to the economy of building a major research facility like the ESS on greenfield is very muddled and built on very loose ideas of how science and technology translates to innovation and economic growth. The end of the postwar economic boom in the 1970s, and the concurrent major growth in expenditure on Big Science, brought expectations and demands for direct economic benefits from these huge public investments in R&D,

and an "economization" (Berman 2014) of science generally, but it was also noted at an early stage of this development that the socio-economic effects of Big Science "are not as easy to prove and quantify as was first believed" (Schmied 1982: 154). Signs abound that expectations are at a reasonable level, albeit misguided and impatiently unsensitive to the inner workings of technology transfer and innovation (Salter and Martin 2001; Jacobsson and Perez Vico 2010), perhaps caught up too much in the catchphrases of the "innovation systems" and "regional economic development" theories that focus way too much on generalizable quantitative measures (Hallonsten 2016a: ch. 6).

There have been surprisingly few attempts to systematize the study of *how* large, publicly funded, hi-tech research facilities can impact the local and global economies. A classic in the area launched a taxonomy of three partly overlapping types of impact: (1) procurement; (2) knowledge and technology transfer; and (3) scientific use (Meusel 1990). The first, effects of procurement, can be easily measured, and it has been shown that a majority of procurement contracts stay in the local economy (Hallonsten 2016a: 201), which speaks in favor of expectations in this regard, that manifest themselves in difficulties in reaching agreement in negotiations over funding solutions for international collaborative projects, and the fact that host countries typically pay by far the largest share. The ESS is a major civil construction project, comparable (in this specific regard, not otherwise) with a bridge or road, which creates jobs and a general boost for local and regional lines of business. On the high-tech side, it has been shown that facilities like the ESS can have both important *spillover effects* and lead to significant learning among vendors, who are pushed to the extremes of technological development by working with the buildup of scientific facilities at the forefront of science and technology (Autio et al. 2004; Schmied 1982; Hallonsten and Christensson 2017a, 2017b). This is close to the issue of knowledge and technology transfer, which occurs not only when the use of facilities contributes to innovation (see below) but also in the developments of technology for the various parts of the facility. Generally, however, it is important to note that lead times can be very long, and innovations ready for introduction on a market may show up in a place and institutional context very remote from that where the original scientific work was done and the innovation can be traced back to. There is much evidence to suggest that the eventual economic benefit of knowledge and technology transfer from the development of instrumentation and scientific research done with this instrumentation at neutron facilities follows unpredictable and complex patterns (Hallonsten and Christensson 2017a: 153).

Which leads us to the issue of the benefit of scientific use of neutrons. The *direct* industrial use of neutrons, meaning that scientists from a corporation come to use a neutron source as part of industrial R&D, is minor. At the state-of-the-art US Spallation Neutron Source (SNS) at Oak Ridge National Lab in Tennessee, which is a direct competitor to the ESS, 16 out of 764 users, or 2%, in the fiscal year 2017 (October 1, 2016 to September 30, 2017) had affiliation in the private sector (Department of Energy 2019). Neutron facilities often boast of having a much larger presence of industrially relevant R&D in their user communities, and claim that this happens through collaborations with academic user groups who apply in the open competitive calls and therefore do not show up in statistics. While the claim is entirely plausible, it is also unclear how large this share of the use actually is. The ILL claim that between 10% and 25% of the use of their instrumentation is "industry-via-academia" experiments, "depending on how this is evaluated," but do not disclose any details on how the figure was established (ILL 2018: 94).

As will be discussed further in other chapters, the campaign to bring the ESS to Lund was very much built on the image of the ESS as a key resource – perhaps even *the* key resource – in efforts to solve grand challenges and make contributions to local and regional economic growth, and global sustainable development. Building the ESS in Lund, its proponents argued, would give a boost to existing capabilities in R&D and also add new ones, extending far beyond basic research and to the forming of hi-tech spinoff companies and scientific clusters. The concluding chapter will pick up on this thread and conduct a deeper discussion, informed by the empirical accounts in chapters 6–10, on what this meant in terms of how the campaign was run, and in terms of how the future ESS facility will harmonize with Swedish science and science policy.

European collaborative Big Science and "research infrastructures"

One of the major puzzles that form a backdrop to the chronicle and analysis of this book, and whose solution also contains several important elements necessary to understand the politics and organization of major European collaborative research facilities like the ESS, is the two-decade time lag between the initial proposal to build the ESS (1993, see chapter 5) and 2014, when a final funding solution was reached between thirteen countries and construction could start (chapter 9). This delay was not due to the inability of European scientists and science administrators to plan and design the ESS – quite the reverse; a first design was published in 1997 – but to political inability. The 1997 ESS design became the groundwork for the planning and construction of both the US and Japanese counterparts, the SNS in Oak Ridge and the J-Parc northeast of Tokyo, which opened in 2007 and 2009 respectively. There, in other words, the time from proposal to start of construction was very much shorter.

The key to understanding this lies in a brief geopolitical and historical summary. The United States has a federal government responsible for planning and funding large national projects including major research facilities and National Labs, of which there are seventeen, and where US Big Science facilities, including neutron sources, typically reside. Japan, although a somewhat smaller country with a somewhat smaller economy, is a scientific powerhouse and has a similar domestic system of large labs and institutes that host, among other things, Big Science. Europe, on the other hand, is still a continent of several smaller countries, and although there has been significant political integration in the past sixty to seventy years, Europe is far from a federation and only has common policy and regulation frameworks in a few specifically designated areas. Science is not one of them, and never has been. When the political integration of Western Europe began in the 1950s, under heavy influence from the United States, nuclear physics and atomic energy were key areas of collaboration, but they were kept separate from the mainstream political integration process of what would become the European Union. CERN, the European Organization for Nuclear Research (originally the European Council for Nuclear Research, *Conseil Européen pour la Recherche Nucléaire*, hence the acronym) was founded in Geneva in 1954, primarily on diplomatic grounds; the purpose was to achieve Western European unity, work together to build up the continent after the devastating war, secure the influence of the United States over Western European science, education, and culture, and, not least, establish a means of subordinating Western Germany under joint control of the other European countries (Krige 2003, 2006; Hermann et al. 1987). Roughly a decade after the founding of CERN, several Western European countries launched the idea of a collaborative ground-based astronomy observatory, which was eventually built in the Andes (the European Southern Observatory, ESO), and partly modelled on the CERN collaboration. When, in 1963, the French Republic and the Federal Republic of Germany, the two key players in Western European politics, embarked on an intensified collaboration with the signing of the Elysée Treaty, this set a new pace for the political integration process, and also established a framework for scientific collaboration within which several large and small projects were launched in the following decades (Cramer 2020). The ILL was the first result of this scientific branch of the "Motor of Europe" that drove the continued political integration of (Western) Europe and eventually produced the Single European Act of 1985, the Maastricht Treaty of 1992, and the adoption of the Euro as common currency at the turn of the millennium (Middlemas 1995). Following the ILL came the European Science Foundation (ESF) (in 1972), the European Molecular Biology Laboratory (EMBL) (in 1973), the European Space Agency (ESA) (in 1975), the Joint European Torus fusion research facility (in 1977), and the ESRF synchrotron radiation facility (in 1984), all in crucial ways pushed through by joint French-German initiatives (Krige 2003: 899). All of these, plus CERN, ESO, and ILL before them and the XFEL and ESS after them, have been founded through agreements completely outside of the EC/EU framework, on an ad hoc basis, and often become reality only through very complicated negotiations where rules are invented along the way. The ESS is a prime example, which will, of course, be returned to in more detail in chapters 5, 8, and 9.

The precedent set for the organizing of European intergovernmental collaboration in such large-scale projects was in a sense *no precedent*. While the (Western) European countries were forced to collaborate in order to remain globally competitive in, for example, particle physics, ground-based astronomy, and eventually synchrotron radiation and neutron scattering, the organization of such collaboration has had to rely on ad hoc solutions and reinvention of legal arrangements and organizational structures for each new project (Hallonsten 2014; Krige 2003). Therefore, existing European Big Science projects have been realized through muddled and opaque political processes of negotiations and horse-trading on several levels, typically ending with the signing of intergovernmental agreements stipulating how the project in question is to be funded, organized and run. The varieties between projects are therefore broad. New shapes and forms have emerged with almost every new collaboration (Papon 2004; Cramer et al. 2020). It has been argued that this way of organizing intergovernmental scientific collaboration in Europe outside of the mainstream EC/EU integration framework has protected the scientific organizations in question against inert EU bureaucracy and wasteful spending (Hoerber 2009: 410;

Gaubert and Lebeau 2009: 38; Hallonsten 2014: 35; Papon 2004), and, in fact, contributed to the scientific successes of the collaborations. This argument certainly seems to have some validity for the ILL (see a previous section) and the ESRF (Hallonsten 2013a). This should be of some comfort for the proponents of the ESS, whose tiresome work of bringing the project from idea to reality perhaps will pay off in a future (retained) global leadership in the area of neutrons. Meanwhile, the collaborations have certainly suffered, not least from the delays and some peculiarities regarding how they are organized that stem from the negotiations and the tendency of European governments to view international scientific collaboration as "the pursuit of national self-interest by other means" (Krige 2003: 914).

"Megascience," which emerged in the 1960s and 1970s, is the root of many of these problems. The two key meanings of this concept were explicated, and their connection established, in chapter 2: First, developments at the forefront of the field of particle physics led to a dramatic increase in the length of experiments, and second, as a consequence, the resources needed to maintain and upgrade particle physics experiments and instrumentation soared and necessitated the pooling of resources on a continent-spanning level, and the concentration of investments to a few places. In Europe, the key particle physics center was CERN in Geneva, an enormously prestigious and diplomatically important project (Krige and Pestre 1987), which was allowed to monopolize the particle physics budgets of its smaller member countries in particular, which led to the closing of much domestic Big Science in these countries. In the wake of the economic downturn, and with memories of the CERN experience, expensive scientific collaborations on a European level became potential competitors to national investments, and so European governments began to protect their national interests and put return for investment in collaborative Big Science above the common good. In practice, this led to significantly stiffer competition for hosting collaborative Big Science facilities, and a default interlacing of the issue of location with the issue of how to share investment costs, because of the generally held (and largely valid) belief that the host country of a collaborative facility reaps great scientific and socio-economic benefits. This was especially obvious in the case of the ESS, since the unspoken starting point for the negotiations over the ESS in 2008–2009 seems to have been that the site contender had to pledge at least 50% of the construction costs to even have a chance in the competition (see chapters 7 and 9). The stiffer competition and increased protection of national self-interest in European scientific collaboration have also led to extensive use of the *fair return* principle (see a previous section) on both procurement of goods and services, and scientific use (especially in the ILL and ESRF), and *in-kind* contributions to the funding of the facilities (especially in the XFEL and ESS).

As noted above, for several decades, the mainstream European political integration process that eventually produced the European Union (EU) did not include collaboration in science and technology, except for the EURATOM agreements (Guzzetti 1995: 1–34). But in the 1970s, the communities made efforts in science policy in order to resurrect European economic competitiveness and revitalize its economies (Middlemas 1995: 112ff; Tindemans 2009: 15–16; Guzzetti 1995: 35–61), with the Framework Programmes for Research and Technological Development as one key mechanism (Middlemas 1995: 251; Papon 2004: 69). The second Framework Programme (FP2), of 1987–1991, entailed earmarked funds (30 million Euro) for facilitating access to experimental resources, databases, and other scientific tools for the entire scientific community of the EC, and in the fifth Framework Programme (1999–2002), this fund was increased to 184 million Euro and named "access to research infrastructures" (Papon 2004: 70).

Since the late 1990s, the European Union and its member states have continuously intensified efforts to stimulate the launch and operation of "research infrastructures," a broad and varied category of resources for scientific research identified as "central" for the competitiveness of the European knowledge-based economy, but a label and concept that is mainly politically motivated and surrounded by political hype (see below and Hallonsten 2020a). The intensified political interest in "research infrastructures" should be seen in light of the innovation policy offensive of the European Union in the early 2000s, with the launch of the European Research Area (ERA) (Borrás 2003; Edler et al. 2003; Delanghe et al. 2009) and renewed attention to the elements of the "innovation system," which includes research infrastructures (Hallonsten 2020a; Chou 2014; Ryan 2015). In 2002, as part of these developments, the European Commission took the initiative to establish an independent advisory body, the European Strategy Forum on Research Infrastructures (ESFRI). As it happened (chapter 7), ESFRI was given a leading role with regard to the ESS project in the low-intensity years between the 2002 Bonn meeting and the 2007 decision of the Swedish Government to support the ESS Scandinavia proposal. The ESFRI is a *forum*, formally independent of the European Union and instructed to facilitate "informal consultations on strategic issues related to Research Infrastructures" (ESFRI 2006: 7). In its first years, ESFRI set up a number of expert groups in a variety of areas, essentially composed of experts in academia and the private sector who could coordinate and prioritize, in peer review-like processes, among research infrastructure projects proposed or launched, to achieve a more comprehensive overview for European politicians to relate to in policy and decision making (Bolliger and Griffiths 2020).

In September 2006, the first ESFRI *roadmap* was published, at the request of the EU Competitiveness Council (the recurrent official gathering of all the ministers for science, and similar, of all EU countries). The list of 35 projects on the roadmap had been put together in a process involving "almost 1,000 high-level experts from all fields of research," who collectively viewed the projects as scientifically strong and of great potential (ESFRI 2006: 10, 15). The 35 projects were very dissimilar and, not least, they varied greatly in size – estimated construction costs ranged from 9 million Euro (the European Social Survey) to 1,186 million Euro (Facility for Antiproton and Ion Research, FAIR) with the ESS the third most expensive (ESFRI 2006: 8). The roadmap emphasized that it was "not a priority list" and that its key aim was "to facilitate discussion and allow for coherent planning." ESFRI chair John Wood commented further on the role of ESFRI and the roadmap, underscoring that while the forum had a coordinating and facilitating role, all political initiative still lay with the member countries (ESFRI 2006: 7). In subsequent updates of 2008, 2010, 2016, and 2018, the ESFRI roadmap has come to include several new projects, and some, including the ESS, have been taken off the list as they have entered "implementation phase."

ESFRI also has the important function of collecting the opinions of EU member states regarding joint European research infrastructure, and in 2007–2008, it formulated an appeal to the European Commission to make an oversight of the legal frameworks in the area in order to facilitate smoother processes to establish collaborative organizations. In July 2008, in response, the Commission proposed a new legal-organizational form, the European Research Infrastructure Consortium (ERIC), which was affirmed by the European Council in a Council regulation one year later (European Council 2009) and then came into force. The first research infrastructures granted ERIC status were databases for health (2011) and linguistics (2012), and in 2015, the ESS organization transitioned into an ERIC (chapter 11). By the end of 2018, twenty ERICs were in place, with one application pending. ERIC is neither a form of company, an association, or a public authority, but an entirely new organizational form. Its legal, political, and organizational uniqueness has been the subject of some research (Hilling et al. 2017; Duclos Lindstrøm and Kropp 2017; Reichel et al. 2014; Ryan 2015; Yu et al. 2017), but it is too early to evaluate if the purpose of the ERIC regulation – to facilitate the process of launching new European research infrastructures – has been fulfilled. An interesting detail that has specific relevance to the ESS is the tax reliefs of the ERIC regulation, including full VAT exemption, that undo several of the alleged advantages for the Swedish economy of hosting the ESS, anticipated by Allan Larsson in his 2005 investigation, which were an important basis for the decision by the Swedish Government to seek to host the facility (chapters 6–7).

The ERIC regulation (European Council 2009) rests legally upon articles 187 and 179 of the Treaty of the Functioning of the European Union (amended to its current form by the 2007 Lisbon Treaty) which give the Union the right to "set up joint undertakings or any other structure necessary for the efficient execution of Union research, technological development and demonstration programmes" (European Union 2012: 131), and to "encourage undertakings" that aim for "the removal of legal and fiscal obstacles" to strengthening the scientific and technological bases of the Union (European Union 2012: 128). The ERIC regulation stands out as an EU policy instrument because its use by member states as a means to facilitate collaboration and joint operation of research infrastructures is conditional upon authorization by the European Commission, which reviews and approves applications for ERIC status, and the Commission's continued monitoring to ensure that the research infrastructure in question operates in accordance with the ERIC regulation (Moskovko et al. 2019; Moskovko 2020). What also makes the ERIC organizational form conspicuous is that it is indeed novel and legally/administratively unprecedented; it is not a form of corporation, nor a part of the EU, or an international treaty organization, but a consortium of member states with the EU as an observer and facilitator (Reichel et al. 2014: 1056). The ERIC regulation establishes a framework that can be used by two or more EU member states that are in the process of establishing a cross-border R&D activity that can qualify as a "research infrastructure." It gives them the opportunity to apply to the European Commission to have their activity (the research infrastructure-designate) established as a legal entity with full legal personality in all EU countries, under the legal name ERIC. Non-EU countries may participate as associate members or observers (Moskovko et al. 2019). The regulation also contains certain mandatory features for all ERICs, and requires all applications for ERIC status to the Commission to be accompanied by *statutes* that, hence, follow a certain template but that are also shaped in accordance with the specificities of the research infrastructure being set up. The purposes of *stimulating* the creation of cross-border "research infrastructures" in Europe and *simplifying* the process whereby this can be done (European Commission 2008), has therefore been fulfilled in a nominal sense, because the ERIC regulation is flexible and easy to use, and new ERICs can be set up relatively quickly (Moskovko et al. 2019).

To conclude, the European Union and many governmental agencies in its member countries have acted decisively in the past two decades to bring order to the policy area of "research infrastructures" in Europe and contribute to the solving of many of the historical challenges to European scientific collaboration; however, these efforts have not necessarily been built on a comprehensive or clear understanding of what "research infrastructures" really are (Hallonsten 2020a) or what types of political solutions are really needed in the area. In particular, the lack of *differentiation* between projects, in terms of their scientific area and size, and also with respect to how they are typically planned and organized on the basis of best practices and tradition in various scientific communities and governmental bureaucracies, is seemingly severe. This knowledge deficit is by no means unusual in politics, and the use of buzzwords and catchphrases significantly more shallow or hollow than "research infrastructures" is commonplace in contemporary political discourse and in management and governance in the public and private sectors alike (Alvesson 2013). In the context of this book, however, it is important to establish that many of the key concepts and categories used by policymakers and campaigners for the ESS and similar projects elsewhere have a limited basis in scholarly work, or, indeed, reality.

4. Swedish research policy and Swedish Big Science

Swedish research policy and the small state dilemma

Like most European countries, Sweden had a research system in the making before World War II, with universities, science academies, innovation-based industrial firms, and a political awareness that science and technology plays an important role in societal development. But World War II, its dramatic end, and the immediate postwar period prompted a buildup of institutions and organizations for science and technology, with regard to research as well as education, and a concerted effort to make science and technological development central to the modernization of society. Spared from the destruction of World War II, Sweden could continue on the path of a remarkable economic growth that took the country from a position as one of the world's poorest in 1870 to one of the world's richest a century later (Schön 2007: 12–15; Bergh 2015: 11). Swedish enterprise had flourished before the World Wars, and built strong industrial capacity on the basis of world-leading innovations, but the Swedish Government had viewed expenditure on research as a liability rather than an investment with long-term growth potential (Stevrin 1978: 81). Before the 1940s, there was hardly any research policy in Sweden at all (Premfors 1986: 11–12), but this changed dramatically after the end of the war. John Desmond Bernal's (1939) appeal to make science into a force of good in broader society, and the influential US policy report *Science, the Endless Frontier* (Bush 1945) that advocated a strong role for the state as funder of science and technological development, became important ideological foundations for the policy changes in many countries, including Sweden (Elzinga 1993, 1997). The examples set by the superpowers in spending and institutional buildup, both within their own borders and in Europe (Graham 1992;

Krige 2006; Westwick 2003), also played a role in setting a new global standard.

The Swedish public science system that developed in the first two post-war decades was characterized largely by expansion within existing structures. The universities expanded as student cohorts grew, and a number of research councils were formed, but little comprehensive reform was undertaken until the 1960s. Industry had demanded a strengthening of the higher education system, especially the engineering schools, and the Government had launched research councils to channel public investment to science, but the old universities with their traditional organization of faculties and chairholder professors maintained influence over the direction and content of publicly funded R&D and over higher education (Premfors 1986: 13). A few semi-public research institutes were founded but were small, branch-specific, and under the control of industry associations (Hallonsten 2018). A governmental investigation chaired by politician and public servant Gösta Malm in 1942–1946 became highly influential, suggesting, among other things, the launch of six area-specific research councils, but most importantly because it underscored the key role of higher education for the critical supply of relevant competence to Swedish industry, and affirmed the alliance between research and education in the universities and engineering schools (Pettersson 2012), in line with the German Humboldt tradition (Östling 2018).

A major governmental investigation of the organization of the university system, undertaken in the late 1950s recommended further expansion and a preserved alliance between education and research, which meant that in the 1960s and 1970s, most of the growth of publicly funded R&D continued to be channeled to the universities (Hallonsten 2018: 638). The "Active Enterprise Policy" of the 1960s and on, and the reforms to the higher education system, were key pieces in the furthering of the modernization of Swedish society through reformist socialization, and solidified the binary structure of the Swedish innovation system: A relatively small number of large industrial firms dominated private sector R&D and were supported by long-term innovation procurement contracts from state monopolies in the energy, telecommunications, transport and, of course, defense sectors (Fridlund 1999). A likewise small number of universities, eventually complemented by regional colleges, all state-run, organized practically all higher education and almost all publicly funded R&D (Holmberg and Hallonsten 2015). The major university reform of 1977, delayed almost a decade due to controversy over its contents and a parliamentary gridlock in 1973–1976, did not change the basic bottom-up governance model of Swedish universities, with its disciplinary compartmentalization and professorial rule matched by line-item funding in the annual governmental budget, which meant practically no room for maneuver or internal prioritization at university or faculty level. The Government's unwillingness to take initiatives in research policy persisted at least until the 1990s, and there was therefore largely an absence of priorities and strategic mobilization in specific areas until the 'new new' research policy (see a later section).

The identification of a "lack of aggregation mechanisms" in the Swedish public R&D system (Benner 2008; Hallonsten 2011) is therefore fitting; it is also possible to speak of particularly strong *path dependence* in the decentralized and democratic Swedish public research system dominated by the universities. The research councils and the agencies in charge of technical development and innovation were relatively weak, and mainly reproduced existing patterns in their programs and funding schemes (Premfors 1986; Benner 2008). One documented exception is the work of the Board of Technical Development (*Styrelsen för Teknisk Utveckling, STU*) in the 1970s and 1980s, which launched sector-specific R&D programs that played an important role in the renewal of some areas of Swedish science, for example the buildup of materials science in Swedish universities in the 1960s and on, which was a purposeful collaborative effort between professors, research councils, and the STU (Gribbe and Hallonsten 2017), and an exception to the relative weakness of public R&D agencies and funders. MAX-lab, the predecessor of MAX IV which was built up gradually over thirty years, was a success story of the system as a whole and is the exception that confirms the rule. But both in the case of materials science generally, and MAX-lab specifically, initiative came from below, and success was achieved mainly through the ingenuity and persistence of individuals, and in no small part *in spite of* rather than *because of* institutional arrangements (Hallonsten and Christensson 2017a; Gribbe and Hallonsten 2017). Resource increases intended to revitalize the system were largely undone by economic downturn and retrenchment, especially in the 1990s (Benner 2001: 162–165).

In 2001, the Government created the Swedish Research Council (*Vetenskapsrådet, VR*) by merging the previous area-specific councils, and the Council essentially maintained the tasks and governance structures of these. A perceived need for coordination and governance of national research facilities and costly instrumentation led to the creation of a Committee for Research Infrastructures (*Kommittén för Forskningens Infrastrukturer, KFI*) within the Swedish Research Council, to coordinate matters of research infrastructure (see next section). In 2001, the Swedish Agency for Innovation Systems (*Verket för Innovationssystem, Vinnova*) was also created to take over the role of supporting research and innovation in universities and the private sector. Despite overlapping slightly with the Swedish Research Council in its capacity of research council-like funder of research in projects and programs at the universities, Vinnova's mandate was to support and trim the "innovation system," a mission statement that has been accused of being ideologically tainted and too vaguely connected to the known best practices of supporting science and technological development for social progress and economic growth (see a later section). Vinnova is, furthermore, an agency under the Ministry of Enterprise, while the universities and research councils are all the responsibility of the Ministry of Education, which creates some dissonance, and the areas are also handled separately in the legislative process. The Government submits a quadrennial research bill to Parliament for approval, where it draws up the main priorities of its research policy and makes legislative reforms, but allocation of funding to universities and research councils is made in the annual budget bills. Although the two research bills introduced by the 2006–2014 center-right coalition government (of four political parties) were called "research and innovation bills," the policies launched in these bills concerned almost exclusively the academic sector and other research activities traditionally in the same realm (international collaborations, national infrastructure). These two research bills introduced the largest ever increase in the public funding of research in Sweden (see a later section), but were also criticized for maintaining too narrow a focus on academic excellence. In contrast, the Swedish National Innovation Strategy, issued in 2012 by the Ministry of Enterprise, outlined broad objectives for the Swedish innovation system and pointed out a number of general goals, including strengthening the national innovative capacity to keep up with international competition, and improvements for small and medium sized enterprises, but launched no new concrete reforms or funding initiatives (Swedish Government 2012a).

Even after the major increases in resources and the reforms to the governance structure outlined above, the Swedish research policy and funding system remains pluralist and, compared to other countries, lacking in "aggregation mechanisms." Of the 27 universities and colleges (6 full-breadth universities, one medical university, three technical universities, one agricultural university, four new universities formed by upgrades of previous regional colleges, twelve non-university regional colleges), all but two are publicly owned and organized as governmental agencies. There is a persistent inequality among them that dates back to the time when the regional colleges were founded: 88% of the direct governmental allocations for research goes to the eleven older universities, with the sixteen new universities and colleges sharing only the remaining 12% (Holmberg and Hallonsten 2015), an imbalance that also shows in their abilities to attract external funding, and other measures of research strength and volume. Until the 2010 "autonomy reform," when the universities were formally put in charge of their own funding (Swedish Government 2010a), little redistribution was made, and money was essentially channeled to areas of research on the basis of tradition. But even after the "autonomy reform" the patterns are largely intact.

The Swedish Research Council and Vinnova are the largest funders of research in Sweden, together allocating approximately SEK 9 billion, or approximately 35%, of the governmental appropriations for R&D (2018) (Swedish Government 2017). In addition to these, several smaller government funding agencies and public research foundations exist. In 1992–1994, the Government made a comprehensive effort to dismantle the Wage Earners' Funds (*Löntagarfonderna*) which had been created in the early 1980s as a policy to begin the socialization of Swedish industry through an additional tax whose revenue was placed in these funds, and decided to use the accumulated capital to create several public research foundations, infusing the research funding system with SEK 20 billion (in 1994 prices), distributed among ten foundations (Holmberg 2012: 47–54). Private capital also plays an important role in Swedish research funding, in particular the various Wallenberg foundations of donations from generations of industrial leaders in the powerful Wallenberg family, among which the Knut and Alice Wallenberg Foundation (KAW) stands out as the largest, and moreover is one of the main contributors to the MAX IV facility (see chapter 7).

Small countries will always face a dilemma in research policy which has to do with a natural absence of institutional critical mass and economies of scale: Should a country of Sweden's size work to spread its resources evenly over a broad spectrum of research areas, so that it can develop the necessary capacity to absorb results and advances from abroad, or should it specialize to increase international visibility and develop niches where domestic research capacities can excel? The former has, historically, not been a viable solution for Sweden, whose geopolitical position as neutral in the Cold War has necessitated a doctrine of self-sufficiency. Yet the alternative, which requires strategic resource prioritization and thus some crowding out of other areas, was also not pursued in active policy until the turn of the millennium (see below). Traditionally, and to some extent still, Swedish research policy has utilized a mix of the strategies, with a wide distribution of resources and a wide delegation of policy initiative; in other words, a pluralistic and decentralized system. Many different policy agents partake in both the formulation and implementation of research policy, but coordination between them is rather weak. With no governmental Ministry of Science, but instead the colocation of research policy with education policy (including the full range from preschool to graduate studies) in the Ministry of Education, and the location of industrial and innovation policy in the Ministry of Enterprise, initiative and authority have historically been delegated to councils and universities, not by any design but by default. Nonetheless, Sweden has managed to develop strong capacity in some highly competitive fields such as materials science (Gribbe and Hallonsten 2017) and the life sciences (Benner 2008: 169ff), although there is much evidence to suggest in both cases that the successes have been achieved in spite of institutional setups rather than because of them. A similar lesson can be learned from the history of MAX-lab (see below). Interestingly, Swedish science has also developed strong capacities in many areas because of its international orientation, and not least Sweden's consistent policy of taking part in international collaborations, especially European.

Sweden and Big Science

Sweden has historically been a reliable but minor partner in European scientific collaborations, typically taking an active part in early planning phases and eventually entering collaborations with a budget contribution of a small percentage of the total. Swedish scientists are also generally good at making use of the resources and facilities provided by the collaborations, thus often making Sweden a net beneficiary (Hallonsten 2009: 245). But in domestic Swedish research policy, participation in international scientific collaborations has not always been an uncontroversial issue, and the default model for the funding of membership fees sometimes leads to disputes and unrest in the scientific communities and their representation in academies and research councils. The model, which has never been articulated or codified but rather become default by historical precedent (see below), essentially puts the responsibility for deciding on large investments in the hands of the scientific community, and the Swedish funding of Big Science is typically an issue of resource prioritization within existing budgetary frameworks, with direct investment by the Government on top of existing appropriations an exception rather than a rule. Interestingly, however, decision making processes have not normally been spared from political influence; quite the reverse, leading Swedish politicians have occasionally put strong pressure on the scientific community to act in certain ways.

The key example is the CERN debate of the 1960s and 1970s, in the wake of the dramatic increase in the CERN investments and operating costs due to the development of the *Megascience* developments discussed in chapter 3. This increase happened largely due to a major upgrade program – CERN II – proposed to enable CERN to keep its globally competitive position in particle physics. Between 1964, when the upgrade was first suggested, and 1976, when the new machine was commissioned, the annual CERN budget increased with a factor of six (CERN 1965, 1977). In Sweden, as in most member countries, CERN membership had been a generally uncontroversial political issue in its first decade of existence, and Sweden's annual contributions to the CERN budget (oscillating between 4% and 5% in 1955–1965) had been on a relatively low level. The upgrade was not going to change the relative sizes of the budgetary contributions of the member countries, but it would unavoidably multiply the overall annual budget and thus, in real terms, significantly increase the Swedish financial commitment (as it turned out, by a factor of five between 1965 and 1975). As Widmalm (1993: 123–126) has shown, this increased financial commitment to CERN meant harsh internal reprioritization for Sweden: Not only were national programs in high energy physics swiftly shut down, but other fields also suffered greatly, including all subfields of physics and many other fields that could have benefited from continuous operation and upgrading of those accelerator labs that were either shut down or saw their funding levels frozen from the late 1960s and on. The connection between a significantly ramped-up CERN contribution and cutbacks in other areas is not only seen in the data compiled by Widmalm (1993: 123–126) but in fact also made explicit in those press releases issued by the Swedish funding bodies responsible at the time of the decision to back the CERN upgrade: the research council responsible, the Swedish Research Council for Atomic Affairs (*Atomforskningsrådet, AFR*) stated in its press release on February 2, 1971 that the size of its commitment to CERN was "near the limit to what a research council can stand" (quoted in Widmalm 1993: 120).

This unforgiving internal reprioritization of resources was, in fact, not only the consequence of the CERN upgrade itself, but also part of the logic by which the decision was made to remain a member of the costlier collaboration. As noted in the previous section, the postwar growth of the science system had occurred predominantly within the frameworks of existing institutions, which led to science policy becoming strongly decentralized and governmental intervention in the affairs of the universities focusing mainly on accomplishing a vast expansion of higher education – the Government seems to have been largely unwilling to engage directly in matters of research policy (Premfors 1986: 15–41). With the CERN II upgrade, this unwillingness was made explicit, but with a twist. Prime Minister Olof Palme referred the decision to join or not to join CERN II back to the scientific community, more specifically the AFR, and forced it to fund membership in CERN II within existing frameworks; but he simultaneously linked CERN II to Sweden's international relations and the concurrent decision by the Swedish Government not to join the European Economic Community (EEC), a decision that had aroused some diplomatic resentment in mainland Europe, which the Prime Minister hoped could be soothed by Sweden backing CERN II. Palme himself complained about what he saw as an inability in the scientific communities to take responsibility and take tough decisions (something which politicians were used to doing "every week") and called the issue of joining or not joining CERN II a "wholesome exercise in determining research priorities" that Swedish scientists themselves had to "take full responsibility for." Meanwhile, "[t]here can be little doubt that the foreign policy situation influenced the Swedish Government's attitude to CERN II" (Widmalm 1993: 121–122). As the decision came closer, several influential academics spoke out, claiming that the project should not be a priority on purely scientific grounds, although politically, it might be a good idea to join. Within the AFR, views were split and the decision was finally taken at an "inflamed meeting" where it was established that "the costs could indeed be met 'within the financial limits laid down for higher education and research'," and simultaneously noted that international political pressure was "one of the most important considerations" that lay behind their eventual decision to join (Widmalm 1993: 117–118).

This decision-making procedure for CERN II set a precedent for Swedish policy regarding Big Science. In all subsequent decisions on whether or not to join European scientific collaborations, decisions have been made at the level of the funding agency that would eventually be responsible for putting up the money, although formal decisions have been taken by the Swedish Parliament. The only comprehensive investigation of Swedish memberships in international scientific collaborations, evaluating these commitments and comparing them in terms of their benefit to Swedish science, was undertaken in 1996–97 on the instructions of the Government, which was searching for a means of cutting its science budget in times of fiscal austerity (Edqvist 2009: 47). The instruction to the investigation was to find a way to save SEK 150 million annually through cutbacks on Swedish memberships in international scientific collaborations, which corresponded exactly with the size of the Swedish membership in CERN at the time. The investigation, entitled "Savings small and large" ("*Besparingar i stort och smått*"), consequently proposed three alternatives: To exit the CERN collaboration, to exit all other collaborations funded in the same appropriations bill (seven collaborations, together costing Sweden a sum of roughly SEK 150 million), or to make minor cuts across the board. The investigation dismissed the first two alternatives on scientific grounds and with reference to research policy considerations, arguing that the collaborations made valuable experimental resources available to Swedish scientists and, interestingly, that a cancellation of the Swedish membership in any of the organizations would harm the image of Sweden as a reliable collaborative partner in international scientific collaborations and thus hurt Sweden diplomatically (Swedish Government 1997b). Meanwhile, as a consequence of the debate, the Government changed the procedure for the allocation of funding to international scientific collaborations, and thus strengthened and codified the existing practice: From 1998 and on, instead of making up line items in each year's governmental budget, funding for memberships in international scientific collaborations would be allocated to the research councils concerned, who would be responsible for making priorities and evaluating the costs and benefits of participation in each international organization (Swedish Government 1997a: 171–172). Ever since, concerning Swedish memberships in international scientific collaborations, line items in the Government instruction to the council contain no numbers but merely instructions to pay, and some expenditures in this column (e.g. the ILL, where the council itself has made the agreement) are not mentioned at all in the instruction but simply handled in the internal budgetary work of the council (Interview: Holmberg).

The Swedish Research Council, since 2001 the only research council in Sweden with responsibility for international collaborations in Big Science, allocates several hundred million Euro each year to international research organizations, among which CERN is by far the most costly, with a Swedish contribution of SEK 286.9 million (approximately 29.8 million Euro) in 2017. The neutron facilities ILL in Grenoble and ISIS (no acronym) in the UK, and the European Synchrotron Radiation Facility (ESRF) in Grenoble, each cost Sweden in the range of SEK 20–25 million (approximately 2.1–2.8 million Euro) per year. A number of collaborations in space science and technology, nuclear physics, molecular biology, and ground-based astronomy received Swedish funding in 2017 on a lower level (a few million Euro or less each) (Swedish Research Council 2018a: 90, 139). Most of these collaborations are regulated in intergovernmental conventions, sometimes signed at ministerial level and sometimes directly by the Council; some are consequently funded by the Council on specific instruction from the Government, whereas the Council itself has decided to participate in, and contribute funding to, others.

Before the MAX IV initiative and the bid to host the ESS, it is fair to say that there had been no Swedish domestic Big Science. Certainly, Sweden had operated a number of national research facilities (and one of them, not counting MAX-lab/MAX IV, is still running), but they have been of minor size compared to those huge accelerator, reactor, and telescope facilities on the international scene that are normally thought of as Big Science, and of which Sweden is a member. At the point of its most intensive operation, MAX-lab, the Swedish national synchrotron radiation facility and the predecessor of MAX IV, ran three storage rings (of which the largest, MAX II, was 90 meters in circumference) and served some thousand users annually (Hallonsten and Christensson 2017a), but it was built in a series of small steps rather than through clear and discontinuous decision making, which makes it very different from its international contemporaries/competitors and also significantly cheaper than most comparable facilities abroad (Hallonsten 2011: 199). Another Big Science facility in Sweden was the Studsvik Nuclear Reactor Facility some 50 km south of Stockholm, which was originally one of the sites of the Swedish nuclear energy development company Atomic Energy Ltd. (*AB Atomenergi*), founded in 1947, and where a series of research reactors were operated from the late 1940s and on (Fjæstad 2010: 74ff). The last of these, the R2, built in 1958–1960, was equipped with some neutron scattering instrumentation, and a Neutron Research Laboratory (*Neutronforskningslaboratoriet, NFL*) was created in the 1970s to host and facilitate experiments. Originally a part of the Natural Science Research Laboratory (*Naturvetenskapliga Forskningslaboratoriet*) under the Natural Sciences Research Council, it was taken over by Uppsala University in 1986, receiving good evaluations in two reviews in the 1990s (Lindgren 1992: 46; Swedish Natural Sciences Research Council 1997b). At the time of the latter review, in 1997, the laboratory had four instruments in operation, and around a hundred annual Swedish neutron users, more than half of whom were from Uppsala University (Swedish Natural Sciences Research Council 1997b: 9, 12–15). In 2003, however, Studsvik Nuclear AB decided to close the reactor, mainly due to its age, which had driven up operating costs (Swedish Research Council 2004: 28–29). To compensate for the loss of a domestic neutron source, the Swedish Research Council decided, in 2005, to join the ILL in order to secure neutron supply for Swedish scientists (see chapter 6). In 2017, a total of 44 Swedish scientists made use of 4.64% of the available experimental time at the ILL, which corresponds to some 60 days (ILL 2018: 94–96).

The original MAX-lab, opened to users in 1986, was a small-scale university project that was built over a period of almost ten years, on the basis of a series of smaller grants and with a great deal of ingenuity and improvisation (Hallonsten 2011). The low budget and very unstable operations in the first years did not prevent scientific output of some dignity (Hallonsten and Christensson 2017a: 114–115), and in the early 1990s, there was a major expansion with the construction of the MAX II ring. Similar to its predecessor, MAX II was built on the basis of several different grants from several different sources; in spite of favorable reviews by international experts, and support from Lund University and the Natural Sciences Research Council, neither the Government nor any of its agencies were willing to take comprehensive responsibility for the project. The MAX II funding issue was instead resolved in a "symptomatically Swedish" way (Hallonsten 2011: 196): Five funders shared the bill for construction and operation of the lab, with constructions costs totaling SEK 172 million (spread over seven grants issued by several funders in the years 1990–1995) and operations costs in the year of MAX II inauguration (1997) totaling SEK 13.85 million and gradually rising to a level of SEK 48.9 million annually some ten years later, which still did not cover the full operating budget of the lab since Lund University contributed significantly by paying for "conventional facilities and utilities" including water and electricity and keeping several MAX-lab staff on their payroll (Hallonsten 2011: 199; Swedish Natural Sciences Research Council 1997a: 3; Swedish Research Council 2009: 97; Swedish Research Council 2010b: 19). Only SEK 62 million of the MAX II investment came directly from the Government (and covered the building that housed the lab). MAX II opened to users in 1997 and grew to become by far the most important Swedish research facility, with an annual budget of SEK 91.6 million (2013) and close to 1,000 users annually, mainly from Sweden but also from the other Nordic countries as well as the rest of the world (Swedish Research Council 2014: 62; Hallonsten and Christensson 2017a: 97–99). In 2007, MAX-lab opened its third storage ring, the MAX III, a prototype for much of the technology later implemented in the MAX IV design (Hallonsten and Christensson 2017a: 107).

Until 2005, MAX-lab was one among four Swedish national facilities receiving recurrent annual funding from the Swedish Research Council for their operation, the others being the Manne Siegbahn Institute (MSI), the Onsala Space Observatory (OSO) and The Svedberg Lab (TSL). In a 2002 evaluation of the four facilities by international expert panels, the judgment was that in order to make the most out of the supposedly limited funding envelope for the national facilities, funding should be concentrated to MAX-lab and OSO at the expense of the other two, and that a National Accelerator Physics Program should be created. In 2003 a decision was made to phase out council support to MSI and TSL, and both MAX-lab and OSO received subsequent funding increases to enable their expansion, due in part to increased user demand but most of all the need to fill some rather urgent holes in their budgets that the 2002 evaluation had identified (Swedish Research Council 2002: 15). The annual council funding allocation to MAX-lab increased every year after that, from SEK 17 million in 2002 to over five times as much (SEK 91.6 million) in 2013 (Swedish Research Council 2002: 9; 2014: 62).

In recent years, some efforts have been made to increase coherence and transparency of Swedish decision-making structures in relation to international and domestic Big Science. In 2005, a Committee for Research Infrastructures (*Kommittén för Forskningens Infrastrukturer, KFI*) was created within the Swedish Research Council to coordinate matters of infrastructure and advise decision making, primarily through the publication of the *Swedish Research Council's Guide to the Infrastructure*, a national roadmap (cf. the ESFRI roadmap, chapter 3) that emerged in a first version in 2006 and has since been updated with some regularity. The guide, or roadmap, lays out a national strategy for investments in "research infrastructure" on the basis of the weighed opinions of the Council and its expert groups, but suffers from similar shortcomings to its European counterpart issued by ESFRI; "research infrastructure" is a very wide concept, and the roadmap consequently contains a broad array of projects of different types and sizes. The Committee for Research Infrastructures was converted into a sub-council of its own in 2010, the Council for Research Infrastructures (*Rådet för Forskningens Infrastrukturer, RFI*) and has managed to increase coherence and coordination in the area. It has established its own procedures for allocation of funding to "research infrastructures" in Sweden of sizes considerably smaller than the ESS and MAX IV, through open calls.

On the basis of the general Swedish research policy history in the previous section, and the examples of CERN II and MAX-lab accounted for above, it can be concluded that Sweden has historically been almost like a mini-cosmos of the European (lack of) structure for prioritizing, funding, and organizing major research facility projects, with no specific mechanism and a history of shuffling responsibility around, or sharing it, between governmental agencies. Sweden, a small country with a history as a reliable partner in European scientific collaborations, has not developed many domestic research facilities of considerable size; in fact, MAX-lab was the largest and most internationally visible long before the proposal to build MAX IV, which is itself a major leap from the old MAX-lab in Lund. Nonetheless, there has been no central initiative for the planning and coordination of large scientific facilities in Sweden of the kind typically found in other countries.

Against this background, the proposal by MAX-lab to undertake the major MAX IV upgrade of the lab at an initial cost estimation of several billion SEK, and the February 2007 announcement by the Swedish Government that it would seek to have the ESS placed in Lund and cover approximately 30% of its construction costs, are rather spectacular events. The MAX IV proposal, though it would prove to be scientifically and technically sound, and was well-supported by Swedish and Nordic scientific communities, suggested a commitment of such unprecedented size in Sweden that its proponents might have been discouraged from suggesting it, given the experience in obtaining funding for MAX II some fifteen years earlier. But while MAX IV could at least be argued to be a natural (yet significantly scaled up) continuation of the rather successful MAX-lab, the Government announcement that it would support the ESS Scandinavia bid marked a clear break with the previous policy of referring decisions to make similar commitments to the scientific community and the relevant research council(s), leaving other countries to host and take the main responsibilities for Big Science, with Sweden as a member. The MAX IV proponents most likely hoped that the Government would come to its senses and acknowledge the impressive track record of MAX-lab highlighted in several evaluations (for a review, see Hallonsten and Christensson 2017a: 80–90), such as the 2002 evaluation which called the lab "very innovative" and a "first class" synchrotron radiation facility "serving fields of science where the Swedish community performs at its best" (Swedish Research Council 2002: 37) and commit to funding MAX IV. Their hopes for such a development were likely increased by the two evaluations of the technical and scientific cases for MAX IV made by international expert groups in 2005 and 2006, respectively, but also somewhat torn by the Government's 2007 decision to bid for the ESS, which came in the midst of strong silence regarding MAX IV in spite of their favorable evaluations (see chapter 7).

The 'new new' research policy

As suggested by the previous sections, the strength of the Swedish research system has historically been its breadth and ability to cultivate a certain level of innovation in a wide range of areas, while perhaps suffering from an inability to take it further and make globally leading achievements.

The latter has been a dominating theme in governmental research policy since the turn of the millennium. In their research policy bills and other policy documents, consecutive governments have emphasized the need for strategic mobilization in certain areas of strength in order to meet the challenges of globalization and retain, and further enhance, the competitiveness of the Swedish knowledge-based economy (Swedish Government 2000b; 2004b; 2005a; 2008c; 2010a). Some governments have gone further in their attempts to launch and implement policies that correspond to these ambitions, but there is a clear and consistent pattern of strategic prioritization and the launch of new funding programs and policies to achieve specialization, a pattern in which the investments in MAX IV and the ESS clearly fit.

Premfors (1986), who wrote one of the first comprehensive historical reviews and analyses of Swedish research policy in the 20th century, distinguishes between the "old" and the "new" research policy, noting a change that began in the mid-1970s and culminated with the issuing of the first governmental research policy bill in 1982. The "new" research policy was the result of pressure on the Government to take a clearer coordinating role – a pressure that was both political (from the opposition parties) and bureaucratic (from the Government's agencies and administration).

The sweeping higher education reform of 1977, a reform of the research councils undertaken in the same year, and a new system of quadrennial governmental research bills, the first of which was issued in 1982, were perhaps the most visible effects of this alleged shift to a "new" research policy of increased coordination and oversight from the Government (Premfors 1986: 28ff).

In light of the two most recent decades of research policy in Sweden, Premfors' (1986) periodization pales slightly. In comparison, the research policy bills of 1982–2000 maintained a *laissez-faire* approach to Swedish science. Major reforms to the system were certainly made, such as the 1993 managerial reform to the higher education system that implemented far-reaching performance-based management; the 1998 reform to doctoral education that sought to increase throughput; and the 1999 abolition of the chairholder professorships and the privileges that came with them; but these were made with comparably little connection to the most important governance tool of research policy, namely funding.

The governance reforms and funding increases made in Swedish research policy since the turn of the millennium have, on the other hand, been monumental enough for the current period – in which the ESS was established in Sweden – to be called the era of a 'new new' research policy in Sweden. In 2000, the research councils were reorganized after some years of preparation and investigatory work, which among other things entailed the launch of the "innovation systems" concept into Swedish science policy (Eklund 2007: 120). In 2001, the Swedish Agency for Innovation Systems (Vinnova) was created on the basis of this idea, simultaneously with the new Swedish Research Council, which was essentially a traditional research council, only broader than its area-specific predecessors, and later criticized for an innate inability to take a proper strategic role in the system of allocating resources according to the needs of the research system (Swedish Government 2008b: 49). The Swedish Research Council certainly was designed as a classic research council with an extra add-on mission to point the way to areas of research of strategic importance and worthy of special attention, and Vinnova became a kind of research council for applied R&D and innovation projects in academia and industry (Swedish Government 2000c). While the launch of the Swedish Research Council and Vinnova were symptoms of the emerging "discursive" shift in Swedish research policy to more emphasis on "excellence" and especially strategic priorities in areas of specific importance for the long-term competitiveness of the Swedish labor force and economy, in the case of the

Swedish Research Council the shift remained "discursive" at least for its first years. Its role was to distribute funding within three areas (natural and technical sciences, humanities and social sciences, and medicine) through open calls and traditional peer review, and it had little room or mandate for resource distribution or occasional discontinuous investment in specific areas due to larger strategic priorities within the overarching research council organization (Benner 2008: 298–299, 382). This situation changed dramatically after 2004–2005, as will be discussed below.

The 2000 research policy bill launched an increase in the total annual governmental allocations to R&D of SEK 1.3 billion over a four-year period, and identified some areas as strategically important, most notably biotechnology, IT, materials science, educational sciences, and environment and sustainable development, which together received nearly half of the funding increase (Swedish Government 2000b: 49ff). The bill emphasized strategic priority, and highlighted that the directed resource increases would lead to substantial gains down the road, through competence enhancement, innovation, and economic growth. The 2005 research bill, delivered six months later than planned, went further down this road and launched an even greater increase – also identified as a "permanent level increase" – in the annual allocations of SEK 2.34 billion over the years 2005–2008. The increase was made across the board, but the Government judged it to be especially urgent to increase the funding for medicine, engineering sciences, and research for sustainable development – these three shared 40% of the total resource increase. Although the previous research bill had noted that "Sweden needs to become better at making priorities in important research areas" (Swedish Government 2000b: 12), it was the 2005 bill that ushered in the 'new new' research policy. The bill proclaimed that in order to meet rapidly changing challenges from "society and the world," Sweden would have to give its national research base better tools to compete on the international stage (Swedish Government 2005a: 10, 80). A 2004 governmental investigation by the Vice-Chancellor of Linköping University, Bertil Andersson, outlined new forms of research funding that could strengthen the international competitiveness of Swedish science in specific areas, and advocated new funding programs structured according to a "triple ten rule" that was based on international role model funding programs: The funding model should consume roughly 10% of Sweden's research council allocations, continue for 10 years, and endow each strong research environment with approximately SEK 10 million yearly (Swedish Government 2004b: 35). The 2005 research bill realized these suggestions, and gave the research councils and Vinnova the joint special task of allocating SEK 300 million to strong research environments (albeit with no specific focus areas identified) (Swedish Government 2005a: 12–13), clearly stating its aims of identifying and supporting Swedish "centers of excellence" with substantial funds, and thus also creating better incentives for the universities to strategically prioritize certain areas (Swedish Government 2005a: 10, 89, 95–96).

The result, implemented by the research councils and Vinnova as instructed in the bill, was four programs: The Linnaeus Grants of SEK 5–10 million annually for 10 years to academic environments, administered by the Swedish Research Council and Formas; the Vinn Excellence Centers of SEK 7 million annually for 10 years, administered by Vinnova; the Berzelii Centers of SEK 10 million annually for 10 years, administered by Vinnova and the Swedish Research Council; and the Institute Excellence Centers of SEK 6 million annually for 6 years, administered by Vinnova together with two public research foundations. In total, almost a hundred grants of this type were distributed to groups at Swedish universities over the years, all in open competition and with no funding earmarked for specific areas, but with some variations between the programs with regard to their focus (fundamental research, strategic research, academy-industry collaboration, etc.) (Hallonsten and Silander 2012).

The 2005 governmental research bill also paid some special attention to research infrastructure. As part of the general resource increase over the years 2005–2008, the Government allocated SEK 42 million to the Swedish Research Council to "develop research infrastructure," and with the instruction that the funding would be distributed based on "assessments of the quality and relevance" of the infrastructures (Swedish Government 2005a: 12–13). Research infrastructures are regarded as "increasingly important," and the instruments used by scientist are "growing bigger in size and often used by many groups" (Swedish Government 2005a: 98–99).

Sweden had been governed by a Social Democratic minority government for ten years when, in 2004, the four opposition parties formed what they called "Alliance for Sweden" ("*Allians för Sverige*") in order to win the upcoming 2006 general election and form a coalition government. The strategy was successful – much of the dominance of the Social Democratic Party of Sweden in the postwar period can be attributed to disunity among the center-right parties, and there was a perceived need for change in the electorate – and in the general election of 2006, held on September 17, the Alliance gained a total of 48.24% of the votes and thus 178 seats (out of

349) in parliament, and formed the first majority government in Sweden since 1981.

The context and preconditions for a majority four-party coalition government is different from that of a minority one-party government, at least in the Swedish context. Composed of three rather small and equal-sized parties and one significantly larger, the incoming center-right Alliance government not only had an ambitious reform agenda and some highly prioritized areas of policy where they wanted to leave a mark; the three smaller coalition parties also had their profile areas – real, or in the views of their voters and the electorate, or both. *Folkpartiet Liberalerna*, whose literal English translation would read "The Liberal People's Party", was the third-largest among the four (with 7.54% of the vote and 28 seats in parliament), and had been particularly active in integration and education and also, in the most recent years leading up to 2006, research. Always paying the required respect to academic freedom in their party program and election platforms, but never going beyond the general phrases and conceptual viewpoints and suggesting specific reforms in the area of research policy (Folkpartiet 2003: 8), the party became the voice for research policy in the Alliance. In his memoirs, the party's chair and incoming Minister for Education and Research in 2006, Lars Leijonborg, writes of having a special relationship to science since the 1970s, which gave him "a rational understanding of the central role of science in meeting the great challenges of humanity, but also an emotional fascination for the search for knowledge itself" (Leijonborg 2018: 324). Ahead of the 2006 election, in various pamphlets and statements, *Folkpartiet* increasingly defined itself as the party of research and innovation, speaking of research as a "very important growth factor" and arguing that the Government "needs to take a greater responsibility for research" which, in practice, means "strong and independent universities and a diversity of research funders." More specifically, the party called for a "radical move to remedy the underfunding of Swedish research infrastructure, most of all through grants to faculties and research councils" (Folkpartiet 2006: 6–7). The political manifestos produced jointly by the Alliance in 2004–2006 largely echo these ideas, and also mention an ambition to strengthen the most prominent research groups and environments in Sweden (Allians för Sverige 2006: 17–18). When, on October 6, 2006, the newly elected Prime Minister, Fredrik Reinfeldt, gave his inaugural address in the Swedish Parliament, he stated that "Sweden will be a strong research nation" and that "the financial frameworks for research will be expanded, beyond earlier parliamentary decisions, in the coming term. A larger share of research funding will be distributed to distinguished research environments" (Swedish Government 2006a). Lars Leijonborg could therefore be said to be destined to take over the helm at the Ministry of Education – "this was a top priority for us" (Interview: Leijonborg) – and as deputy, Leijonborg recruited long time director of administration at Lund University Peter Honeth, who had unmatched experience in university politics, research and education funding and policy, and the functioning of the Swedish research system, and, moreover, was a long-standing member (and previous operative) of *Folkpartiet*. Leijonborg and Honeth wasted no time in pressing ahead with reforms and major decisions at a pace not seen in many years. The ESS project is a good example – core proponents of the project at the time speak of a "game change" (see chapter 7) – but this was certainly not the only new governmental initiative in research policy.

The first thing on the agenda for an incoming government after an election in Sweden is to present a budget for the coming year. In 2006, like any other election year, the new government had a mere ten days from taking office before the budget was to be delivered to parliament, which it was on October 16, 2006. Nonetheless, the budget included several new policies and funding initiatives, including in research. The total block grants for research to the universities were increased by SEK 200 million in 2007, in addition to the increases that had been envisaged in the previous government's research bill one and a half years earlier. In addition, the budget mentioned the plan to abolish a specific tax (8%) on external research grants, paid by the universities, which would mean another resource increase of several hundred million once the reform was made (this commenced on January 1, 2009). The budget also revealed the new government's ambition to increase the total public expenditure on research by several hundred million SEK in 2008 and 2009 (Swedish Government 2006b: 131, 156). A year later, the Governmental budget for 2008 reiterated the key principles of the Government's research policy, and launched the plan to use various policy measures, not least including new funding programs, to increase strategic profiling of universities and a strengthening of key research environments in order to enhance Swedish competitiveness in research and achieve a more efficient resource utilization (Swedish Government 2007a: 120–122). In the economic policy bill of spring 2008, the Government continued to lay out the details of its research policy orientation. In a long discussion about the role of research and innovation for national competitiveness in a globalized economy, the Government gave voice to the opinion that while Swedish research was strong in an international perspective, "there are signs of a lowered quality and competitiveness" and "there are flaws in the mobilization of governmental resources for research and the profiling of research." The bill hence notes work in progress by the Government to develop new models for allocation of research funding to the universities, and that "governmental research funding shall, to a significantly greater degree, be allocated on basis of scientific quality" (Swedish Government 2008a: 18, 54).

In the fall of 2008, it was time for the new government's first research bill. In light of what had been stated in preceding budget bills (above), anticipation was great. Indeed, the bill contained vast general increases in the Governmental R&D appropriations for several coming years, announcing a general permanent increase of the annual R&D budget of SEK 2.4 billion (≈ USD 370 million) in 2009, followed by permanent increases of SEK 1 billion (≈ USD 150 million) in 2010 and SEK 500 million (≈ USD 75 million) in 2011, in all gradually increasing the annual governmental R&D appropriations by no less than 18% between 2008 and 2011 (Swedish Government 2008c: 1). Later, these promises were exceeded by the increases launched in the budget bills and in actual allocations. Leijonborg claims in his memoirs that "no other governmental expenditure post had a larger permanent increase during the eight years of the Alliance government" (Leijonborg 2018: 325).

The 2008 research policy bill argued strongly for the necessity of its targeted policies and new funding programs. Citing the "grand challenges of humanity" which "cannot be met without new knowledge," and a need for the Swedish economy to enhance the "knowledge content" in its export products to retain and strengthen competitive advantage in times of globalization, the Government's dramatically increased funding levels to research in the academic sector were presented in the bill as the only reasonable and responsible road ahead for Swedish research policy. But the bill also appealed to Swedish history and the world-renowned achievements of Carl Linnaeus, Svante Arrhenius, Alfred Nobel, and Eli Heckscher, as well as commercial successes such as safety matches, ball bearings, refrigerators, and the drug Losec, to argue indirectly that Sweden also has a responsibility to excel in research and innovation for the benefit of humanity, according to the pattern of "invention of tradition" (chapter 2). The conclusion of the Swedish history of science drawn in the bill is that academic science, with no immediate application in view, is the best breeding ground for innovation, but that simultaneously, new resources should be concentrated to a smaller number of excellent environments and allocated according to some predefined priorities. So far, argued the Government, too much funding had been allocated without competition or quality criteria, and fragmentation and shortsightedness had also stymied many promising developments and inhibited commercialization. Most of all, Swedish research policy and funding had suffered from a lack of cross-disciplinary, long-term, coordinated funding to independent and excellent research environments (Swedish Government 2008c: 14–19).

The bill launched several specific policy initiatives to live up to these ambitions, including a new formula for allocation of a small part of the block grant funding to the universities that would be based on aggregated publication and citation performance and the ability to attract external grants; increased funding to the research councils, which was presented in the bill as unfettered and supposed to strengthen the general mission of the councils, but most of which went to predefined and targeted programs (see below); and funding for "innovation offices" at some universities to stimulate and simplify commercialization of academic research (Swedish Government 2008c: 19–27). Most importantly, however, the bill launched the Strategic Research Areas (*Strategiska Forskningsområden*) grants that would be issued to research in six major areas that fulfilled three criteria: "research that can contribute to finding solutions to urgent global problems"; areas where Sweden already had "world class research"; and areas where Swedish enterprises were active with their own R&D and where the Governmental research funding would hence strengthen Swedish commercial competitiveness. These areas were to be funded with a total of SEK 2.6 billion over five years, through a model that "combines the external quality control of the research council funding model with the long-term orientation of the university block grants." Just like the preceding Linnaeus Grants (see above) that were launched in the 2005 research bill, the Strategic Research Areas program "aims to build up a number of new world class research environments." but unlike the Linnaeus Grants, the funding would go to areas predefined by the Government.

As seen in figure 1 below, the Governmental appropriations to research increased on all accounts from 2009 and on, at a somewhat higher pace than in preceding years. The most dramatic increases in this period took place between 2007 and 2008 (a 10% increase), between 2008 and 2009 (11.4%), and between 2011 and 2012 (10.6%). Adjusted for inflation, the increases are somewhat smaller, but the total increase in the annual appropriations for R&D between 2006 and 2014 (the eight years with the four-party Alliance government in power) was, nonetheless, more than 50%.

Figure 1: Governmental appropriations for R&D in Sweden, 2001–2014, billion SEK, not adjusted for inflation

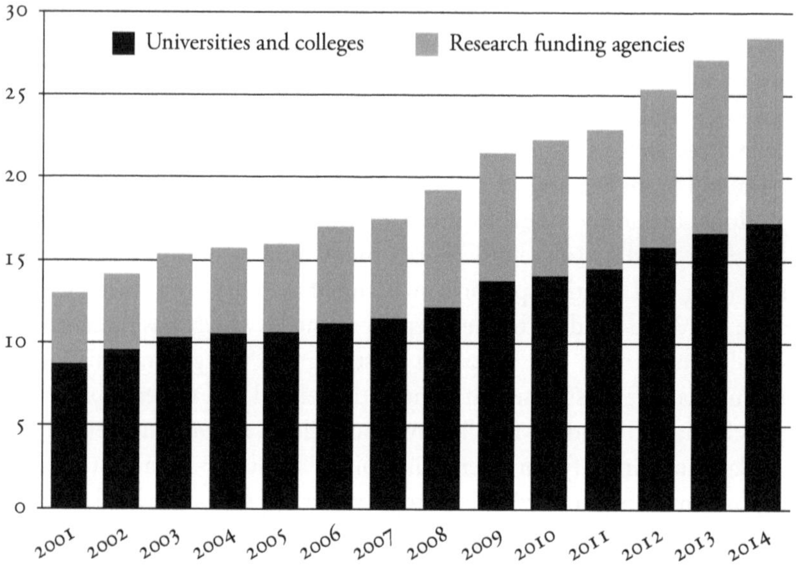

Note: The figures include appropriations to universities and colleges (only grants for research and doctoral education), research councils (not including specific grants for administration), other research institutes and funders under the Ministry of Education (including the Swedish National Space Agency), but not including other special budget posts like the Royal National Library, the Central Ethical Review Board, etc.
Source: Governmental budget bills for the noted years.

While it is clear that the research system in Sweden received a generous increase in resources in the time period studied, primarily as the result of the research policy and funding offensive of the 2006–2014 center-right government, it has also been argued that the increase was unevenly distributed in the system, with more money pouring in to the old universities and to "strategic" research areas specifically chosen by the Government, leaving both the smaller regional colleges and the less fashionable research areas (and the general and unfettered funding streams) with less growth (Benner 2008; Holmberg and Hallonsten 2015). What is clear is that the Government had increased its role in deciding where money was to be directed; a prime example of this is the dramatically altered role of the Swedish Research Council in the past fifteen-year period. The Governmental instruction to the Council, issued annually in the same way as for any governmental agency, details the allocated funding and its conditions.

From its establishment in 2001, the Council received funding through six different appropriations, one for each of the subcouncils (humanities and social sciences, medicine, natural and technical sciences) plus one for education sciences, one for "expensive scientific equipment etc." ("*dyrbar vetenskaplig utrustning m.m.*"), and one for "other research funding etc." ("*övrig forskningsfinansiering m.m.*") (Swedish Government 2004c). From 2005 and on, the appropriations have numbered five: one for each subcouncil, one for educational sciences, and one for "other research funding and research infrastructure" (Swedish Government 2005d). This is when "research infrastructure" entered the Governmental regulation of the Council, and although it had been used in reviews of the research landscape in Sweden and abroad in the Council's annual reports since 2002, it was not until the forming of the Committee for Research Infrastructures and the change in the appropriations that the concept was given special attention, as manifested by highlights in the annual reports of the Council from 2005 and on (Swedish Research Council 2005). As seen in figure 2 below, it was the fifth appropriation, for "other research funding and research infrastructure," that grew most dramatically in the ten years between 2005 and 2014 when the 'new new' research policy doctrine had its breakthrough and implementation. From 2008 and on, adjusting for inflation, the four appropriations to the subcouncils and to educational sciences hardly increased at all, whereas the fifth appropriation (including ESS) almost doubled.

But it is the level of detail in the line-item funding of the fifth appropriation that says the most about the implementation of the 'new new' research policy as seen from the perspective of the Swedish Research Council. In 2004, the instruction contained six specified pieces of line-item funding: three in natural and technical sciences, amounting to SEK 70 million, and three in the "other" appropriation, amounting to SEK 30 million. The rest of the instruction contained only general guidelines of how to use the total budget of roughly SEK 2.5 billion, including, of course, how this sum was divided among the six appropriations (Swedish Government 2004c). Ten years later, the Council's total budget had more than doubled, and totaled SEK 5.6 billion, and the line-items had a significantly more prominent role. The four subcouncils had the liberty to spend their funding in accordance with general guidelines, but for the two additional appropriations, the instruction contained considerably more detail. SEK 900 million were earmarked for funding of the ESS, and no less than 24 other line-items were detailed in the instruction, with a total sum of

SEK 887 million (Swedish Government 2014d). The change had occurred gradually in the years between, and figure 2 gives a rough image of how: The subcouncils received a modest increase, but the "other research funding and research infrastructure" appropriation tripled in the period, especially the Government's funding for the ESS, for reasons of convenience channeled through the Council (Interview: Holmberg), which accounted for almost a billion in 2014. As noted, this appropriation contained a lot of funding with sums and recipients specified by the Government, as well as a large sum of money to be used for special initiatives by the Council itself, or by the Government but funded by the Council. The diagram itself is a testimony both to a shift of power in Swedish research funding, and a shift in the role of the Swedish Research Council.

Figure 2: Appropriations to the Swedish Research Council, 2005–2014, billion SEK, not adjusted for inflation

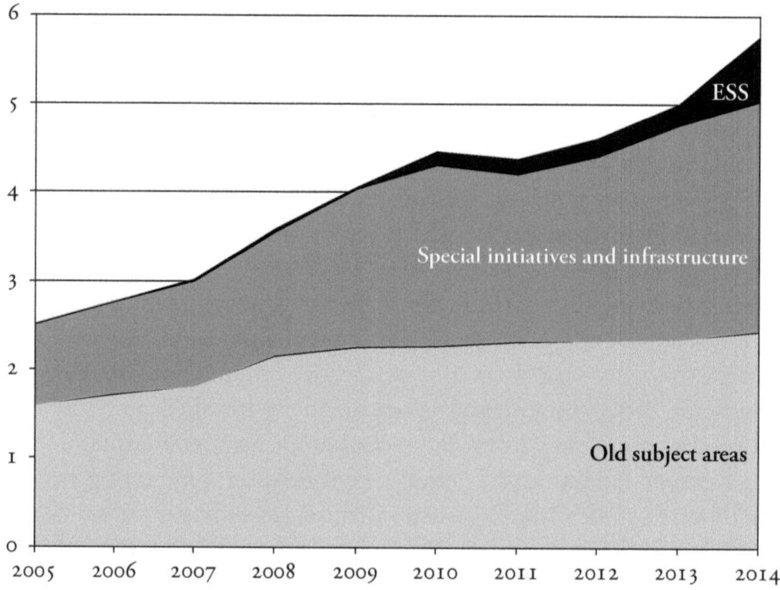

Source: Governmental budget bills for the noted years.

Nowadays, the Council finds itself not only funding (and thus, in a sense, owning) the MAX IV facility and channeling money to the ESS, but also carrying out all kinds of funding programs devised by the state. Given the basic structure of the Council's organization, and the foundations for its governance (see above; Swedish Government 2000a), it is probably safe to say that it was only partially prepared for taking on the great battery of missions and tasks it was assigned by the Government, and not prepared at all to take on the responsibility of strategically and comprehensively staking out the road forward for a small and relatively unprepared country blessed or cursed by the success of bringing the ESS facility to Lund. The Council is the Swedish Government's default agency for carrying out its research policy, yet it was founded to allocate competitive funding for basic research to areas defined both by tradition and on the basis of strategic considerations, but with representatives of the research community in charge, and thus with peer review as key governance principle.

Somewhat simplified, the 'new new' research policy, initiated by the Social Democratic government in the first years of the new millennium but clearly accelerated by the 2006–2014 center-right Alliance government, entailed a shift away from curiosity-driven, "basic" research (although a considerable amount of R&D of "strategic" character and also directly "applied" research takes place in this sector and with this funding) and toward more directed or earmarked funds, allocated to specific programs and areas, and most often distributed on the basis of a peer review assessment of applications, albeit targeted to specific areas or programs such as the 2005–2015 Linnaeus Grants or the program for recruitment of internationally leading research leaders, both of which were administered by the Swedish Research Council on the direct instructions of the Government.

According to Lars Leijonborg, one of the key people behind the 'new new' research policy, the ESS was a perfect symbol of the ambitions of his party and the Alliance government – a "moon lander" project (Interview: Leijonborg). But also with regard to its consequences, which were most likely not anticipated (at least not in full) at the time the decision was made in 2006 and on, the ESS is a key piece in the doctrinal shift in research policy in Sweden in the two first decades of the new millennium. We will return to this discussion in chapter 11. At this point, it suffices to conclude that the Swedish national ESS campaign as initiated by Peter Honeth upon his appointment to the position of highest-level advisor to the Minister for Education and Research in 2006, and vigorously carried on by the Government and minister Lars Leijonborg, had a broader context and fitted quite well in the 'new new' research policy. Put differently, the ESS was a way for the Government to live up to its own words in the 2008 research and innovation bill, that "in order to be able to conduct research of the highest quality, Swedish science must be granted access to the most advanced research infrastructure" (Swedish Government 2008c: 2).

5. Bridging the European neutron gap (1993–2008)

A pan-European facility

As outlined in chapter 3, the experimental use of neutron scattering has been almost entirely dependent on the operation of nuclear reactors. Early on, as long as reactor construction and development was growing, this meant a near abundance of neutrons, and until the late 1970s, the European neutron user community had not had to plan, design, and seek funding for its own infrastructure, with the exception of the ILL. But nuclear reactors have limited lifetimes, and as the 1980s approached, many reactors were taken out of operation and the neutron communities of Europe saw their supply of experimental opportunities fall. The identification of the "neutron gap" between supply and demand for neutrons, however, was not only a threat but also an opportunity, if the gap could be bridged by the design and construction of new, purpose-built neutron scattering facilities with higher performance as well as more stable and safer operation.

In 1986, the European Commission convened a panel to draft a plan for a future European neutron source, and four years later, the panel delivered a report outlining a spallation-based state-of-the-art facility, organized as an intergovernmental collaboration (Kaiserfeld 2013: 29). Two of the British and German neutron strongholds, the Rutherford Appleton Lab in Harwell, UK (host of ISIS) and the Jülich Research Centre (*Forschungszentrum Jülich*) took the initiative for a series of workshops and meetings in 1992 and 1993, where a basic design for a spallation facility was drafted. In 1993, the ESS Council was formed, with representatives from Austria, Denmark, Germany, Italy, the Netherlands, Sweden, Switzerland, and the UK, and observers from France and Spain (Berggren and Hallonsten 2012: 22–23; Kaiserfeld 2013: 29). Three years later, a consortium of institutes from these countries and with support from the European Commission delivered a "site-independent" technical study of the ESS, including cost estimations made "with 20% accuracy." Both the European Commission and the European Science Foundation (ESF) had endorsed the work of the ESS Council in 1994. In 1996, the ESF carried out an independent study of the scientific prospects for neutron scattering, reaching the conclusion, and recommending, that a next generation neutron source should be built in Europe, and that it should become operational within 15 years (ESS R&D Council 1998a).

A parallel organizing effort by European neutron users was the 1994 forming of the European Neutron Scattering Association (ENSA), an umbrella organization for national neutron user associations. The ESS plan was a major reason for the forming of ENSA, which saw as one of its key missions to coordinate support and lobby for the ESS. A 1998 ENSA report, which collected the opinions of its estimated 4,400 European members (organized through national user associations), stated that the association "strongly and unequivocally supports a positive commitment" to the ESS, while also advocating upgrades of both the ILL and ISIS in order "to fully utilise the potential of their capital investment, to maximise scientific productivity and to prototype instruments for ESS" (ENSA 1998).

But in spite of all these endorsements, not much progress was made on a political level. The comprehensive three-volume ESS *Reference Design* report published by the ESS Council in 1997 outlined a very ambitious and costly facility with world-leading capabilities and a performance significantly higher than the ILL, the ISIS, and other comparable facilities globally. The design had been developed with involvement from the RAL in the UK, the University of Frankfurt, the Jülich Research Centre, and the Hahn Meitner Institute (HMI) in Germany, the University of Aarhus in Denmark, The Svedberg Laboratory (TSL) in Sweden, the National Institute for Nuclear Physics (*Istituto Nazionale di Fisica Nucleare, INFN*) in Italy, and the Eindhoven University of Technology in the Netherlands (ESS Council 1997). But instead of starting a political decision-making process on the funding and location of the facility, the ESS Council began an R&D phase driven by scientific interest and with politics remaining absent from the process. Thomasson and Carlile (2017: 17) seem to suggest that Germany, Britain, and France were unwilling to press ahead with the ESS project in the 1990s, although they were the three European neutron strongholds and thus natural champions of the ESS, and traditionally the biggest investors in European collaborative Big Science, and that their vested interests as owners of the ILL trumped the opportunities they saw in the ESS. Germany had also embarked on the construction of a reactor-based neutron facility in Munich, the *Forschungsreaktor München II* (*FRM-II*) which made another major German investment in neutron sources unlikely; in addition, Germany had recently proposed the building of the international TESLA facility for particle physics at the federal research lab DESY in Hamburg (Heinze et al. 2017: 433). France apparently had similar reasons, as hosts of the ILL and also operating a reactor at CEA Saclay (Interview: Tindemans).

The situation in the United Kingdom was complicated, and remained so for a long time, even after the 2009 site decision (see also chapters 7, 8 and 9). Some claim that an informal agreement had been made in the early 1970s, when the ILL was established with UK support, that the next major collaborative neutron facility in Europe was to be built in Britain (Interview: Tindemans). The United Kingdom has always been hesitant about European Big Science collaboration (Hallonsten 2014; Cramer 2017), but in the case of the ESS, there are also testimonies of British ambitions to merge the ESS project with a future major upgrade of ISIS by trying to block a political decision on the ESS and gambling that resources would eventually be redirected to ISIS (Interview: Tindemans). Others claim that the site issue itself was enough to make those directly involved in the ESS unwilling to lift the issue to a political level; in order to make the ESS reality, "both expertise and funding from all of Europe" was needed, and "the day someone would have stated that 'It's going to be Jülich,' or 'It's going to be Rutherford,' the dynamics would have changed completely, and they didn't want that change of dynamics in the project" (Interview: Börjesson).

The aforementioned 1998 ENSA report conveyed that the Austrian, Belgian, Czech, Danish, German, Hungarian, Italian, Dutch, Polish, Russian, Spanish, Swiss, and British neutron user organizations were very positive to the ESS, almost without reservation. The French community was cautiously positive but emphasized – unsurprisingly, given their interest in keeping the ILL running – that spallation sources are only complementary to reactor sources. Interestingly, the Swedish Neutron Scattering Society (SNSS), represented by Lars Börjesson, noted that "ESS is THE answer to the future demand of neutron scattering" but that it should be "supported by a network of smaller facilities" (ENSA 1998: 6). Swedish researchers had been involved in the ESS design work since the early 1990s, and took an active part in ENSA through its domestic association, the

SNSS (Interview: Börjesson). The controversy over CERN in 1997 and on, which produced a governmental investigation on Swedish participation in European collaborative Big Science (see chapter 4), seems to have put some brakes on Swedish contributions to the ESS R&D in the late 1990s, if only temporarily (ESS R&D Council 1997).

The ESS R&D phase was initiated by the forming of an ESS R&D Council with five founding members: the Paul Scherrer Institute (PSI) in Switzerland, the CEA (*Commissariat à l'énergie atomique et aux énergies alternatives*) in France, the Risø National Laboratory in Denmark, the Jülich Research Centre (*Forschungszentrum Jülich, FZJ*) in Germany, and the Council for the Central Laboratory of the Research Councils (CCLRC) in the UK. Jørgen Kjems of Risø was elected chair and Andrew Taylor of the UK CCLRC was elected secretary. The Jülich Research Centre continued to host and financially support an ESS Office, with Hans Ullmaier as director (ESS R&D Council 1997). The following year, the five institutes signed an MoU concerning the ESS R&D phase, where they specified the goals of this work, namely

> to collaborate [...] in order to prepare for a decision about the construction of the ESS as a world class next generation neutron spallation source to the benefit of European science, technology, and industry. [...] ESS shall be an internationally co-owned and operated user facility for fundamental and applied research [...] [and] shall be regarded as a regional European facility with a significant role on the global scene in parallel to similar next generation facilities planned in USA and Japan. (ESS R&D Council 1998a.)

Later the same year, an additional eight laboratories, universities, and other organizations joined the ESS R&D Council, including Uppsala University in Sweden, the National Institute of Materials Physics (*Istituto Nazionale per la Fisica della Materia, INFM*) in Italy, and the Center for Energy, Environmental and Technological Research (*Centro de Investigaciones Energeticas, Medioambientales y Technolggicas, CIEMAT*) in Spain (ESS R&D Council 1998b; ESS R&D Council 1998c). In 2000, the members of the R&D phase signed an extension of its MoU, where they stated explicitly their goal of completing the design process and delivering a "fully costed proposal to put to potential funders for either an ESS that is part of a multipurpose facility or a stand-alone ESS by May 2003" (ESS R&D Council 2000b).

In 1999, the OECD Megascience Forum made an announcement that has come to be regarded as decisive not only for the eventual successful launch of the ESS, but for the construction of spallation sources elsewhere. Formed in 1992 to foster cooperation among OECD countries in the planning, construction, and operation of large research facilities, the OECD Megascience Forum had no formal decision-making powers but played an important role as a center of discussion for participating countries (Papon 2004: 63). In the late 1990s, the forum drafted a global neutron strategy to bridge the "neutron gap" by developing spallation sources to eventually replace reactors, which essentially meant a continuation of the operation of the ILL and ISIS as long as these were competitive, including necessary upgrades, and then the launch of "three high performing next generation neutron spallation sources, one in the US, one in Japan, and one in Europe" (Interview: Tindemans). At this time, the plan in the United States was to build a very powerful reactor-based neutron source, the Advanced Neutron Source (ANS), but the efforts were redirected to spallation, in part due to the OECD Megascience Forum's strategy. In 1997, the Oak Ridge National Laboratory in Tennessee won the competition to host the new US neutron spallation facility, and construction started in 1999 (Rush 2015: 147–149). In addition, the Japanese government took the recommendations of the OECD Megascience Forum seriously, and began planning its own spallation source. Both of these, the Spallation Neutron Source (SNS) at Oak Ridge and the Japan Proton Accelerator Research Complex (J-Parc) north of Tokyo, were designed and built on the basis of the 1997 ESS reference design. They opened to scientific use in 2007 and 2009, respectively (Berggren and Hallonsten 2012: 23; Kaiserfeld 2013: 30).

The ESS R&D Council had placed a lot of faith in the Megascience Forum's neutron strategy work (ESS R&D Council 1997), and was not disappointed. On the recommendation of the Megascience Forum, the OECD Ministerial Conference of March 1999 endorsed the plan for three new neutron sources in Europe, Japan, and the United States, which seems to have given the ESS R&D Council renewed confidence and energy. In November 1999, the ESS R&D Council appointed both an Executive Team, including a project leader and a chair, and a Science Advisory Committee (SAC), and drafted a new MoU (ESS R&D Council 1999). In February 2000, the ESS R&D Council proposed the creation of a Central Team with more direct executive powers, to be based at the Jülich Research Centre, which would oversee an update of the technical design. But council members were reluctant to press ahead (ESS R&D Council 2000a). Technically and scientifically, and with regard to budget and timetable, the

ESS project was well-equipped, but the main hurdle was to get it through the national and European politics (Interview: Tindemans).

In May 2000, Jørgen Kjems was succeeded by Peter Tindemans as chair of the ESS R&D Council (ESS R&D Council 2000b). Tindemans, a Dutch national science policy advisor and administrator, and chair of the OECD Megascience Forum (1992–1999), had been personally approached by Kjems with the request to take over the chair of the ESS R&D Council, partly in order to give the project a spokesperson with political skill and experience. Tindemans was, of course, well aware of the difficulty of reaching agreements on scientific collaborations in Europe, and saw how scientists had been working for eight years without any formal planning schedule for the next steps towards realization of the facility. One of his first initiatives as chair was therefore to establish a deadline for a political decision on the ESS, in order to improve the standing of the project in political circles, well aware of the risks involved but determined to make a difference (Interview: Tindemans). In May 2000, the members of the ESS R&D Council declared the initial *R&D phase* of the ESS project concluded, and moved on to a *project proposal phase*, aiming to reach a multilateral European political agreement regarding the construction of the ESS in late 2003 or early 2004, and the eventual start of operation of the facility in 2010 (Kaiserfeld 2013: 37). As part of the project proposal phase, a new and ambitious conceptual design was adopted by the ESS R&D Council in 2001 that outlined a European neutron source significantly more powerful than both the Japanese and the US sources. The deadline for making a decision on a site was set to the May 2002 meeting of the ENSA in Bonn. On this occasion, the ESS would be "presented to the scientific and political public" with the help of "professional material for the media and the press." The candidates were invited to present their site proposals at the meeting (ESS R&D Council 2001a).

The Bonn meeting and after

The ENSA users' meeting took place on May 16–17, 2002, in Bonn. The ESS R&D Council presented its project proposal for the ESS: an updated scientific case, and a new cost estimate of 1.5 billion Euro. In the views of most pundits, the proposal was solid and the technical design convincing (Interview: Tindemans). The four-volume *ESS Project* documentation, totaling 907 pages and drawing on the expertise of some 70 scientists organized in the ESS R&D Council, Project Directorate, Task Leaders, Central

Project Team, Scientific Advisory Committee, and Technical Advisory Committee, was a solid piece of documentation on the basis of which evaluations of the technical feasibility and scientific viability of the proposed ESS project, and subsequent decisions for funding and location, were to be made. Specifically, the increase of available neutron source power by one to two orders of magnitude above what any existing or currently proposed neutron facility could deliver, as shown in the report, enabled the outlining of an ambitious and broad science case with several new areas of application (ESS R&D Council 2002a). The Bonn meeting was expected to be a success for the ESS proposal. But politics, it became clear, was not on the same page as science.

Held in the *Bundeshaus* in Bonn, the former seat of the (West) German parliament, the meeting attracted 900 participants from 20 European countries, including 750 scientists, as well as politicians, industrialists, and journalists. A press conference, attended by some 50 journalists, was held to make a public announcement about the ESS and its possible future locations. On the podium were Peter Tindemans (chair of the ESS R&D Council), Dieter Richter (ESS Science Director), Robert Cywinski (chair of ENSA), Jean-Louis Laclare (ESS Project Director), Helmut Krebs (Secretary of State for Nord-Rhine Westphalia), and representatives from the five candidate sites for the ESS. The press conference reportedly ended with Peter Tindemans triumphantly declaring that "the ESS is entirely feasible and necessary!" (Carsughi and Clausen 2002).

The five candidates that presented their *Expressions of Interest* in hosting the future ESS were the Jülich Research Centre (backed by the German Land State of North Rhine-Westphalia), the Rutherford Appleton Laboratory (backed by the UK governmental Council for the Central Laboratory of the Research Councils, CCLRC, and the existing ISIS neutron spallation facility), the Halle/Leipzig Inter-state Consortium (backed by the German states of Saxony and Saxony-Anhalt as well as the cities of Halle and Leipzig and a number of universities and research institutions in the region), the Yorkshire-ESS (YESS)/White Rose University Consortium (backed by the universities of Leeds, Sheffield, and York) and the ESS Scandinavia Consortium (Berggren and Hallonsten 2012: 24–25). But none of these proposals could demonstrate any support from their national governments, as the ESS R&D council had hoped. Thus, while the Bonn meeting was a "great meeting for neutrons," it was "not good for the ESS" (Interview: Carlile). A "huge meeting, but no high-level meeting, [...] nobody was there who knew how to push this politically, [...] they hadn't anchored it on the level of ministers anywhere" (Interview: Matic). Some claim that politicians from France, Germany, and the UK had already decided jointly in 2001 that they would not pursue the ESS, and that the meeting and Peter Tindemans' efforts to reach a decision pushed them even further in this direction (Thomasson and Carlile 2017: 18–19).

Even more devastating for the ESS project, however, was the result of Germany's thorough inventory of future investments in large research facilities in the summer of 2002. Germany was generally expected to take a leading role in the realization of the ESS, given its strong history in neutron scattering (in Munich and Jülich, among other places), and the Jülich Research Centre had been a central actor in the preparatory work ever since the early 1990s. But Germany had a battery of big projects on the drawing board, and an impressive system of fifteen national laboratories within the Helmholtz Association, several of which were long-time hosts of Big Science facilities in continuous need of renewed missions in this area, to keep activities running (Hallonsten and Heinze 2016). Most of all, the German particle physics and photon science laboratory DESY in Hamburg had proposed building the gargantuan TESLA facility, a 33 km-long linear collider for particle physics with a free electron laser extension, in all at an estimated total cost of 3.4 billion Euro. Although it was an international collaboration, Germany knew it would have to bear a large share of the costs and, not least, take the political initiative towards its realization (Heinze et al. 2017: 439). Together with several other proposed projects, including the ESS, the TESLA proposal was reviewed by the German Council for Science and Humanities (the *Wissenschaftsrat*), on the instructions of the German federal government, which published its report of the investigation in July 2002, giving no fewer than six other projects (including TESLA) higher priority than the ESS (Wissenschaftsrat 2002).

The message was clear: Germany was not going to take a leading role in bringing the ESS from concept to reality, let alone seek to host it. The reasons, as noted in the report, were that the project could be pursued with other European countries in leading roles and hence did not require a German initiative; that the neutron scattering technique used in the ESS was likely to be replaced by cheaper techniques within only a decade or two; and that in comparison with synchrotron radiation, neutrons were judged to be of lesser scientific potential. The scientific case for the ESS had not, in the eyes of the Council, been sufficiently developed (Wissenschaftsrat 2002: 56). The ESS R&D Council objected to the assessment, arguing that the evaluation was based on the wrong premises, and that the four-volume ESS project report, published in connection with the May 2002 Bonn meeting, "provides the solid foundation, as requested by the Wissenschaftsrat, upon which a definitive assessment and priority allocation of ESS can be formulated" (ESS R&D Council 2002c). Some controversy also surfaced on the news pages of *Nature*, where members of the expert group enrolled by the Wissenschaftsrat claimed that their positive views of the ESS project had been neglected in the final assessment (*Nature* 2002a; 2002b). The result was, nonetheless, that Germany clearly and openly withdrew from the ESS project. Commentators agree that internal German priorities, including the recent opening of a new reactor in Munich (see above) and the decision to go ahead with the x-ray free electron laser, the European XFEL, at DESY in Hamburg, a project of size and cost comparable to the ESS and focusing on a complementary experimental technique (Heinze et al. 2017: 440) were the key reasons for this course of action (Börjesson 2003: 98; Interviews: Tindemans, Börjesson).

The German withdrawal was monumental, and caused a chain reaction of negative decisions for the ESS. First, the ESS R&D Council decided to close down its ongoing work and not spend any more money on the project. Second, France, represented by the CEA, formally withdrew its support (Interview: Tindemans). Some efforts to keep the R&D project alive in some form continued throughout 2002 (ESS R&D Council 2002d), but at the January 22, 2003 meeting of the ESS R&D Council, the final ax fell. People present speak of a meeting that was "ill-tempered" (Thomasson and Carlile 2017: 19) and "not very pleasant" (Interview: Berggren). The German delegate, M. Steiner, advocated termination of the ESS project and a prioritization of other options for neutrons in Europe, on a longer term, and this position received support from other delegates (ESS R&D Council 2003a). This was the first European ESS meeting for Karl-Fredrik Berggren, newly appointed project leader for the ESS Scandinavia Consortium (see chapter 6), and he was taken aback: "All these people had worked for years and years on this, and the meeting just threw it under the bus" (Interview: Berggren). The decision to terminate the ESS R&D Council, effective September 1, 2003, was taken at a meeting in Leipzig in July 2003 (ESS R&D Council 2003b).

A glimmer of hope nonetheless remained. At its founding in 2002, ESFRI had put together several expert panels, and a working group on neutrons had explicitly been charged with the task to "carry out a comparative study of different scenarios for the development of facilities for neutron based science in Europe." Its report, published in January 2003, noted that action was required to maintain and develop Europe's global lead in neutron use, and suggested three alternative scenarios: (1) that the ESS, as detailed in the 1997 and 2002 design studies, be built according to plan, which would be the highest priority given the aims; (2) that the ESS be built partially, with only the long pulse target station, which would mean a retained European lead "in some fields"; and (3) that only a smaller short pulse source be built, which would make Europe "competitive but not leading" in the long run (ESFRI 2003: 4). Although the report was carefully written so as not to make a clear recommendation for any of the three scenarios, the general sentiment inside the working group and in ESFRI, which seems to have spread widely and swiftly (see below), was that the first scenario was unrealistic and that scenario two was the viable way forward (Interview: Tindemans).

The spring of 2003 was to be decisive for the future of the ESS project, for many reasons. Peter Tindemans, newly appointed chair of the ESS R&D Council, decided to keep the project alive in spite of its termination, and formed the *ESS Initiative* with the interested partner countries, a smaller organization with a secretariat at the ILL, and with ENSA, the larger neutron labs, and the official site contenders as members, each of which was expected to contribute with approximately SEK 200,000 annually (ESS Scandinavia 2003n). The stated purpose of the organization, as codified in its statutes of October 11, 2004, was to "gather all those interested in building at the earliest possible occasion a next generation top tier neutron source in Europe". In a first version, the ESS Initiative would exist for two years, until the end of 2005, "unless the members deem a continuation useful" (ESS Initiative 2004b). The ESFRI report of January 2003 contributed strongly to the abandoning of the original and very ambitious ESS design, and the common conclusion that if the ESS had a future at all, it would be in a scaled down and significantly cheaper version (Interview: Tindemans). Some ten years later, when the *ESS Technical Design Report* (*TDR*) was published as the central document for the future facility and its construction (see chapter 8), it identified 2003 as a turning point for the ESS, not least technically, as it introduced a "new concept [...] that involved descoping the whole facility and a fundamental change of technical orientation" without dramatically lowering the scientific ambitions (ESS 2013b: 3). Politically, the strategy of keeping scientific ambitions high, but lowering the costs, obviously made sense. One of the first decisions of the ESS Initiative, taken to simplify the "lobbying" effort that was its main purpose, was to reduce the ambitions of the ESS and go forward with the smaller versions (Interview: Tindemans).

But what is most interesting and significant in the context of this book was the *window of opportunity* for smaller countries that opened in the spring of 2003, as Germany, France, and the UK essentially withdrew from the project. This window, open all the way until the 2009 meeting in Brussels (see chapter 7), was probably a major reason that the ESS ended up in Sweden. Lars Börjesson, a key actor throughout the whole history, noted in 2003 that "[t]he big opportunity for Scandinavia to attract a world-leading research facility like the ESS lies in an early initiative, while larger countries are occupied with other business. When the three big ones start acting, it will be too late for Scandinavia" (Börjesson 2003: 101).

The United Kingdom, however, had not bowed out completely. In 2003, its Council for the Central Laboratory of the Research Councils (CCLRC), which oversees large facilities in the UK, invited representatives from France, Germany, Italy, and Spain to round table discussions on the ESS. An internal ESS Scandinavia memo from May 2003 indicates that "UK authorities [...] believe that the ESS should be located in the UK" in the medium to long term, and that the Rutherford Appleton Laboratory (RAL) seemed to view the recently decided upgrade of ISIS as a stepping stone toward locating the ESS to the UK on a somewhat longer term, which the UK government apparently "is ready to work hard for" when that time comes (ESS Scandinavia 2003k). The UK Minister for Science, Lord Sainsbury, reportedly expressed the expectation that the UK government would issue a bid in 2005, which would enable start of construction in 2008 (ESS Scandinavia 2003n). The Halle/Leipzig initiative in Germany also remained in the race, despite the lack of federal support, and a Hungarian initiative to seek to host the ESS was underway but with no formal declaration yet made (ESS Scandinavia 2004b). Minutes from 2003 meetings show clearly that the decisions in 2002 by Germany and France to drop out did not mean that these larger countries would refuse to participate in the project altogether. The course of events in 2002 and 2003 should, therefore, be interpreted as positive for Sweden in the long run, as it left the field open for smaller countries to apply to become hosts, while still keeping the larger countries in the group of future contributors (ESS Scandinavia 2003d).

To a significant extent, the situation after 2003, with the ESS Initiative in place and with the major site contenders out (except the United Kingdom, whose credibility was weaker and whose real position was unclear), was a new game. ESS Scandinavia continued its efforts, although the delay of any decision on the European stage impacted its organization negatively and caused powerful sponsors to back out (see chapter 6). In 2004, Hungary entered as a site contender (ESS Scandinavia 2004b), and Spain joined the game for real in 2006, apparently with initial French support (ESS Scandinavia 2006d). Proponents of a British ESS made their voices heard again in 2005, when John Wood, then director of the CCLRC, said he "pleaded with ministers" to make the UK government submit a formal site bid for the ESS (*Research Europe* 2007). At the end of 2006, the Halle/Leipzig Inter-state Consortium formally dropped out.

Although the role of the ESFRI working group on neutron facilities of 2002–2003 could well have been interpreted, at the time, as an assassin of the ESS project, the benefit of hindsight gives way to another interpretation, namely that it was ESFRI and the working group that came to the rescue of the ESS project, by seizing the opportunity to present the scaled-down version as a realistic alternative. The ESFRI working group turned out to be "the anchor for the ESS to become alive again" (Interview: Tindemans), in part because it connected the ESS to the work of ESFRI and its roadmaps, where only projects of some priority were taken up. By including the ESS on its "List of Opportunities" (preceding the first roadmap), ESFRI declared that the project fulfilled a number of important criteria, including being "of pan-European interest"; "relevant at international level"; "timely and mature"; and "technologically feasible" (ESFRI 2005: 11). A year later, the first edition of the ESFRI roadmap listed the ESS among six projects in the Materials Science category, together with an upgrade of the ILL (ESFRI 2006: 24). The fact that the descriptions of the ESS project remained largely unchanged through ESFRI's 2005 List of Opportunities and its 2006 and 2008 roadmaps can be taken as an indication that the years 2003 to 2008 brought little, if any, changes to the ESS project on European level, politically or in terms of technical design and scientific aims. The ILL upgrade, the ESRF upgrade, and the European X-ray Free Electron Laser (XFEL) are all identified in the 2008 ESFRI roadmap as "well advanced in their negotiation phases or have even entered the construction phase," but not the ESS (ESFRI 2008a: 5).

What is quite clear is that political initiative from prospective host countries, and their determined efforts to move the project forward, was the only thing that could really resurrect the ESS and bring it closer to realization.

6. The ESS Scandinavia Initiative (2000–2006)

Origins and organization

Like most smaller European countries, Denmark and Sweden had begun exploration of peaceful uses of nuclear energy in the mid-1950s, in the wake of the Atoms for Peace program. In Denmark, the first research reactor at the Atomic Energy Commission's Research Facility Risø (*Atomenergikommisionens Forsøgsanlæg Risø*) north of Roskilde began operation in 1956, and enabled Danish scientists to establish small-scale and parasitic research with neutron scattering (Nielsen and Knudsen 2010). Over the decades, the Danish neutron scattering community grew to some prominence and from the 1970s and on, formed part of the user community of ILL in Grenoble. In the 1990s, the Danish Neutron Scattering Society (DANSSK) and the Swedish Neutron Scattering Society (SNSS) began to engage directly in the planning of the ESS on European level, in order to secure some influence over the scientific and technical design of the facility and its instruments (Interview: Börjesson). The work on the 1997 ESS technical design had some Danish involvement, and prominent Danish neutron users Jørgen Kjems and Kurt Clausen reportedly worked to mobilize Swedish support to make the Scandinavian voice stronger in the ESS collaboration (Interview: Johansson).

The origin of the ESS Scandinavia Initiative was, therefore, not an effort to bring the ESS to Scandinavia, but to have a strong Scandinavian presence in the forums where the ESS was planned and promoted. Lars Börjesson, one of the key figures behind the initiative, describes the idea of locating the ESS in Scandinavia as an "add-on" – the early meetings between the Danish and Swedish neutron scattering societies were mainly about declaring official support for the ESS project, and about securing a seat at the table when the ESS was discussed on European level. It was only after some time that the idea that Scandinavia could also seek to host it came up (Interview: Börjesson). Denmark has a long history of ambivalence and public resistance to nuclear energy and most things associated with it (Nielsen and Knudsen 2010), and while the original idea to propose that the ESS be built in Scandinavia seems to have come from Denmark, it was also clear from the beginning that Danish scientists favored a location in Sweden. The Øresund Bridge, connecting the Copenhagen area with Southern Sweden, was under construction, to be inaugurated in July 2000, and the very first idea was reportedly to place the ESS right at the Swedish bridgehead – as close as possible to Denmark but with no risk of the project being stuck in a Danish national debate over nuclear energy (Interview: Berggren).

On June 23, 2000, the Danish ESS initiative was formalized at a meeting of the Danish Neutron Scattering Society (DANSSK), as a response to the plans to close the Risø research reactor some years into the new millennium. Later the same summer, the Swedish Neutron Scattering Society (SNSS) was contacted with the proposal that a joint meeting be held to explore the possibilities of a Scandinavian (Danish-Swedish) initiative, and on October 3, 2000, the inaugural meeting of what was already being called the ESS Scandinavia Initiative was held at Lund University, with invitees from the Danish and Swedish neutron scattering societies, and with MAX-lab and Risø National Laboratory as sponsors. The meeting dealt, in turn, with the scientific case for the ESS and the status of the ESS project on the European stage, the activities relating to the ESS in Sweden and Denmark, and discussions on how to form and structure ESS Scandinavia. The minutes of the meeting are fairly detailed, and reveal that some key people took upon themselves the role of promoting a Scandinavian initiative to host the ESS; Martin Vigild, neutron user at Risø and associate professor at the Technical University of Denmark (DTU) at the time, emphasized early on that "time is short" and that hard work was needed to lift the initiative onto the political agenda quickly, in order to be able to submit a site proposal at the Bonn meeting in May of 2002. According to Lars Börjesson, "there was a positive attitude amongst Swedish science policy decision makers toward bringing a large-scale facility to Sweden," which should increase the chances (ESS Scandinavia 2000).

The October 3, 2000 meeting appears to have reached a consensus that a joint Danish-Swedish initiative should be formed, first of all to demonstrate and organize Scandinavian involvement in the ESS (regardless of its eventual location), and then also to work toward a site proposal. The ESS Scandinavia initiative was thereby officially inaugurated, and a board was elected, consisting of Lars Börjesson (chair of the SNSS and Professor at Chalmers University of Technology), Börje Johansson (Professor of Theoretical Physics at Uppsala University), Kim Lefmann (Senior Scientist at Risø National Laboratory), Robert McGreevy (Director of the Studsvik Neutron Research Laboratory), Kell Mortensen (Professor and Head of Research Programme at Risø National Laboratory), and Martin E. Vigild (Assistant Professor at the Technical University of Denmark) (ESS Scandinavia 2000).

The work of the initiative in the first year consisted mainly of small-scale lobbying towards organizations in the region that could be engaged in the work and persuaded to form a consortium to push the issue politically. In Sweden, Lars Börjesson, Börje Johansson, and Aleksandar Matic approached university management and met surprisingly positive attitudes to the idea, especially when a Scandinavian bid to host the future ESS was discussed – this seems to have awakened the interest of both academic and political leaders in the region (Interviews: Börjesson, Johansson, Matic).

Aleksandar Matic, newly graduated PhD in physics from Chalmers University of Technology in Gothenburg, was enrolled almost immediately after the October 2000 meeting in Lund, as ESS Scandinavia's first operative. In collaboration with a Danish counterpart, Lise Arleth of the University of Copenhagen, Matic coordinated meetings and began drafting a proposal with the 2002 Bonn meeting as a goal. In the first year, before the consortium was formed (see below), Matic had a temporary position at Chalmers, funded by Lund University. His work was largely to prepare documentation to present to local and national policymakers, decision makers, and other potential stakeholders, including municipalities in Skåne, universities across Sweden, and the Swedish Research Council (see a later section). A small scale operation, in this period ESS Scandinavia ran on the enthusiasm of the scientists involved, who had limited experience of political work, as illustrated by the following anecdote told by Aleksandar Matic:

> Just to put this in perspective, [...] we had a meeting with the director of city planning in Lund, Anders Tingvar, very early. I don't know what the purpose of the meeting was, perhaps we were to convince them of something or just inform them. [...] So we come there, present ourselves, and they say like, "OK, you're bringing this facility here, for 15 billion. It's like the Olympic Games. What's your marketing budget?" We had no money at all. So I just said a figure that I thought was big,

> compared with my private finances, I said "50,000." That's a lot, right? He didn't even laugh. He was annoyed. Next meeting, I said 250,000. (Interview: Matic.)

Nonetheless, encounters like this one did not turn off professionals like Tingvar. People in power were apparently attracted by the idea of a major European research facility in Lund.

In early 2002, the lobbying had borne fruit and a number of organizations with a common interest in a Scandinavian ESS entered into a consortium agreement. On March 22, 2002, the ESS Scandinavia Consortium was formally inaugurated in Copenhagen by the Vice-Chancellors of the Universities of Copenhagen and Lund, Linda Nielsen and Boel Flodgren (*Sydsvenskan* 2002d). The same day, the first ESS Scandinavia Consortium agreement was signed by the Universities of Lund, Copenhagen, Oslo, and Linköping, the Technical University of Denmark, Chalmers University of Technology in Gothenburg, the Royal Institute of Technology in Stockholm, the Norwegian University of Science and Technology in Trondheim, the Danish, Norwegian, and Swedish neutron scattering associations, MAX-lab, Risø National Laboratory, the Norwegian and Swedish national nuclear energy research laboratories in Kjeller and Studsvik, the national Swedish accelerator laboratory The Svedberg Lab in Uppsala, the City of Lund, Region Skåne, and cross-border associations of universities and high-tech industry including Øresund University, the Medicon Valley Academy, the Øresund Science Region, the Øresund Committee, and Copenhagen Capacity (ESS Scandinavia 2002c).

With the exception of Studsvik, all the 23 organizations listed above would remain consortium members, jointly renewing the consortium agreement every year until 2007, when the consortium was perceived to have outlived its purpose and was dissolved. The consortium agreement from 2002 obliged every member organization to "formally back the realisation of ESS in Scandinavia, in the Øresund region." The consortium would function as the "formal body presenting the expression of interest for localisation of ESS to the Øresund region," with the goal to "present an expression of interest to host the ESS in the Øresund region at the ESS European Conference in Bonn, 16 May 2002." More specifically, the work of the consortium included drafting a feasibility study "for hosting ESS in one or more physical sites in the Øresund region on the Swedish side"; "to promote the Øresund region as a scientific, technological, industrial and cultural centre in the perspective of ESS"; "to inform decision makers, industry, scientists and the public about ESS and the advantages of hosting the ESS in the Øresund region"; and "to work for support for the ESS-Scandinavian initiative from the Scandinavian scientific community." The consortium council, with representatives from each member organization, was the governing body of the consortium and initiative, complemented by project management and a secretariat (ESS Scandinavia 2002c). After one year, when the consortium agreement was renewed for the first time, a slightly more detailed organizational structure was outlined, with the Council identified as responsible for overall strategy, the appointment of the Management Board and the project secretariat, deciding on new members, and making changes to the MoU. The Management Board had the executive role and responsibility for financial matters. Any member organization was free to leave the consortium at any time, and the financial contribution of each member organization was to be settled in individual agreements (ESS Scandinavia 2002n).

The project secretariat was installed as a subunit to the planning unit of the Lund University central management, with two people working full time in 2002 and another person hired late the same year (Interview: Wickberg). The first two recruitments were Patrik Carlsson, neutron user and former doctoral student of Lars Börjesson at Chalmers University of Technology in Gothenburg, replacing Aleksandar Matic who went on to do a postdoc (Interview: Matic), and Anders Björhammar, who had worked with the Swedish bid to host the 2004 Summer Olympics, and who became acting project director in 2002 (*Sydsvenskan* 2002q). Patrik Carlsson remained employed in the ESS Scandinavia (and later ESS) organization until 2015. In the fall of 2002, Carina Wickberg (then Johansson) was hired as project coordinator, on full time (ESS Scandinavia 2002k). All staff of the project secretariat were formally employed by Lund University, and costs were shared by consortium members with Lund University and Region Skåne contributing the largest parts.

The first few years of the ESS Scandinavia operations were characterized by striking enthusiasm from local and regional politicians. In the spring of 2002, a group of political supporters of ESS Scandinavia was formed under the name "the OS group." OS means "*Offentliga Skåne*" which translates to "Public Scania," but the acronym was also intended as a play of words since "OS" is the Swedish abbreviation for Olympic Games, and many of the local and regional politicians viewed the ESS campaign as similar to a campaign to host the Olympics (ESS Scandinavia 2002e). The OS group was formed as a discussion forum for politicians, where the political strategy for promoting ESS Scandinavia on a local, regional, and national level was worked out, with several members from Region Skåne, as well as representatives from regional authorities, municipalities, and Lund University, many of whom came from the highest political leadership in the respective organizations, such as the mayor of Malmö, Ilmar Reepalu, and the governor of Skåne, Bengt Holgersson (ESS Scandinavia 2002i). Several of the local politicians represented in the OS Group joined the ESS Scandinavia delegation to the meeting in Bonn in 2002 (Interviews: Börjesson, Matic).

The project secretariat was originally run by Anders Björhammar and Patrik Carlsson, but after the Bonn meeting in particular, it became clear that a more senior person with more political and scientific weight was needed (Interview: Matic). Börje Johansson, professor of theoretical physics at Uppsala University, was asked by Lars Börjesson to take on the part-time job of scientific project leader, but declined, and argued instead for the recruitment of Karl-Fredrik Berggren, "who was just about to retire and who had run the super computer center and was generally interested in physics" (Interview: Johansson). Berggren was apparently easy to convince (Interview: Börjesson), and recalls being quite excited by the idea right from the start: "neutrons is a really important technique" and "it was time for Sweden to become host of one of the large facilities" (Interview: Berggren).

The work of the consortium gained a structure and momentum with Berggren as scientific project leader, Björhammar and Carlsson as project staff, and Carina Wickberg as coordinator and administrator. Funding for the activities, in 2002, came from Copenhagen University (SEK 50,000), the City of Lund (SEK 350,000), Lund University (SEK 600,000), Region Skåne (SEK 1,000,000) and Øresund University (SEK 50,000), making SEK 2,050,000 in total (ESS Scandinavia 2002l). But financial commitments only took the project through 2002, and the majority of the applications for funding submitted in the course of 2002 were turned down (ESS Scandinavia 2002m). In the summer of 2003, *Sydsvenskan* reported that the implosion of the ESS project at European level meant that interest in Sweden was also waning, and so the ESS Scandinavia Consortium and secretariat needed to scale down to keep operations running longer. The position funded by Region Skåne, held by Anders Björhammar, was abolished due to the decision by the regional authorities to withdraw funding, and Björhammar lost his job (*Sydsvenskan* 2003d). The work of the project secretariat was thus unavoidably reduced and streamlined, and focus was placed on the work to anchor the initiative on national governmental level and in Europe, as well as preparations to participate in the work to revise the ESS design (ESS Scandinavia 2003m). In June 2003, the most urgent cash shortage was resolved by a grant from the private Knut and Alice Wallenberg Foundation (KAW) of SEK 1.84 million (ESS Scandinavia 2003n). "The money from Wallenberg kept us going for another year. Without it we wouldn't have survived at all" (Interview: Berggren). The City of Lund, which had extended its commitment through 2003 with a further SEK 350,000 (*Sydsvenskan* 2003a), announced in May 2003 that it would continue to fund ESS Scandinavia for another year, until the fall of 2004, but that it then expected the Swedish national government to take over. Local politician Lennart Prytz (of the Social Democrats) complained over the passive stance taken by the national government (*Sydsvenskan* 2003c). Nonetheless, the city continued to fund ESS Scandinavia for at least another couple of years (*Sydsvenskan* 2005a). Region Skåne continued to make substantial contributions (several million SEK) to the operating budget of the consortium (ESS Scandinavia 2002d), and also pledged a major financial contribution to the eventual ESS facility, should it be built in Lund (see a later section and chapter 9). But after some time with little visible progress, the interest faded among other consortium members, and the project secretariat had to look for other sources of funding. In the summer of 2004, the financial situation was still strained. Funds from the Swedish Research Council (SEK 750,000 in 2004 and SEK 1,500,000 in 2005) (Swedish Research Council 2005: 40; Swedish Research Council 2006c: 16) as well as the aforementioned Wallenberg grant kept the activities running until mid-2005, but the expected cash shortage led to further downscaling, most notably the cancellation of the collaboration with lobbying firm Kreab (see below) (ESS Scandinavia 2004a). The Swedish Research Council's grants to ESS Scandinavia in these years were minor compared to its total budget and turnover, but made a huge difference for the very small and fragile ESS Scandinavia organization in 2003–2005 (Interview: Börjesson), especially after the Knut and Alice Wallenberg Foundation turned down a follow-up application from ESS Scandinavia, in June 2004, for continued funding amounting to SEK 1.5 million (Berggren 2004). In 2005, the ESS Scandinavia Consortium had a total budget of SEK 2.4 million, shared by the City of Lund (SEK 225,000), Lund University (SEK 450,000), Region Skåne (SEK 225,000) and the Swedish Research Council (SEK 1.5 million) (*Sydsvenskan* 2005a). In 2006, Region Skåne paid an additional SEK 1.2 million to the consortium to fund consultancy services on environmental issues (*Sydsvenskan* 2006a).

Supporters and antagonists

To go forward with a project like ESS Scandinavia, working to realize a "golden opportunity" but with slim chances of success (Interview: Börjesson), several different competences are certainly required, and several different allies in powerful positions. Later in this chapter, and in chapter 7, the role of leading politicians and public administrators, and the importance of a shift in government that put these people in powerful positions, will be discussed. From 2004 and on, a former politician with an extensive network and a determination that few could match, played a decisive role, especially in the later stages of the negotiations around 2008–2009. But in the early years, it was the enthusiasts that counted. They were found not only in scientific/academic circles but also in university administrations and local government.

A key person who was active from the very start of the ESS Scandinavia Initiative in October 2000, and who remained chair of the highest governing body of the ESS (the Council) until the summer of 2019, is Lars Börjesson. A physicist by training, and professor of physics at Chalmers University of Technology in Gothenburg since 1995, Börjesson was a frequent neutron user in the 1990s, a founding member (and first chair, 1995–2000) of the Swedish Neutron Scattering Society (SNSS), and hence also one of the people behind the original initiative to bring the Danish and Swedish neutron communities together and form the ESS Scandinavia Initiative in 2000. Lars Börjesson acted as the project leader and chair of ESS Scandinavia in its first years and was also nominated by the Swedish Government to be the scientist representative in the ESFRI on its formation in 2002. In early 2003, Börjesson was appointed deputy secretary general of the natural sciences council of the Swedish Research Council, with specific responsibility for research infrastructures (Interview: Börjesson). With this appointment, Börjesson had to step down as chair of ESS Scandinavia and was no longer able to work actively for the consortium. Shortly thereafter, the Swedish Research Council's Committee for Research Infrastructures (*Kommittén för Forskningens Infrastrukturer, KFI*) was formed, with Börjesson as secretary general. Through this turn of events, ESS Scandinavia had one of its leading supporters and champions placed in two important policy fora: ESFRI and the KFI. It is hard to pinpoint exactly the importance of the role of Börjesson in these functions. As an informal promoter of ESS Scandinavia, behind the scenes, his influence in eventually bringing the proposal to the attention of the Swedish Government, and in the series of events from

2007 to 2014 that, in stages, realized a Scandinavian ESS, was probably significant. Others took more front-stage positions and wrote themselves into the history of the ESS campaign much more actively. Clearly, however, Börjesson is an *éminence grise* of ESS Scandinavia and one of the key persons behind the success of the campaign.

Another key person in the early days was Uppsala physics professor and theorist Börje Johansson, whose enthusiasm for the ESS came not from a scientific interest in neutrons but from a general commitment to bringing a larger international scientific project to Sweden (Interview: Johansson). As a person of high repute in Swedish physics, Johansson was an important ally for ESS Scandinavia in the early work to raise awareness at universities and gain the support of scientists and science administrators, perhaps especially in combination with Lars Börjesson, whose reputation was also important. "When Börje and Lars said something, people listened," says Aleksandar Matic, "so Börje was in that sphere, and could open all these doors" (Interview: Matic).

An important organizational ally for ESS Scandinavia, right from the start, was Region Skåne. Although not represented at the inaugural meeting in Lund in October 2000, which was more of a scientific meeting and workshop, the regional authority was among the first to be contacted by ESS Scandinavia when it started working in 2001, probably by Lund University director of administration Peter Honeth (see below), who had both extensive knowledge and political instinct, and a wide network. Carl Sonesson, chair of the regional executive council, and Rolf Tufvesson, a member of the regional executive council, invited ESS Scandinavia to a hearing, which led to the creation of the OS Group (Interview: Kinhult). The region itself was a new construct, with a partly new mandate, which embodied a regional euphoria for which the ESS Scandinavia plan was an extraordinary good fit. The ancient geographical area of *Skåne* in Southern Sweden, with its own flag and heraldry, has a history of cultural independence and strategic importance in economy and trade, which had made it subject to several wars between Denmark and Sweden. It has belonged to Sweden since the signing of the last peace treaty of 1660 and was fully integrated into the Kingdom of Sweden in 1719 (Wetterberg 2017: 497ff). In 1997, the two counties that made up the traditional geographic area of Skåne, Malmöhus County and Kristianstad County, were merged, and in 1999, the newly created Skåne County became a *Region*, which meant that during a trial period, it would have a broader responsibility than a county, complementing its tasks of organizing health care and public transport with the coordination of regional development, including infrastructural planning, promotion of economic and regional growth, environmental planning, and interregional collaboration (Wetterberg 2018: 643ff). In Skåne, and among politicians in particular, the regional trial was a major change with huge symbolic value. The Øresund Bridge that connects Skåne with Denmark had opened in July 2000, after a record-breaking construction period, ahead of schedule and with a significantly lower total cost overrun than is commonplace in major transport infrastructure projects around the world (Flyvbjerg et al. 2003: 14). A fixed link, over or under the Øresund Strait, had been discussed for a century, and its materialization was a symbolic event both for Skåne and the eastern parts of Denmark, around Copenhagen, which together form the Øresund Region. Euphoria over these achievements characterized much of the political work in Skåne at the time, and spilled over on the ESS Scandinavia Initiative: "Against this background, people were probably prepared to believe in the impossible to an extent that they wouldn't have just ten to fifteen years earlier" (Interview: Kinhult). All the chairs of the regional executive committee in the years covered by this history – Carl Sonesson (1998–2002), Uno Aldegren (2002–2006), Jerker Swanstein (2006–2010), and Pia Kinhult (2010–2014), have shown great enthusiasm towards the ESS, and the officials in the Region Skåne organization had been active in promoting it (Interview: Lindström). The commitment of the regional authority went beyond enthusiasm, representation in the OS Group and the ESS Scandinavia council, co-funding of the work of the secretariat (see above), and lobbying of different kinds. The region's work to secure the land for the ESS, in 2009–10, is an interesting episode in its own right, that will be returned to in chapter 8.

Another supporter and enthusiast of the ESS Scandinavia Initiative, who was in the game for almost as long as Lars Börjesson, is Peter Honeth, long time director of administration at Lund University and a former politician. Honeth would eventually rise to become deputy (or state secretary, *statssekreterare* in Swedish) to the Minister for Education, Lars Leijonborg, after the election of the center-right government in 2006, and brought the ESS to his attention, which turned out to be absolutely decisive for the campaign (see chapter 7). But he was also very important to ESS Scandinavia in the early years:

> He called me, when I was on my way to a meeting with the smaller ESS group in Lund, and I had no idea who he was but he told me he was the director of administration at Lund University and said our project was very interesting and asked if he

> could come to our meeting. And he brought someone from the business development division of the regional authorities and they asked us how we were planning on developing the project, and we had only vague ideas and so he offered to find the money to hire some people to work part-time with this. And we welcomed it, of course. (Interview: Börjesson.)

Honeth himself recalls that he became very interested and decided to act to help the project achieve some stability in its organization – "the scientists were a bit unclear on this point" – and also argued strongly, early on, to bring the ESS to Lund. "I had no trouble seeing that this was big enough to be very interesting" (Interview: Honeth). "He realized that this was a good project to bring to Sweden," and that it "would reflect well on Lund University, but it would also tie the Øresund Region together" (Interview: Berggren). And the project seemed to be in need of this kind of help: "It became my role to work with the enthusiasm of the scientists involved and start getting some work done" (Interview: Honeth). Early on, Peter Honeth seems to have strongly advocated the mobilization of support in relevant scientific communities in Sweden and Denmark before working to attract political attention to the project, not only to build strong coalitions but also to obtain active support from leading scientists in the shape of letters and proposals to the Government. "He connected us to Denmark, and to Region Skåne, and the City of Lund" (Interview: Matic). Honeth also highlighted the risk that individual scientists would see the ESS as a threat to their own research activities in the competition for resources, and that the ambition therefore should be to present the ESS as a new investment outside of the existing funding frameworks.

Also going beyond the personal enthusiasm of Peter Honeth, Lund University engaged strongly in the ESS Scandinavia proposal at an early stage and also took some risk by employing the officials of the secretariat – "they showed that they were prepared to gamble on this" (Interview: Börjesson) – and thus investing some money. Honeth himself argues that this kind of risk-taking and willingness to gamble on good projects has been in the DNA of the University for some time. A recent impact evaluation of MAX-lab confirms this: the lab had formidable support from a succession of University Vice-Chancellors, both in practice and in rhetoric, especially in comparison with other national research facilities in Sweden whose host universities have acted less forcefully in their support (Hallonsten and Christensson 2017a). Honeth explains that he learned early on, when taking up the post of director of administration in 1990, that the spirit of the central university management and its Vice-Chancellors was to take the initiative and push for things that lay outside of the regular faculty structure but where they saw potential (Interview: Honeth). When ESS Scandinavia emerged in the early 2000s, Boel Flodgren was the Vice-Chancellor of Lund University and became involved in the project, taking part in the Swedish delegation to the 2002 meeting in Bonn (see below) and promoting the ESS Scandinavia candidature. "She was very active. She probably had no idea what the ESS was, but she did what she was supposed to do as Vice-Chancellor, namely to say that if it's good for Lund, you can count on my support. And we could" (Interview: Johansson). "When we needed it, she would work the phones, meet people, go to meetings. So, yes, she was also a guarantee" (Interview: Matic). Flodgren's successor as Vice-Chancellor, Göran Bexell, was "less active, but certainly not negative" (Interview: Berggren) and kept the funding commitments to ESS Scandinavia, also actively supporting and approving the recruitments of Colin Carlile and Christian Vettier as guest professors in 2006 and 2007 (see below). Internal discontent – highly predictable given the traditional structure of Swedish universities (chapter 4) and summarized by Honeth as "the classic one, from the faculties and down: We shouldn't invest in anything but our own activities, to put it bluntly" (Interview: Honeth) – never surfaced other than in an occasional letter to the editor of the local newspaper *Sydsvenskan* (see below). But this was not so much connected to the University and the fears for displacement effects in the local academic environment, as more focused on environmental concerns and issues of local land area use.

Early on, local critics of the ESS established a connection between the project and the long-standing issue of the local nuclear power plant in Barsebäck, some 10 km west of Lund. This issue had caused severe problems in the bilateral contacts between Sweden and Denmark for several years; these also affected ESS Scandinavia, and had had a symbolic importance in the campaign leading up to the Swedish national referendum on nuclear energy in 1980 as well as indirectly playing an important role for the eventual founding of the Swedish Green Party. The intellectual left-leaning environmentalist groups of Lund were instrumental in this campaign, and while several decades had passed since, and the national government had moreover pledged to close the second Barsebäck nuclear power reactor in 2005 (the first had been closed in 1999), a small but vocal group of activists in Lund connected the plans for an ESS in Lund with the Barsebäck issue and feared that the latter would be kept in operation to serve the former with power, a suspicion that apparently also spread to the Danish Government (see chapter 7) (Larsson 2019: 89).

At a public information meeting in November 2002, representatives of ESS Scandinavia made it clear that the ESS would not make itself dependent on a single power source like Barsebäck, especially since its closing was imminent (ESS Scandinavia 2002k). In order to further placate the local opinion, in early 2003 the ESS Scandinavia Consortium ordered an investigation from the consultancy firm *ÅF Energi och Miljö* regarding the future energy use of the ESS (see next section), and emphasized its lack of connection to the Barsebäck power plant, which was cited in several documents and agreements of the consortium afterwards, as a means to preclude any further discussion on the matter (ESS Scandinavia 2003e, 2003g).

At local level, the City of Lund became involved in ESS Scandinavia early on, with politicians across the spectrum endorsing the ESS Scandinavia plans and supporting them in all possible ways, with the exception of the Green Party and the Left Party. Ahead of the presentation of the Scandinavian site candidature in Bonn in May 2002, the City of Lund commissioned a study from consultancy firm SWECO that investigated the preconditions for establishing the ESS in the designated area at Brunnshög northeast of the city. This provided ESS Scandinavia with the substantial documentation needed to put together a formal expression of interest, and also laid the groundwork for the local planning that began later the same year (see next section). The city planning office worked actively to help ESS Scandinavia prepare for the permit process, with the blessing of politicians and officials from five of the seven parties represented on the City Council (*Kommunfullmäktige*) and representing close to 80% of the voters. The Green Party and the Left Party, representing the remaining 20%, voted against, and accused the director of city planning, Anders Tingvar, of not staying neutral in the process of reviewing the ESS plans in accordance with normal procedure (*Sydsvenskan* 2005c). The Left Party and Green Party also made their voices heard by writing several opinion letters to the Lund section of *Sydsvenskan* (see below), and, of course, in the City Council. The issue of mercury – at this time, the ESS design still entailed a target station of mercury – seems to have been the most provocative aspect for the Left Party, whose local representative, Katrine Arvidsson, said it was a "stupid idea to drag 30 tons of mercury into the town" (*Sydsvenskan* 2002e). The Green Party concurred, and also expressed their resistance and resentment toward the idea of using premium farm land for the ESS facility. Both parties also criticized the heavy engagement of the City of Lund in ESS Scandinavia, arguing that the whole ESS issue had been "handled very undemocratically." Green Party representative Rolf Englesson argued in the fall of 2002 that the city's executive committee had entered into the ESS Scandinavia Consortium "[w]ithout any public debate, and without any processing of the issue in the City Council" (*Sydsvenskan* 2002m). The Left Party expressed reservations with regard to the City of Lund's continued participation in the ESS Scandinavia Consortium in February 2004, calling for an investigation of environmental aspects before giving the project their support (*Sydsvenskan* 2004b). The Green Party also advocated a local referendum on the ESS, an idea that other parties did not rule out, but which didn't muster enough support to become reality (*Sydsvenskan* 2002h, 2002i). In its place, *Sydsvenskan* undertook an opinion poll in collaboration with the opinion analyst company Temo, in September 2002, asking 500 people in Lund if they were "for or against" Lund as the site of the proposed ESS. 66% answered "for," 12% answered "against," and 21% were undecided. Younger people, men, and voters leaning to the right were the most positive. Sympathizers of the Green Party were the most negative (*Sydsvenskan* 2002n).

Both at local level (in Lund) and on the national stage, the Green Party continued to criticize the ESS project after the Government's announcement of its active support in February 2007 (*Sydsvenskan* 2007b, 2007c). The local branch of the national NGO the Swedish Society for Nature Conservation (*Naturskyddsföreningen*) (see below) also maintained their position of resistance to the ESS until 2008–2009 (*Sydsvenskan* 2009c).

A lively debate over the ESS raged on the local Lund section of *Sydsvenskan* in 2002–2004, focusing mainly on the mercury issue, the use of farm land, and radioactivity, and to some extent the (lack of) suitability for a small town like Lund to host such a major international research center, not to mention the alleged undemocratic processes whereby the city had thrown its support behind ESS Scandinavia. Two neighbors of the future ESS facility, Bo Wennergren and Bengt Jönsson, were the most active in the debate, writing eight and nine letters respectively, in 2002–2004. Other people occasionally joined in, including (but not limited to) Left Party and Green Party operatives. ESS representatives and scientists at Lund University responded in writing, promising that the facility would be built in a safe and sound manner, and local politicians in favor of the project also gave their counterarguments, as did, in rare cases, ordinary citizens with a generally positive attitude to the ESS.

Some local NGOs also voiced their resistance and undertook more or less organized campaigns against the ESS. A regional branch of the Federation of Swedish Farmers (*Lantbrukarnas Riksförbund, LRF*) spoke out against the ESS plans in June 2002, citing the premium farm land that the ESS was to be built on (*Sydsvenskan* 2002g). A working group was formed by members of the environmentalist activist group Action Skåne Environment (*Aktion Skåne-Miljö*), and the local branches of the National Swedish Association of Tenants (*Hyresgästföreningen*) and the Swedish Society for Nature Conservation, SSNC (*Naturskyddsföreningen*) in August 2002, to protest the ESS plans. In a letter to the executive committee of the City of Lund, the group "demanded more information about the project and questioned the appropriateness of locating a facility of this type close to urban environments and on premium farm land" (*Sydsvenskan* 2002k). The local SSNC branch also wrote to the Minister for Education, Leif Pagrotsky, in 2003, asking if the ESS was really compatible with the aims of Swedish environmental legislation of "promoting long term sustainable development" (*Sydsvenskan* 2003g). Other groups that formed an active opposition to ESS Scandinavia, mainly citing the planned use of mercury in the target station, the energy consumption, and the lack of a thorough investigation of the environmental impact of the planned ESS were the Swedish Environmental Movement's Nuclear Waste Secretariat (*Miljörörelsens Kärnavfallssekretariat*), the Rio Group (*Rio-gruppen*), and an organization calling itself the Working Group Against the ESS in Lund (*Arbetsgruppen mot ESS i Lund*), apparently consisting of one person, Bo Wennergren (above).

It was the local branch of the national NGO the Swedish Society for Nature Conservation, called the Society for Nature Conservation in the Lund District (*Lundabygdens Naturskyddsförening, LNF* hereafter) that undertook the most energetic campaign against the ESS. Apparently unable to mount broader or greater resistance with the help of more actors, and, surprisingly, lacking the support of its national parent organization, the LNF and its ESS working group "failed, by its own account, to generate a local resistance to the ESS" (Stenborg and Klintman 2012: 189), but it cannot be ruled out that its efforts contributed to the rather ambitious work of the ESS organization to make the future ESS facility more environmentally friendly. But other groups also contributed, and remained active as late as 2007–2009. In October 2007, a meeting arranged by four organizations (the local SSNC branch, the Rio Group, the People's Campaign against Nuclear Power, and the Nuclear Waste Secretariat of the Environmental Movement) took place in central Lund, but gathered only "some 20 people" (*Sydsvenskan* 2007f). In June 2009, when the decision in favor of Lund had been made, the LNF made the case in an article in

Sydsvenskan that the ESS should be located some ten kilometers east of the intended site, where the land was not as valuable. The official standpoint of the LNF was that it would still prefer the facility not to end up in Lund at all, but given that Europe's governments had apparently decided to situate the ESS in Lund, it now suggested a compromise solution to save outstanding farm land (*Sydsvenskan* 2009m).

ESS Scandinavia took the issue of public relations seriously right from the start, and worked purposefully to interact with the local population and the critics within it. Two information meetings were held, in October and November 2002, with Q&A sessions where representatives of the ESS and Lund University responded to queries from anyone interested. The meetings attracted some 80 people each (*Sydsvenskan* 2002p; ESS Scandinavia 2002k). Later the same year, a meeting was organized in Copenhagen by opponents to ESS Scandinavia, primarily the Danish opposition to nuclear energy, and gathered high-profile environmentalists including Danish lawyer Niels Henrik Hooge, Member of the European Parliament Inger Schörling (Swedish Green Party), French anti-nuclear activist Xavier Coeytaux, German environmentalist Bernd Fieboese, and local opponents such as Kim Ejlertsen from the Danish environmentalist organization Noah, and Brunnshög resident Bo Wennergren. The proponents of ESS in Sweden and Denmark, and the ESS Scandinavia consortium, decided to cancel their planned participation, claiming that the meeting was "not organized in a serious way," since it was "based on a number of misinterpretations" (*Sydsvenskan* 2002t). Patrik Carlsson of ESS Scandinavia told *Sydsvenskan* that the consortium had provided the meeting with clarifying information which they had ignored (*Sydsvenskan* 2002s). Several similar meetings, organized both by ESS Scandinavia and the organizations taking a stand against the ESS, were organized in the years to come. The issues of conflict seem to have remained roughly the same – mercury, land use, energy consumption, an undemocratic process (Stenborg and Klintman 2012). At a public meeting in April 2005, representatives of ESS Scandinavia described the ESS as "heaven for scientists and industry," whereafter critical voices in the audience called the project "hell for the environment" (*Sydsvenskan* 2005b). Allan Larsson's investigation report, published in the summer of 2005 (see a later section), was criticized for not discussing environmental aspects at all (*Sydsvenskan* 2005e, 2005f, 2005g) but Larsson defended this choice by claiming that it was never part of his mission and that the proper authorities in Sweden would deal with these issues through due process (*Sydsvenskan* 2005d).

An important and interesting aspect of the history of ESS Scandinavia and its campaigning efforts is the relatively weak local resistance towards the ESS plans, and the lack of a clear resistance from the national environmentalist movement in Sweden (for a thorough analysis, see Stenborg and Klintman 2012). While it cannot be ruled out that the Green Party, which occupied 17 seats in the Swedish Parliament between 2002 and 2006 and formed part of the parliamentary base of the Social Democratic government, helped block the ESS from becoming a national political priority in 2003 and on (see below), the environmentalist resistance was minuscule, especially compared to the support from a broad variety of organizations and actors, locally, regionally, and nationally. Some debate would ensue in the Swedish Parliament after the Government announcement in February 2007 that it would support ESS Scandinavia and offer to host the ESS in Sweden; and contrary to the apparent opposition behind the scenes during the period of Social Democratic of government between 2002–2006, when the Green Party formed part of the parliamentary support for the Government, after the change of government in the fall of 2006 its opposition was more open and vocal: "We believe that the Government has thrown itself into a project that it has no real idea of the scope and character of," Green Party member of parliament Per Bolund argued in a parliamentary debate in March 2007 (Swedish Parliament 2007b).

At national level, outside of the environmentalist movement, the main opposition to ESS Scandinavia's plans to mobilize the Swedish science and science policy system in favor of Lund as host candidate for the ESS came from academic environments in Stockholm and Uppsala. This is hardly surprising, given the petty rivalry between the two oldest universities in Sweden – Uppsala (founded 1477) and Lund (founded 1666) – and Sweden's general centralization of political and cultural power to Stockholm. Several key people in the early phase of national campaigning by ESS Scandinavia – Börje Johansson, Lars Börjesson, Karl-Fredrik Berggren – bear witness to the enthusiasm of local and regional political and academic leaders and figureheads around Lund, a cautiously positive but never outright negative attitude among people in the rest of Sweden (including the university towns of Gothenburg, Linköping, Umeå, and Karlstad), and a rather blunt and arrogant opposition from Stockholm and Uppsala. Representatives of the regional authorities confirm this view and note that Skåne is viewed as a peripheral region of Sweden and too close to Denmark, which invokes a rivalry between neighboring countries (Interviews: Kinhult, Lindström). Efforts to alert the national government of the ESS Scandinavia plans in

2000–2004 were generally fruitless and signaled a strange lack of interest on the part of cabinet-level officials (see a later section). The campaign was viewed by politicians and decision makers in Stockholm as a regional, or even local, affair, and not something for the national government or any of its agencies to concern themselves with (Interviews: Wickberg, Kinhult). Stories of outright negative responses to ESS Scandinavia's plans from the Vice-Chancellors of Uppsala University and the internationally renowned Karolinska Institute in Stockholm abound among informants involved in the campaign in the early years (Interviews: Johansson, Matic, Berggren) and are confirmed in part by the absence of these universities on the list of ESS Scandinavia consortium members in 2002 and on, and their comparably negative referral responses to the Larsson investigation in 2005 (see a later section).

The relationship with Denmark had its own very special dynamic, the details of which can probably not be mapped in full within the scope of this book. But in the course of the ten-year period from the forming of the ESS Scandinavia Initiative with Swedish and Danish participation in 2000, to the official Danish entry into partnership in the ESS company in 2010 (see chapter 9), the relations went through several cycles of trust, distrust, agreement, and disagreement.

> It was always two steps ahead and two steps back. All the time. They are really smart negotiators. [...] They always signed the consortium agreements, every year, but someone always also called me up and said they weren't going to participate because the chances of an ESS in Lund were as slim as a snowball in hell. (Interview: Wickberg.)

A likely root of the problem, according to Lars Börjesson, was Denmark's ambivalence towards neutrons; a scientific stronghold in the use of neutrons for various experiments, manifested not least by the strength of the research environment at Risø National Laboratory, was countered by a very strong national opposition to nuclear power. The ability to neutralize the latter while still building on the former was probably what led Danish neutron scientists to enter into the ESS Scandinavia collaboration and advocate Lund as a site for the ESS, but there are signs that leading Danish scientists and science policymakers never completely abandoned the dream of a Danish ESS (Interview: Börjesson), a dream that was also bolstered by the previous history of having campaigned for, and lost, a candidacy of the European Synchrotron Radiation Facility (ESRF) in the 1980s (Cramer 2017: 20). Similar ambivalence on the part of Norway can also be seen in the early years, although this was likely due rather to the relatively small size of the Norwegian neutron scattering community, and the junior position of Norway in European research collaborations (ESS Scandinavia 2003q). Finland, for its part, always remained outside ESS Scandinavia.

Danish support and involvement were clearly more crucial. The inaugural meeting of ESS Scandinavia in October 2000 was attended by several Danish scientists and leading science administrators. Several Danish universities had been members of the ESS Scandinavia Consortium since its founding in 2002, and the ceremony marking the launch of the consortium collaboration was held in Copenhagen, hosted by the Vice-Chancellors of the Universities of Copenhagen and Lund, Linda Nielsen and Boel Flodgren (*Sydsvenskan* 2002d). But the first sign of Danish relinquishment also appeared in 2002, when the Danish representative to the ESS Scandinavia Consortium, Robert Feidenhans'l, suddenly announced that Denmark would not be able to contribute financially to the ESS (ESS Scandinavia 2002g). A strategy to make Denmark adopt a more positive stance was discussed in the OS Group in the spring of 2003, with the focus on emphasizing the positive socio-economic effects of an ESS in Lund for the whole Øresund Region, and the possibility of locating some key part of the future ESS facility, such as the administration, to Copenhagen (ESS Scandinavia 2003h). In 2003, the importance of Danish support and direct Danish involvement was stressed in internal discussions, but their low degree of commitment was simultaneously acknowledged as a problem in need of a solution (ESS Scandinavia 2003b).

On January 7, 2003, representatives from ESS Scandinavia Henning Christophersen, Lars Börjesson and Kell Mortensen managed to put the ESS on the agenda for the coming meeting of the Nordic Ministers for Education and Research in March the same year, only to see this removed at the request of the Danish delegation (ESS Scandinavia 2003m). When the Danish Minister for Education, Helge Sander, was approached by his Swedish colleague Thomas Östros about the ESS in 2003, he seems to have dodged the question completely, expecting the ESS to end up in Britain (*Sydsvenskan* 2003f). An internal memo in the Offices of the Swedish Government from December 2003, summarizing the current state of the ESS project, the situation on the European stage, and the Scandinavian researcher-led initiative, states that the Danish Minister for Science, Helge Sander, "has declined Danish participation in hosting the ESS in Scandinavia in writing, during 2003" (Swedish Government 2003b). Meanwhile,

Mats Johnsson of the Swedish Ministry of Education noted in an interview in late 2003 that "Danish support is crucial, because if the ESS facility is to become reality, it will have to be through Swedish-Danish collaboration" (*Sydsvenskan* 2003e).

A key person in Denmark was Jørgen Kjems, who had represented Denmark and its national laboratory in Risø in the ESS R&D Council since its founding in 1997 (ESS R&D Council 1997), serving as its chair until 2000 (ESS R&D Council 2000a), and also participated in the inaugural meeting of the ESS Scandinavia Initiative in October 2003. Kjems, a former director of Risø National Laboratory (1997–2000), a physicist and long time neutron user, was a member of the ESFRI working group on neutrons of 2002–2003 (chapter 5), a 2005 working group on "future research infrastructure in Denmark" that took a positive attitude towards the ESS (Forskningsstyrelsen 2005), and in the expert panels advising ESFRI in its work on the consecutive infrastructure roadmaps in 2006 and forward, among many other things. In the first years of the 2000s, Kjems was very positive to ESS Scandinavia, but reportedly "turned against it" later on, with no explanation given (Interview: Berggren). Another person with influence in the Danish neutron community was Robert Feidenhans'l, a scientist at Risø National Laboratory and later professor of x-ray physics at the Niels Bohr Institute of Copenhagen University and director general of the European XFEL in Hamburg, who also participated in ESS Scandinavia Consortium meetings in 2002–2004. In 2003, Feidenhans'l authored a report entitled *Large Scale Facilities for Synchrotron Radiation and Neutrons – New Possibilities for Denmark* which mentioned the plans for an ESS in Southern Sweden but kept a strangely neutral position in relation to the initiative, advocating instead upgrades of the ILL and ISIS, and a greater presence of European scientists at the future SNS in Oak Ridge (Feidenhans'l 2003: 7, 13). The report caused surprise and resentment in Lund (Interview: Berggren).

The Vice-Chancellor of the Technical University of Denmark (*Danmarks Tekniske Universitet, DTU*), Lars Pallesen, although early on a pronounced supporter (ESS Scandinavia 2002b), later "was not so enthusiastic, to put it mildly" (Interview: Matic). Some claim Pallesen feared that the ESS would make Lund University more attractive for the most prominent natural and technical scientists, to the disadvantage of his university (Larsson 2019: 135), which others claim was a general trend among Danish university leaders, who simply felt envy towards their Swedish neighbors, and fear that they "might take the limelight from the Danish universities" (Thomasson and

Carlile 2017: 26) and that "money will flow not to them but to the ESS, on the Swedish side" (Interview: Börjesson).

At ministry level, however, the attitudes seem to have been mainly positive, and policymakers are said to have been "well aware" of the "tensions that existed in Denmark" but nonetheless "stayed committed to the project" (Thomasson and Carlile 2017: 26). In 2010, through the purchase of stocks in the Swedish limited liability company ESS AB, which was in charge of the design update and planning for the facility in Lund, Denmark became an equal partner in ESS Scandinavia and the future co-host of the facility, after years of negotiations (see chapters 7 and 9).

Campaigning, 2000–2003

The minutes of the inaugural meeting of ESS Scandinavia in October 2000 give a fairly detailed view of the prospects for the initiative as interpreted by the founding members, and their ambitions. As noted, the primary purpose of ESS Scandinavia was to gather and organize competence and interest in neutrons in Scandinavia so that Denmark, Norway, and Sweden would be represented in the continued work at European level towards the realization of the ESS. Locating the ESS in Scandinavia seemed quite improbable given that European neutron strongholds such as the Jülich Research Centre and the Rutherford Appleton Laboratory (RAL) were still site contenders, at least officially, but the idea was not out of the question – Lars Börjesson reportedly called it "a golden opportunity with a slim chance." Some participants at the October 2000 meeting, most notably Risø laboratory director Jørgen Kjems, voiced serious doubts over the prospects of a successful Scandinavian site proposal, calling an ESS Scandinavia initiative a "high-risk project." But the Øresund Region had some lure, and its attractiveness appeared as a most obvious selling point of an ESS Scandinavia campaign. The meeting highlighted several favorable features of the region: an international airport, several cities and towns with cultural activities (Lund, Malmö, Copenhagen, Lyngby, Roskilde), several universities, laboratories, and industries, and good infrastructure (ESS Scandinavia 2000). "We realized that one small country would not succeed here, that there would have to be several countries behind it, and then a border region is an easier sell" (Interview: Börjesson).

Fairly soon, as the consortium started working and Lars Börjesson, Börje Johansson and Aleksandar Matic started lobbying local and regional politicians, university managers, and university boards, it became clear that the most attractive option was to work to establish the ESS in Scandinavia, and, more specifically, in the Swedish part of the Øresund Region. "Most of them were interested in the general case for the ESS, but they were really interested in our idea to try to bring the ESS to Sweden. So based on these experiences, and as part of this work, we adjusted our arguments" (Interview: Börjesson). "Sure, it was a Danish thing. They had the competence in neutrons, but we had Lund" (Interview: Johansson). Lars Börjesson is quite frank in his recall of the sentiment back in these days: "I didn't think it was possible for it to end up in Sweden" (Interview: Börjesson). But Börje Johansson, especially, "was really good at seeing things strategically" and a "why not?" sentiment spread among the people involved (Interview: Matic).

Once it was clear that a campaign to situate the ESS in Southern Sweden would probably win the support of local and regional politicians, and other important supporters had emerged, such as Peter Honeth, the director of administration at Lund University, a local site selection process ensued, with more than half of the 33 municipalities of Skåne declaring willingness to host the facility (*Helsingborgs Dagblad* 2002), including Ängelholm (*Helsingborgs Dagblad* 2001b), Svalöv, Bjuv (*Helsingborgs Dagblad* 2001a), Trelleborg (*Sydsvenskan* 2002a), Landskrona, Svedala, Höör (*Sydsvenskan* 2002b), and of course, northeast of Lund, the area called Brunnshög (*Sydsvenskan* 2002c), where the ESS is now being built. The proposals were assessed by ESS Scandinavia on the basis of several criteria, including the quality of land, accessibility by road and public transport, proximity to urban areas and university environments, and to an international airport. The proposals from Lund, Landskrona, Höör, and Svedala were judged to fulfill the criteria set up, and ESS Scandinavia decided to continue working with Lund as the preferred option, with other options remaining in the game until the aptness of the proposed site in Lund had been thoroughly investigated (SWECO FFNS 2002: 5). Lund was, in many ways, the obvious location. "We were very clear that it should be located close to existing scientific environments" (Interview: Matic), and Lund University was already Sweden's largest and broadest university. "And the Øresund Bridge had just opened, with a rail connection from Copenhagen Airport to Lund. We realized, 'wow, you just hop on the train and you're there'" (Interview: Matic).

After the SWECO study of 2002 (see below), the Brunnshög location was considered to have been properly assessed and it became the designated site for a Scandinavian ESS. The City of Lund, and its planning office, began the local work of planning for the new Brunnshög area in late 2002, including city planning and initiation of the process to obtain the necessary environmental permits (*Sydsvenskan* 2002r), and Lund University shared the responsibility for local preparations, starting the applications for a local trial process of the type that all major construction projects need to go through ahead of start of construction. The City of Lund also put together a "host package" jointly with Region Skåne, including support infrastructure and housing in the city and region (ESS Scandinavia 2002k).

In 2002, the ESS Scandinavia initiative had achieved a maturity that allowed the compilation of a formal expression of interest and hence the ability to present the Scandinavian candidature at the May 17–19, 2002, meeting in Bonn. As noted in chapter 5, the meeting and the events in its aftermath were disastrous for the ESS project but meant the opening of a window of opportunity for smaller countries. The ESS Scandinavia Expression of Interest (EoI) competed for attention in Bonn with the candidatures of Jülich, the RAL in Oxfordshire, the Halle/Leipzig inter-state consortium, and the Yorkshire-ESS (YESS) (Berggren and Hallonsten 2012: 24–25). Lars Börjesson had presented the ESS Scandinavia proposal to the ESS R&D Council at its meeting in January the same year, highlighting the strong support from Swedish and Danish universities and "high-tech industries" in the region and pointing out the favorable local conditions, including rich cultural life, good transport, and a dynamic local knowledge economy (ESS R&D Council 2002b). The EoI echoed these descriptions of Lund and the Øresund Region. It was written and edited almost completely by Aleksandar Matic, who used a lot of material from the ESS R&D Council, a range of arguments compiled as part of the work of ESS Scandinavia in the one and a half years that had passed since the inaugural meeting of October 2000, and not least the results of a study done by consultancy firm SWECO commissioned by the City of Lund, which "gave a superb foundation for all the technical stuff" and "made the EoI look really good" (Interview: Matic). The SWECO study had made a basic investigation of many of the physical and technical aspects of a site proposal that had to be in place (see below), with the proper references to applicable legislation, and a lot of figures and diagrams that supported the argumentation. The EoI itself, a sixty-page document, cited a range of advantages of the designated site northeast of Lund, including "A rich scientific environment," "Excellent transport infrastructure," "Perfect technical conditions at the site," "A high quality of life," and a "High-tech industry environment" (ESS Scandinavia 2002a: 5, 10, 29–31). It placed particular emphasis on the scientific and industrial environments, including not least MAX-lab in Lund and Risø National Laboratory some 20 km west of Copenhagen, "with a tradition of almost 40 years in neutron scattering"; 12 universities within a radius of 100 km, with over 130,000 students and 10,000 researchers in total; and over 500 companies within the life sciences, with some 32,000 employees, all of which, ESS Scandinavia contends, will mean that "considerable synergy effects can be expected between the scientific institutions in the region and the ESS" (ESS Scandinavia 2002a: 7, 13, 19, 21). The cross-border region's recent experience of large construction projects is also mentioned, most notably the Øresund Bridge with adjoining tunnel and railways/motorways, the planned City Tunnel that would connect the bridge with an underground railway through Malmö, the Copenhagen metro, the housing fair in Malmö (opening 2001), the hospitals in Lund and Malmö, the Biomedical Center in Lund, and MAX-lab (ESS Scandinavia 2002a: 9). Competence in accelerator technology on both sides of the border, with MAX-lab in Lund, Swedish national laboratories in Stockholm and Uppsala, the University of Aarhus in Denmark, and private companies Scanditronix in Sweden and Danfysik in Denmark, is also mentioned (ESS Scandinavia 2002a: 10). The EoI also contained extensive descriptions of the excellent "social infrastructure," with several nearby major cities with an affordable housing market and good accommodation for visitors, and a rich cultural environment (ESS Scandinavia 2002a: 29–31), as well as the scientific landscape in Scandinavia (ESS Scandinavia 2002a: 33–38), and details on the site and the anticipated licensing and regulations procedure (ESS Scandinavia 2002a: 39–53).

As noted above, the latter in particular relied strongly on the SWECO study commissioned by the City of Lund in 2001–2002. Entitled "An overview of the preconditions for locating a research facility for neutron scattering, European Spallation Source (ESS) to the Brunnshög Area in Lund," the study was undertaken in collaboration with a working group from the Lund city planning office, Lund University, and ESS Scandinavia, and entailed specialized substudies on "climate," "geology, geotechnics, and hydrology," "technical supply," and "energy supply," (SWECO FFNS 2002: 2, 5) much in line with what was required of a site proposal, as specified by the general *Guidelines on how to submit an expression of interest to host the European Spallation Source*, issued by the ESS R&D Council in 2001 (ESS R&D Council 2001b). These specifications stated that the area proposed as a site for the ESS should be "ideal" from the perspective of construction of the facility, and for its eventual operation and activities, covering an approximate area of 1 square kilometer with expansion opportunities beyond this area; moreover, it should be made available so that the future ESS organization has full ownership and so that no fees or taxes on the land are applicable. The area should also be well-served by roads, power, water, drainage, phone, and data wires, and none of these conditions shall be subject to change over time (ESS R&D Council 2001b).

The SWECO report does a very thorough job of outlining the basic structure and function of the ESS facility, including accelerator, target, neutron guides and instruments, and support infrastructure, including lab buildings and office space ("conventional facilities"), also noting that the facility will be in operation 24/7 and all year long, host several hundred permanent employees, and facilitate temporary visits of several thousand external users annually. With regard to safety and operations, the report is also detailed in its descriptions of the different parts of the facility and how they will be constructed and interrelated, with different safety aspects and different modes of operation, and also lists probable environmental impacts and risks (SWECO FFNS 2002: 8–12). The report investigates the pros and cons of different locations within the designated area at Brunnshög, taking into account geology, topography, and climate, as well as both natural and cultural values, and making a thorough assessment of the availability of communications and power and water supply infrastructure, matching these with the specific demands for the site as detailed by the ESS R&D Council (SWECO FFNS 2002: 13–39). A creative detail in the report, which was also used in the EoI, was the proposal to use excess heat from the future ESS facility for district heating, which is relatively extensively built-out with a network reaching 90% of all buildings in the town, to which the ESS could be connected (SWECO FFNS 2002: 41–42). The conclusion of the SWECO report is that "no reasons to completely rule out the location of the facility within any part of the designated area have emerged, pertaining to those aspects that have been studied," and the report rather points at a number of advantages of situating the ESS in Brunnshög and Lund, including several "especially favorable conditions" in terms of physical planning, infrastructure, and embedment in a scientific and cultural environment (SWECO FFNS 2002: 51).

At the Bonn meeting, discussed in chapter 5, ESS Scandinavia was represented by Lars Börjesson, Aleksandar Matic, and Patrik Carlsson, complemented by a delegation of, among others, Lund University Vice-

Chancellor Boel Flodgren, Malmö Mayor Ilmar Reepalu, and local Lund politicians Christine Jönsson and Lennart Prytz (*Sydsvenskan* 2002e). In total, 40–50 delegates were present to back up the Scandinavian candidature (*Sydsvenskan* 2002f).

> I think we fared well in the competition. The SWECO report gave a firm foundation, and we had worked a lot on the environment in Lund, the universities, and the Øresund region. We had collected a lot of material, so we could just push it. And we had an ad agency in Copenhagen called Kiberg Gormsen, who designed the cover [of the EoI], [...] and signs, so we had a really nice showcase stand. (Interview: Matic.)

Another study of the effects of placing the ESS in Lund, more academically oriented and carried out by Gunnar Törnqvist, professor of economic geography at Lund University, had been procured by a newly created Committee for Research and Development of the Øresund Region, called *Øforsk*, in 2001. The study, published in early 2002 as a book with the title *Science at the Cutting Edge: The Future of the Øresund Region*, had the purpose of investigating the "effects upon the Øresund region if such a facility [the ESS] was located to the region" (Törnqvist 2002: 5). *Øforsk* had been formed in 1998, with funds from the Danish and Swedish governments, to support initiatives and promote Danish-Swedish cooperation in research, mainly through project funding (Olshov 2013: 187). Törnqvist had been very active in research on the economic geography of regions, and especially the Øresund Region (e.g. Maskell and Törnqvist 1999), and could draw on this experience when compensating for the fact that the study was carried out in a "relatively short period of time." A disclaimer, noting that the book "is primarily intended to provide a basis for discussion and future research," is found in the first chapter, where the author also states that "several of the ideas presented in the report will require to be further developed" (Törnqvist 2002: 11). On the basis of an extensive historical account of the close contact and association between Sweden and Denmark across the Øresund Strait and how it had recently been manifested physically in the shape of the Øresund Bridge (inaugurated in the summer of 2000), Törnqvist argues that the region has great potential in terms of attractiveness and creativity, and appeals to policy and decision makers to continue the work of integrating the region through the harmonizing of institutions and the lowering of other barriers (Törnqvist 2002: 14–23). The ESS is briefly presented and placed in a context of the most popular contemporary theories of socio-economic impacts and dynamic growth effects of a major research facility in an already well-equipped region, including "the knowledge-based economy," "the post-industrial society," "the Information Age," "a Europe of regions," "the new economic geography," and "cultures of creativity," complemented by a comprehensive summary of empirical studies into similar matters. Törnqvist evidently builds strongly on his expertise in the area and gives a convincing summary of regional development perspectives on the location of a major international research facility in the Øresund Region, including both theoretically relevant conclusions and empirical observations of sharpness and depth. The concluding chapter is very practically oriented and discusses both challenges and choices that lie ahead for decision makers, should the ESS become closer to reality in the Øresund Region, and realistic expectations for the development that would follow from it. Allan Larsson did not cite Törnqvist in his 2005 report (see a later section), but the 2010 TITA prestudy (see chapter 10) did – not explicitly as a source for any of its claims, but generally, as part of its reference list. It should therefore not be ruled out that the visionary (or even grandiose) message of Törnqvist's (2002) book, albeit expressed with considerable care and based on solid scholarly credentials, through hundreds of pages of analysis and discussion, had a role to play in the buildup of unrealistic expectations in Skåne around the years 2008–2011.

A window of opportunity, but waning domestic support

As he recalls, Börje Johansson was just about to enter the meeting venue in Bonn on May 17 when he noticed a news item in *Süddeutsche Zeitung* that contained the following final sentence: "Joachim Treusch, chair of the board of the Research Center in Jülich, is convinced that in the end, Schröder, Blair, and Chirac will give the neutron source away over breakfast" (*Süddeutsche Zeitung* 2002).

Börje Johansson's reaction was determination not to let this be a game played way above his head and on an international stage where scientists and local/regional politicians, no matter how enthusiastically they pushed their campaign, would not have a say. "We mustn't give up," was his conclusion (Interview: Johansson), and it was timely – the window of opportunity that opened for smaller countries in 2002–2003 gave new hope to those prepared to stay in the game and fight.

From 2003 and on, with the ESS R&D Council terminated and the initiative more in the hands of the respective site candidates, and with Germany effectively out of the game, the ESS Scandinavia initiative had some breathing space and could start working more efficiently on cost estimates, environmental investigations, socio-economic studies, and all the things that could back up the claims that Lund was the ideal site for the ESS (Interview: Berggren). A "Scientific Strategy" for ESS Scandinavia was adopted in the second half of 2002, and entailed plans for information about the proposal to build the ESS on local, regional, national, Scandinavian and European level, in order to build a strong support in the scientific communities that could be mobilized when the Government was approached with a proposal. The strategy also entailed plans for targeting scientific communities with weaker existing ties to neutron scattering technologies, and to inform neutron users and convince them of the excellence of the ESS Scandinavia proposal, as well as to build contacts in foundations and academies in Sweden such as the Royal Academy of Sciences, the Royal Academy of Engineering Sciences, the Wallenberg foundations, and Danish counterparts (ESS Scandinavia 2002h). In the spring of 2003, ESS Scandinavia was also in contact with the Nordic Investment Bank to gauge the possibilities of financing a future ESS facility in Lund in part with bank loans, and received tentatively positive responses (Nordic Investment Bank 2003). A 2004 study on energy consumption of the future ESS facility by *ÅF-Energi & Miljö AB*, a major Swedish consultancy firm in energy, industry, and infrastructure, concluded that the ESS would render hundreds of million SEK every year in electricity tax revenue to the Swedish state, and that the investment could thus even be a net gain for the Swedish state, depending, of course, on several details with regard to funding, costs, and technical design (ÅF 2004). Karl-Fredrik Berggren commented on the study in *Sydsvenskan*, saying, "It's a shame we haven't come up with this brilliant argument before," and adding that he hoped that the Ministry of Enterprise would show more interest now that the tax argument had been brought into the mix (*Sydsvenskan* 2004a). Another 2004 study, undertaken by the Research Policy Institute of Lund University, provided a general overview of possible socio-economic effects of siting the ESS in Lund (Hallonsten et al. 2004). The study seems to have had some significance, together with a Danish counterpart (Valentin et al. 2005), for the investigatory work of Allan Larsson which began the same year (see next section).

But regardless of how well ESS Scandinavia prepared the ground, and regardless of how much documentation and investigation it procured, the biggest challenge was to obtain the necessary political support above the local (Lund and Malmö) and regional (Region Skåne) levels. Partly for this purpose, contacts had already been established in the fall of 2002 with the communications and public relations firm Kreab, which specialized in political lobbying, to assist the consortium in communicating with policy and decision makers in Denmark and Norway and on a European level, although not in Sweden, where the consortium's executive group and project secretariat would still run the campaign (ESS Scandinavia 2002f). On October 14, 2002, an agreement was made to hire Kreab for six months, at Region Skåne's expense, amounting to approximately SEK 2 million until the summer of 2003, when apparently Lund University took over the bill and paid Kreab approximately SEK 100,000 per month for a year (*Sydsvenskan* 2003d; *Sydsvenskan* 2004c). This financial commitment by Lund University was subject to some internal criticism, and scrutiny by *Sydsvenskan*, which tried unsuccessfully to obtain the contract between the University and the PR firm after the Swedish Administrative Court of Appeals judged that the confidentiality of the agreement was important and that the University did not have to make it public (*Sydsvenskan* 2004c).

In 2002–2003, Kreab held a number of meetings with Danish and Norwegian cabinet ministers, members of the European Parliament from Denmark, Norway, and Sweden, and EU commissioners, which established contacts that the operatives of ESS Scandinavia would not otherwise have been able to achieve (ESS Scandinavia 2002p; Interview: Matic). In late 2003, Kreab's work was aimed more directly at Swedish politicians, in parallel with continued work to lobby Danish politicians and enterprise in both countries (ESS Scandinavia 2003p). In one of the last efforts made by Kreab before the collaboration was terminated in the summer of 2004 due to ESS Scandinavia's budgetary limits (see a previous section), the former Danish Minister for Finance and Kreab Brussels' partner and lobbyist, Henning Christophersen, approached several Danish politicians, industrial leaders, and public servants, including Prime Minister Anders Fogh Rasmussen and Deputy Prime Minister Bendt Bendtsen, to alert them to the new momentum ESS Scandinavia was gaining through the appointment of Allan Larsson as the Swedish Government's investigator (see next section), and to reestablish contact in order to again raise the issue to relevant decision making levels in Sweden and Denmark (Christophersen 2004).

ESS Scandinavia also conducted their own lobbying campaign, holding several meetings with the representatives of influential organizations in the Swedish academic system, research policy system, and industry, as well as Nordic and European counterparts (ESS Scandinavia 2002m; 2003a; 2003m; 2003q). Karl-Fredrik Berggren recalls that it was fairly easy to have access to industrial leaders, but that their interest was generally lukewarm and that they could not be counted on as allies in the national campaign. "It led nowhere. They gave moral support, but nothing more" (Interview: Berggren). Exceptions exist though, as when the Danish and Swedish industry confederations jointly approached several cabinet ministers in Denmark and Sweden in November 2003 (ESS Scandinavia 2003q).

Generally, it seems the window of opportunity that had opened for Sweden (and, presumably, other smaller countries) did not correlate with intensified support from politicians in Sweden, but that the absence of a decision at European level, the termination of the ESS R&D Council in 2003, and the apparent general sentiment at European level that the ESS was not going to materialize, at least not in the near future, instead caused some disappointment in Lund and Skåne, which led to waning interest and a significant weakening of the enthusiasm that had characterized the work of ESS Scandinavia in the first two years (see an earlier section). With Lars Börjesson appointed deputy secretary general of the natural sciences council of the Swedish Research Council, and then secretary general of the Committee for Research Infrastructures, and thus formally disconnected from ESS Scandinavia, with Aleksandar Matic starting a postdoc position at Chalmers University of Technology, and with Börje Johansson largely returning to his duties as professor at Uppsala University, the ESS Scandinavia secretariat at Lund University was diminished numerically, and to some extent also competence-wise. But Karl-Fredrik Berggren, Patrik Carlsson, and Carina Wickberg kept operations going in 2003–2004 and on, impressing people in the greater ESS Scandinavia network with their great stamina and enthusiasm. "They were really convincing, and I believe it was genuine. I don't think you would be able to keep up that energy unless it was genuine" (Interview: Kinhult).

The lack of support from Uppsala and Stockholm was discussed in a previous section, and it played out in a very specific way in the attempts of ESS Scandinavia to win the support of the national government. Individual members of parliament, especially from Skåne, acted to draw the attention of their peers and the Government to ESS Scandinavia. A delegation from ESS Scandinavia visited the Swedish Parliament in April 2003 and informed MPs from Skåne about the ESS Scandinavia plans, with a "good response" (ESS Scandinavia 2003i). The first action on the part of individual MPs was in the fall of 2003, when Lars Lindblad of the Moderate Party (*Moderaterna*, the main opposition party) proposed a motion for a governmental initiative regarding the ESS (Lindblad 2003). From the fall of 2003 and until the Government's declaration of intent in February of 2007, nine motions from individual MPs or groups of MPs, sometimes from single party groups and sometimes gathering members of different parties, were filed calling for a governmental initiative to support the ESS Scandinavia initiative and work toward establishing the ESS in Lund. Before the shift in government of 2006, most motions came from members of the opposition parties (Nylander et al. 2004; Wahlgren 2005; Lindblad 2005; Lindblad et al. 2005), but a further two came from members of the governing Social Democratic party (Bernhardsson et al. 2004, 2005). In the fall of 2006, members of the four parties in the newly installed coalition government (see chapter 7) and members of the new opposition filed motions calling for the Government to act (Olander et al. 2006; Jönsson et al. 2006). In 2005 and 2006, motions were also filed from MPs representing the Green Party, calling for the Government not to support the ESS Scandinavia initiative and not to work toward locating the ESS in Lund, citing environmental concerns and costs (Holm et al. 2005, 2006).

When ESS Scandinavia was established, in October 2000, Sweden had had a Social Democratic minority government for six years, with the parliamentary support of the Left Party (and, from 1998, also the Green Party), and with four opposition parties. Since 1996, the Prime Minister had been Göran Persson, who was known for his autocratic leadership style, and enjoyed strong domestic support, especially after the Swedish presidency of the European Council in the spring of 2001, although he had experienced many setbacks from 2003 and on (Svenning 2005: 305, 330). Persson and the Social Democrats had stayed in power after the election of 2002, with support intact from the Left and Green parties, but towards the end of the term, and as the 2006 election drew closer, the Government and the Prime Minister appeared both exhausted and passive, an impression later confirmed by Göran Persson himself (Persson 2007: 456–457). The tighter center-right collaboration between the four opposition parties under the banner "Alliance for Sweden" ("*Allians för Sverige*") consequently produced a decisive electoral defeat for the Social Democrats in 2006, which turned out to be key to the ESS Scandinavia campaign (see chapter 7).

On January 22, 2003, Swedish daily newspaper *Svenska Dagbladet* reported that the ESS was a topic of discussion in the Government offices

(*Svenska Dagbladet* 2003). Meetings were taking place with the state secretaries of several ministries (presumably including education, enterprise and finance), to discuss ESS and ESS Scandinavia, (ESS Scandinavia 2003f), and on January 28, Karl-Fredrik Berggren and Patrik Carlsson met with the Assistant Minister for Finance, Gunnar Lund, who was "interested and positive," and seemingly unperturbed by the estimated costs of the ESS (ESS Scandinavia 2003f), reportedly promising to raise the matter with Minister for Education and Research, Thomas Östros (ESS Scandinavia 2003c). It was apparently not yet possible to arrange a meeting with Östros himself, in spite of signals from ministry aid Mats Johnson that Östros was "positive to the ESS and ESS Scandinavia" (ESS Scandinavia 2003g). In the spring of 2003, it became clear to the consortium and its members that a government decision regarding the ESS was not to be expected within the coming year. The Swedish Government seemingly interpreted the German decision to pull out (see chapter 5) as a reason not to rush any decision domestically. In March, Ministry of Education official Mats Johnsson was quoted in *Sydsvenskan*, saying that "[t]here's no hurry now that Germany has pulled out," and that while the issue had been discussed with other ministries concerned, and "everyone thinks this is very exciting and interesting," the Government offices had "not yet had reason to consider the issue of funding" (*Sydsvenskan* 2003b).

But ESS Scandinavia continued to push for action in Stockholm. The ongoing "constructive" dialogue with the Swedish Ministries of Education and Enterprise continued, and several leading politicians in Malmö and Lund also acted to convince their peers on national level to pay closer attention to the ESS project (ESS Scandinavia 2003p). Several letters to Swedish politicians in support of ESS Scandinavia, from scientists, officials of Lund University, and industrial leaders, produced follow-up meetings, but only at deputy level: "We always got stuck before the ministerial level" (Interview: Berggren). Representatives of ESS Scandinavia visited the Ministry of Enterprise and met with state secretary (deputy) Claes Ånstrand on May 16, 2003, asking for the Government's support in their campaign. After the meeting, discussions between the Ministries of Enterprise and Education ensued, but no further action was taken, and state secretary Ånstrand voiced the opinion a month later that the possibilities of achieving a European funding solution for ESS were meager, and the project's future, therefore, "uncertain" (Swedish Government 2003a). At the meeting of the Nordic Council of Ministers in October 2003, secretary general Per Unckel brought up ESS Scandinavia in his opening remarks and suggested a pan-Nordic investigation of the prospects of hosting the ESS in Scandinavia. Thomas Östros responded that he would return to the issue "within two weeks," but seems to have ignored it thereafter (ESS Scandinavia 2003q).

The most important move towards a Scandinavian and Swedish ESS taken by the social-democratic government of 2002–2006 was to appoint Allan Larsson to investigate the matter – indeed, involving Larsson in ESS Scandinavia might very well have been a decisive action and one of the most fortunate turns of events in the whole campaign (see below), but at the time, it was probably viewed as a disappointment on the part of ESS Scandinavia, who had lobbied for years for a governmental intervention and had to see its cause put on hold for another period of time. It should be noted that an investigation was the normal response of the national government to a proposal and campaign like ESS Scandinavia: Any Swedish government would likely have had to conduct something similar as part of the preparations to make a decision on whether to invest on this scale. Therefore, the appointment of Allan Larsson to investigate the matter, in July 2004, can be taken as sign that the discussions in and between ministries had reached the point where the issue could only be taken further by a thorough investigation. On the other hand, the Government was considerably passive and there are many things to suggest that the investigation was put in place not on the initiative of the Government offices or a specific minister, but by Allan Larsson himself (see below). Also, in the view of the consortium members and representatives of ESS Scandinavia, the Government had dragged its feet for several years, and the frustration among the campaigners at the time is quite clear from meeting minutes and interviews.

There are several possible reasons that the Government did not act more forcefully on the ESS Scandinavia proposal in 2002 and on, and the safest bet is that all these reasons contributed. First of all, the ESS was a very expensive, high risk project. Another likely important factor was timing; the issue simply had to mature. Meanwhile, there is also reason to suspect that the ESS Scandinavia proponents displayed political naivety and did not fully understand how a campaign directed at politicians, in Sweden or elsewhere, should be conducted (Interview: Börjesson). There are signs that the collaboration with Kreab, a leading actor in political lobbying, yielded some results, although the main outcomes seem to have been better contacts in Brussels, which helped little given that the ESS (like all large scientific collaborative projects in Europe) was, and is, detached from EU politics, and in Denmark, which remained hesitant anyway, and whose involvement in the ESS was settled only after the Swedish Government had decided to act. Another sign that political naivety characterized the ESS Scandinavia campaign in the early years is that once a politically skilled person like Allan Larsson was recruited to the ESS cause, things started moving.

What is clear, however, is that the ESS Scandinavia campaign gained little sympathy from the Government and its Minister for Education, Thomas Östros, in 2002–2004 (Interview: Berggren). "Östros was cautious. He probably hadn't made up his mind. But I don't think he believed in it. When I talked to him, he had no real opinion, but he didn't seem to believe in it" (Interview: Johnsson). On 21 October 2004, Thomas Östros became Minister for Enterprise, and the Ministry of Education was merged with the Ministry of Culture, under Leif Pagrotsky, who had little or no experience in education and research. "Leif [Pagrotsky] had no real knowledge about this, no real knowledge about science or higher education. So he had to learn it all, and I can't remember that he had any opinion about the ESS" (Interview: Eliasson). "Pagrotsky had no real opinion" (Interview: Johnson). One question concerns Prime Minister Göran Persson, known for his autocratic leadership style and for paying attention to detail. Some claim that Persson undoubtedly knew about the ESS and was negative toward it, which might explain in part why the Social Democratic government did not act. Some sources seem to indicate that the Prime Minister had indeed taken ownership of the issue in the Government: At a meeting of the Nordic Council of Ministers in Copenhagen in April 2005, Leif Pagrotsky explained that the Swedish Government was discussing a possible Swedish bid but that it was awaiting the results of Allan Larsson's investigation and that "the decision will be made by the Prime Minister" (Nordic Council of Ministers 2005). Others confirm this, claiming that Persson clearly was informed about the issue (Interviews: Berggren, Matic), and that he was against it (Interview: Johnsson). Rumor has it that the ESS was to be included in the spring 2006 financial bill, but that the Prime Minister personally removed it (Interview: Berggren).

The question is why. An interesting connection can be established between the Vice-Chancellor of Karolinska Institute Hans Wigzell, who allegedly was personally against the ESS (Interview: Berggren) and represented one of very few organizations in the Swedish science system who openly and directly opposed the ESS (see next section). Wigzell was informal research policy advisor to the Prime Minister in the early 2000s (Interview: Johnsson) and could possibly have convinced Göran Persson not to act on it. But the timing should also not be underestimated. As noted above, the obvious thing for any Swedish government to do when something like the ESS comes up is to put an investigation in place, and in 2004, when the dust had settled on the European stage, and discussions among ministers and deputies in Sweden had possibly run out of steam, a governmental investigation was the natural next step, and needed proper time. Once it had delivered its report, and the report had gone through the obligatory referral round (see next section), it was the fall of 2005 and too close to the 2006 general election. "You don't enter an election campaign with such a thing" (Interview: Kinhult). "The signal from the Ministry of Finance was that we'd do this after the election" (Interview: Eliasson). "I have been told that if they would have won the 2006 election, then the Social Democrats would probably have moved on the ESS" (Interview: Berggren). Connected to this is the role of the Green Party, which was clearly against the ESS, and which was part of the parliamentary base for the 2002–2006 Social Democratic government. Some claim their influence over the issue was significant enough to prevent the Government from acting on it (Leijonborg 2018: 338–339; Interviews: Börjesson, Berggren).

The Larsson Investigation

On July 8, 2004, the Government appointed Allan Larsson "special negotiator" (note the title, not "special investigator," see below) to investigate the prospects of establishing the ESS in Lund and Sweden. Larsson was a well-known Social Democratic politician and public servant, among other things Minister for Finance for a year and a half in 1990–91, Member of Parliament 1991–1995, and Director General for Employment, Social Affairs and Equal Opportunities in the European Commission under commissioner Pádraig Flynn 1995–2000, and was appointed chair of the board of Lund University in 2003, succeeding industrialist Lennart Nilsson. Assuming the role on January 1, 2004, in his own account, Larsson immediately started looking for "big things" to get involved in; "as chair, one should do something, not just sit there with gavel in hand" (Interview: Larsson). Carina Wickberg, ESS Scandinavia project coordinator and employed in the planning unit of the University's central management, made sure to include the ESS plans in the material used to brief Larsson upon his arrival. "And Allan was totally hooked" (Interview: Wickberg).

A meeting with Allan Larsson, Karl-Fredrik Berggren, Patrik Carlsson, and Peter Honeth was swiftly called, and a comprehensive presentation made. Larsson himself apparently understood immediately that the ESS project was in a political limbo, with the initiative being shuffled between Lund University and the Danish and Swedish ministries of education and science, "and nobody really had any grip on it" (Interview: Larsson). While the project was clearly in the interest of Lund University and, hence, appropriate for its chair to promote, Larsson was also attracted by the prospects of challenging the big European countries and bringing a megaproject to Scandinavia. What was missing was the political momentum, but Larsson had the right contacts and was able to arrange a meeting with Thomas Östros right away (Interview: Larsson). This took place in mid-March 2004, and the minister reportedly "promised to work for an initiative by the Government and begin the necessary preparations," which in practice meant that Larsson was to be given the assignment to conduct an investigation (Larsson 2019: 84–85). In July 2004, the Government decided "to appoint Allan Larsson special negotiator for the Swedish Government, to investigate the possibilities of locating the ESS in Sweden" (Swedish Government 2004d). The assignment specifically included clarifying the interest from the scientific communities and the private sector, analyzing the expected long-term growth effects, clarifying the possibilities of private sector involvement in the financing of an ESS facility in Sweden, and investigating regional, Nordic and European prospective funders and their interest in participating, as well as the interest of European countries and the European Commission. The task was to be concluded and the results presented on July 1, 2005, at the latest, after which a possible decision by the Government regarding a Swedish bid to host the ESS could be expected (Swedish Government 2004d). The Swedish Research Council was instructed to "assist the negotiator in scientific assessments of the project and in initial contacts with enterprise," and "relevant embassies shall also assist in the negotiation assignment" (Swedish Government 2004d). The expertise of the ESS Scandinavia consortium and secretariat, most importantly Karl-Fredrik Berggren and Patrik Carlsson, also assisted Larsson throughout the work on the investigation (Larsson 2019: 87; Interview: Berggren).

It has been suggested that it was thanks to Larsson that the Government finally acted to have the prospects for a Swedish ESS investigated. "He was interested, and convinced Östros to appoint him to investigate the issue, not the other way around" (Interview: Berggren). "Allan wanted to look closely at the matter, and persuaded the Government to give him the task" (Interview: Johnsson). Interestingly, commentators also agree that when Allan Larsson is truly involved in something, he will push it very hard: "I used to say: If you want something done, ask Allan Larsson" (Interview: Eliasson). Note also the title *special negotiator for the Swedish Government*. It is not entirely clear whether this title really reflected the assignment, and why it was chosen. But it is clear that it gave a general impression that Larsson had a government mandate to probe the terrain in Europe on behalf of the Government, not merely to investigate the situation. And Larsson clearly undertook his first tentative negotiations in 2004–2005. In a presentation to the Nordic Council of Ministers on March 29, 2005, Larsson reported on his plans for negotiations with Swedish companies "about cofounding the facility within the framework of a PPP solution (Public-Private Partnership)," and on deliberations already held with representatives of the governments of Denmark, Norway, Finland, Iceland, Estonia, Latvia, Lithuania, Poland, and Germany, and with the European Commission and the OECD. He also reported on results: The Baltic countries were reportedly interested in participating in and supporting a Scandinavian ESS; the Polish government was "open to continued dialogue"; and "leading representatives of the German government" were reportedly interested in a Swedish hosting of the ESS, "on the condition that Sweden can fulfill certain financial conditions." Deliberations with the European Commission had apparently yielded a tentative pledge to fund 10% of the construction costs for the future ESS (Larsson 2005b). Clearly, Larsson's role went above and beyond investigating. "Everyone was convinced that the Swedish Government was actively pushing this. That's what he wanted, of course" (Interview: Johnsson). The title *special negotiator*, in combination with Larsson's prior assignments in the Swedish Government and parliament, and the European Commission (see above), certainly opened many doors. In his memoirs, Larsson himself defines his work in 2004–2005 as "negotiations" rather than investigation (Larsson 2019: 89).

The content of these negotiations, and their outcomes, will be returned to in chapter 7, but it can be noted here that Allan Larsson visited some thirteen countries to gauge their interest in the ESS project and the Scandinavian proposal, and made a number of other considerations that are summarized in his report "*Svenskt värdskap för ESS*" ("Swedish hosting of the ESS"), delivered in June 2005 (Larsson 2005a). Discussing costs, financial effects, socio-economic impacts, and environmental concerns, along with a general assessment of the prospects of establishing the ESS in Lund, Larsson strongly recommended the Government to act and "within 2005 give a declaration of intent that Sweden is preparing an offer to host the ESS, in order to thereby clarify the preconditions for Sweden to offer to become the host" (Larsson 2005a: 9). The recommendation was, however, conditional. The investigation envisioned that Sweden, if it became the host of the ESS, would have to commit "two types of funding" to the project's realization, namely (1) a "basic funding, which follows from Sweden's GDP share among the participating countries and which Sweden will pay regardless of whether the ESS is built in Sweden or another European country" (if Sweden is to participate in the ESS, that is, which Larsson views as self-evident), and (2) a "site premium," i.e. an "extra allocation, motivated by economic and other advantages that hosting countries will reap" (Larsson 2005a: 10).

The first, Larsson argued, should be funded "just like in the case of other European research facilities [that Sweden is a member of], with research policy funding," by which is meant ordinary governmental R&D appropriations. This part of the funding "can be expected to amount to approximately 3 per cent if the participation in the ESS is on the same level as other European research facilities," which Larsson estimated at SEK 40–60 million annually and concluded was on the level of the existing commitment to the Swedish national research reactor in Studsvik (see chapter 4). The second type of funding, on the other hand, "should in its entirety be funded by industrial support funding, from the Government and the private sector" (Larsson 2005a: 33–34). In its "discussion" section, the report acknowledges the arguments against a Swedish bid, stating that "if a Swedish hosting of the ESS would be modeled only as a research policy project," the extra allocation of money to construction and operation of the ESS that would be required from the host country would mean significant displacement effects in the national R&D budget, such as the "creation of other strong research environments" in Sweden (which had been a new theme in the 2004 research bill and would be a major theme in national research policy in coming years; see chapter 4). "Although the studies of growth effects and the effects of ESS on the Governmental budget are very positive," the report continues, "the income generated by the ESS will not be of direct benefit for the Governmental R&D budget" and, therefore, it will "be hard to win the necessary support for a Swedish bid to host the ESS purely on research policy grounds." A prerequisite for achieving a broad support for Sweden to host the ESS is therefore "that the initiative has a strong industrial policy and growth policy motivation and that it receives corresponding financial support." Larsson estimates the preconditions for such a public-private-partnership (PPP) covering "those 20–25 per cent [of the total construction and operations funds] that Sweden will have to commit in addition to its GDP share in order to have support for its bid" to be "good" (Larsson 2005a: 29–30). No details are given on what could be expected from the Swedish private sector, but fourteen years later, Larsson (2019: 172) explains "what he had in mind," namely a contribution from industry of 5%. This figure is, however, not mentioned in the 2005 report. Quite a lot of text in the report is, on the other hand, devoted to motivating an industrial commitment to the ESS, by discussing potential benefits for local business activities and regional economic growth, from the breeding ground for technological development and innovation that the ESS would make up, and from the direct effects of the large investment and the inflow of a high-skilled workforce and several thousands of annual visiting users to the facility (see below). In this sense, the proposed PPP solution was well-motivated, but seen from another perspective, it was peculiar, and it is unclear what gave Larsson the idea that it would be viable. It had very meager support among those companies and enterprise organizations that Larsson consulted as part of the investigation, who were generally positive to the ESS Scandinavia proposal (Larsson 2005a: 75–77) but who apparently had been very clear in meetings with Larsson during his negotiations that they would not be ready to invest in the ESS (Larsson 2019: 94). Karl-Fredrik Berggren remembers a meeting that Larsson had called in Stockholm in 2005, with several industrialists, who demonstrated clearly that they would not be prepared to make the kind of long-term investments that the ESS project required, and that this was the moment when he "stopped believing in the PPP solution" (Interview: Berggren).

There is much to suggest that Larsson had political motives for proposing a PPP solution. Anders Granberg (2012), who has made a thorough review of the Larsson investigation and its responses in the academic community (see below), notes that "Larsson is clearly at pains to remove, or at least substantially reduce, the risk (and fears) that the ESS would encroach on the national research budget," and places strong emphasis on "the distinction between research-policy motives and economic-political motives in justifying a Swedish candidacy," with the main point being that "both types of justification are needed." "In other words, trade-offs and crowding-out effects can be avoided only if financial support can be mobilized outside the research-policy sphere; hence the need for a PPP solution" (Granberg 2012: 130). Larsson confirms this in his memoirs, writing that "with an enterprise and regional policy funding, it would be easier to explain to the scientific community that the ESS would not be a cuckoo in the scientific nest, that would displace other urgent projects" (Larsson 2019: 172).

But while this argument is completely understandable from a political point of view, it makes no sense when looking at historical and contemporary evidence. Comparisons abound across Europe: Both the ILL and ESRF received over 25% of their funding from the host country (France), and the intergovernmental negotiations over the European XFEL in Hamburg (a facility of similar size and scope to the ESS) reached a Memorandum of Understanding between eight countries in 2004 only after Germany, the host, had pledged to fund approximately half of the construction costs (Hallonsten 2014). In none of the cases was a PPP solution even on the table in the negotiations (Cramer 2020). The same is generally true for national large-scale facilities, in Europe and North America alike, although, of course, most of these are built and funded by larger countries than Sweden, which makes the comparison slightly unfair. But even with this in mind, PPP solutions have never been used for scientific projects comparable to the ESS, and were consequently not used to fund the Swedish commitment to the ESS: As discussed at length in chapters 7 and 9, the full Swedish contribution to the construction costs was funded within the Governmental R&D budget, and no private funding was ever solicited. Obviously, in 2005 nobody could have foreseen the dramatic increases in governmental outlays on R&D from 2007 and on, that both contained the ESS commitment and increased every other science appropriation (see chapter 4), so the comment in a previous analysis that Larsson's conclusion that the Swedish ESS commitment could not be funded within ordinary frameworks was "erroneous" is unfair, although correct (Hallonsten 2015a: 419).

But the PPP solution is not the only unrealistic feature of Larsson's investigation. Appended to the report is a study by the Institute for Growth Policy Studies (*Institutet för Tillväxtpolitiska Studier, ITPS*), a governmental research institute, where attempts are made at quantifying not only the expected direct socio-economic effects of locating the ESS in Lund but also the indirect effects which, according to the report itself, are "difficult to measure," a caveat that the report repeats several times. The study claims that "[d]irect effects on income and employment are of significant importance and will diminish the cost of the investment considerably or completely neutralize it," due primarily to multiplication effects. Indirect effects will come through "increased investments in Sweden" and a growth in knowledge and technological development that raises the "total factor productivity (TFP)"; the latter is calculated with formula from Guellec and van Pottelsberghe de la Potterie (2004), data from some OECD countries (not including Sweden) in the years 1980–1998, a series of estimations and extrapolations, and the assumption that the ESS will mean a net increase of R&D investments in Sweden. The report calculates a likely Swedish GDP growth as a result of locating the ESS in Sweden of SEK 4.3 billion, corresponding to 0.17%, and creating roughly 6,000 new jobs (Larsson 2005a: 116–118). A somewhat more qualitatively oriented analysis of regional enterprise in Skåne leads to the conclusion that productivity increases and the founding of new businesses in the region will have "potentially significant growth effects." An interesting note towards the end of the study, repeated by Larsson in the main report, is that "the benefits won't materialize by themselves" but require that the ESS is "incorporated into a suitable innovation system" (Larsson 2005a: 72, 125). Very little effort is made, either by ITPS or by Larsson, to discuss whether such a "suitable innovation system" is in place in and around Lund, whether it can be built up at the adequate levels as part of a governmental investment in the ESS, or what such an innovation system should consist of.

Larsson repeated several of the conclusions of the ITPS study in his own recommendations to the Government, although omitting to mention the many method-related uncertainties and the repeated acknowledgement that effects are difficult to measure. Thus, while the ITPS study repeatedly nuances its own calculations and results with statements of their uncertainty and arbitrariness, which makes the claims ambiguous and indefinite, their use by Larsson in the main report contains no such caveats but merely repeats the very optimistic inferences. Thus, hiding, in a sense, behind the ITPS, and exaggerating the expected positive effects of an ESS in Lund, Larsson manages to paint a very attractive picture that almost completely neglects the issue of whether Sweden is scientifically equipped to reap the benefits of the ESS, and points almost solely to socio-economic benefits: "There are, according to ITPS, grounds for asserting that the ESS would open possibilities for a dynamic economic development at regional and national level" (Larsson 2005a: 19). A senior official in the Government offices notes:

> I worked with the final editing of the report, so I have read it thoroughly. I was very critical of the appendix [the ITPS report], because Allan needed someone

> who could point to an economic gain from building the ESS, and then he went straight to Sture Öberg at ITPS. I don't know what happened there, but he apparently appointed someone who later had to present the study to us, and I felt sorry for him for having to do that job, because it meant vouching for something that was shaky. And therefore Allan put it in the appendix and just referred to it, "ITPS has said that…" And so he hasn't said it himself but let someone else say it. […] I experienced Allan's dishonesty then, he lets someone else do the dirty work and he can just refer to it, "Look, this is how much we will gain…" (Interview: Johnsson.)

It is hard to say whether Larsson knew of the lack of scientific strength in the related areas, or if he did not understand the complexity of this matter, or if he simply bought into the marketing efforts by the ESS campaign. What is clear, though, is that he ignored the fact that Sweden was one of the least prepared countries in Europe, *scientifically*, to host the ESS. Overall, it is quite clear that it was not in Larsson's interest to discuss the scientific case for the ESS, and what the facility would actually do and how. He viewed the ESS as a generic megaproject, which was expected to produce dynamic effects and economic growth in the region. Years later, when asked about the skew of the report in this regard, Larsson claimed that he was not competent to make a scientific evaluation (Interview: Larsson), which seems humble but is a rather cheap excuse. Having consulted ITPS about regional growth scenarios, Larsson could have consulted, say, the Royal Swedish Academy of Sciences about the strength of the Swedish neutron user communities. Not least since the Government's decision to assign Larsson to the investigative task clearly stated that the assignment included "to clarify the interest from the scientific community" (Swedish Government 2004d).

This part of the assignment was clearly undertaken with an imbalance of attention to the arguments of the ESS Scandinavia campaign, and probably also on the basis of a positive bias on the part of the investigator/negotiator himself. For example, Larsson repeats the conclusions of the 2002 ESS Scandinavia EoI in chapter 4 of his report, including the claims that Lund provides a very "rich scientific environment," "outstanding transport conditions," "perfect technical preconditions on the site," a "high quality of life," and a "high-tech industrial environment," completely uncritically (Larsson 2005a: 55–59). The report also relies strongly on consultations with the OECD Global Science Forum, the European Science Foundation, the European Neutron Scattering Association (ENSA), and the European Strategy Forum for Research Infrastructures (ESFRI) (Larsson 2005a: 20), all of which were supportive of the ESS (Larsson 2005a: 42, 73), although hardly any of them were representative of a Swedish or Scandinavian scientific community. Citing the judgment by ESFRI that the ESS is an important project, not least for continuing Europe's global lead in neutron scattering (Larsson 2005a: 74), is a mere repetition of the largely uncontroversial opinion that Europe is in need of a new neutron source to close the looming "neutron gap" (chapter 5) and take up competition with Japan and the United States (where spallation neutron sources were under construction at the time, based largely on the 1997 ESS reference design). Larsson makes no assessment, and collects no balanced opinions, about the scientific need for an ESS *in Sweden*, the fit (or lack thereof) between the experimental capacities offered by a future ESS and what the Swedish scientific communities need and/or would benefit from, or the appropriateness of prioritizing an investment in the ESS in the Swedish science budget. Of course, the latter is unnecessary given that the PPP solution is used as premise for the whole investigation, and all its conclusions.

Overall, it can be concluded that Larsson made very little effort to interact with scientific communities, other than with the organizations mentioned and the tight collaboration with ESS Scandinavia. The members and representatives of the ESS Scandinavia consortium most likely comprised the group of scientists and science administrators in Sweden most biased in favor of a Swedish ESS bid, and their support for the idea of a Scandinavian ESS, as noted by Larsson (2005a: 74) should, at best, be read as a tautology. The Swedish Research Council's proposal to bring a major European research facility to Sweden in order to strengthen the international position of Sweden in science, and the initiative from the Swedish Research Council and the Swedish Agency for Innovation Systems (Vinnova) to create an ESS Innovation Forum cannot reasonably be the reliable signs of "commitments from the scientific community" as presented by Larsson (Larsson 2005a: 20–21, 42, 75). The "scenario for developing, building, and operating the ESS in Scandinavia" presented in chapter 9 of the Larsson report was developed after consultations "with a number of Swedish companies with broad experience and high-level competence in organizing and executing large international facility and technology projects" (Larsson 2005a: 25); yet no representatives from the scientific communities with experience of developing and building large scientific facilities were involved. Such consultations could have made Larsson aware of many important features of the process of planning and building international research facilities like the ESS, including the growing importance of *in-kind* contributions in their financing solutions (chapter 3), which would also prove important in the case of the ESS (chapter 8), the patterns by which scientists use similar facilities (chapter 3), and the challenges of a small country like Sweden in making the complementary investments needed to reap the benefits (see below and chapter 11). The latter in particular would probably have made Larsson reevaluate some of his conclusions about the socio-economic effects for Sweden and the local environment, and the prospects for Sweden, including its existing scientific activities, to reap the benefits of locating the ESS in Sweden, in comparison with the costs it would incur to be able to properly do so.

The report includes a chapter about permit issues and the environment, although it is clear that Larsson wants to avoid these – "it has not been explicitly part of my negotiation assignment" (Larsson 2005a: 81) – and the analysis in the chapter is based entirely on information provided by ESS Scandinavia. Obviously, Larsson refers to the fact that any facility similar to the ESS would have to go through due process and fulfill all the prerequisites in national legislation and European regulation, and that it "will fall on local and national authorities to decide if the ESS fulfills these demands" (Larsson 2005a: 81). The chapter in question, therefore, mainly contains a comprehensive review of the typical procedure that the ESS will have to go through in terms of permits, and a section on the environmental aspects based on information from ESS Scandinavia, where the ESS is compared primarily with nuclear reactor-based neutron sources of the kind common in Europe, which is a comparison that clearly comes down in favor of the ESS, a review of the investigations procured by ESS Scandinavia concerning energy and land use, radioactivity, and safety, and some highly speculative arguments about all the positive contributions to developments in the area of environmentally friendly technology that the ESS will make (Larsson 2005a: 84–89). The final chapter contains a scenario for the planning, construction, and operation of the ESS (Larsson 2005a: 91–97) that, with the benefit of hindsight, can be judged somewhat optimistic but not completely unrealistic.

The Larsson report was sent to a large number of *referral bodies* (*remissinstanser*) for review, as is always done with Swedish governmental investigations. The list of addressees includes 83 government agencies, regional and local authorities, business enterprise organizations, universities and university colleges, research councils, and labor organizations. As in all similar cases, anyone (organizations and private citizens alike) was free to spontaneously express an opinion about the investigation report, in written form, and their letters must also be attached to the file of the investigation and considered in the continued work on the issue. Of the 68 responses collected, 27 were from universities, university colleges, and subunits of these; five were from non-university academic organizations such as the Swedish Neutron Scattering Society and the Royal Academy of Sciences; six were from governmental research funding agencies including the Swedish Research Council and Vinnova; eleven were from other governmental agencies including the Swedish Nuclear Power Inspectorate and the Swedish Radiation Protection Authority; three were from municipalities, including the cities of Lund and Malmö; 13 were from NGOs, interest organizations, and private companies, including the environmental protection action groups mentioned earlier in this chapter as well as the Association of Swedish Engineering Industries, the Swedish Trade Union Confederation, and *Studsvik Nuclear AB*; one came from a political party, namely the Left Party in Malmö; and two were spontaneous responses from private citizens, namely Bengt Jönsson and Bo Wennergren (Granberg 2012). Ten responses contained no comments at all. While Larsson's own memory is that the report "had a very broad support from the referral bodies" (Larsson 2019: 36), and the same assessment was the message of ESS Scandinavia's summary of the referral responses in November 2005 (ESS Scandinavia 2005c), a thorough review gives a somewhat more nuanced picture.

More than half of the 68 responses came from academic organizations and research funders, and a clear majority of these, writes Granberg (2012: 136) in his comprehensive analysis of the referral round, "express a positive valuation or endorsement." Seven of these positive responses came from members of the ESS Scandinavia Consortium, which, of course, is "exactly what was to be expected." Only two of the "academic" respondents rejected the proposal, Karolinska Institute and Uppsala University, and the Royal Academy of Sciences accepted it only with reservations (see below). But most of the positive responses and the endorsements came with similar very clear reservations: the financing of the facility should not be allowed to interfere with the ordinary governmental R&D appropriations. In other words, the model suggested by the Larsson report, that the "site premium" be paid by industrial support funding both from the Government and from the private sector, was seen as a prerequisite by the respondents. Interestingly, however, the respondents' expectations of benefits are almost exclusively in the scientific area – few of them discuss the socio-economic benefits that are so prominent in the Larsson report (Granberg 2012: 136–138).

In a sense, what Granberg (2012: 141) identifies as a rather unselfish attitude towards the ESS project – that is, the tendency of representatives of the Swedish research community to view the future benefits not so much from their own perspective but rather generally, and for Swedish research more broadly – is a sign that the anticipated benefits of a Swedish ESS were quite vague. It is important to note that in 2005, when the responses were collected, the ESS was very sketchily defined in terms of technical design, scope, and future scientific use. In addition, the funding model proposed by Allan Larsson likely worked to calm fears in the community: Facing the question of whether a Swedish bid to host the ESS was a good idea or not, the respondents could only rely on very sketchy (and naturally somewhat over-positive) descriptions of the future capabilities and areas of use of the ESS, and on the PPP funding model proposed by Allan Larsson, to make their judgment. In this perspective, the positive attitudes towards the ESS found in the responses are less interesting than those fears and cautions communicated. In addition to the rather clear signal by most of the respondents that "crowding-out effects are not acceptable and must be avoided" (Granberg 2012: 142), two specific points deserve mentioning.

Many of the respondents underscore the need for a national scientific and technological mobilization in fields close to the ESS, in order for Sweden to build the necessary competences both to take the leadership role in design, construction, and operation of the facility, and so that the domestic scientific community is able to reap the potential benefits of having the ESS in their back yard. This mobilization, the respondents make clear, supersedes the calculated costs for construction and operation of the ESS facility as outlined by Allan Larsson in his investigation report, which in effect means that the Swedish financial commitment will be even greater, its exact size naturally hard to estimate (Granberg 2012: 139–142). An analysis on the same topic, some years later, concluded that "nobody should really expect the ESS not to have any displacement effect on Swedish science as a whole," because "not mobilizing in favor of such scientific fields when there is the prospect of having a very advanced neutron source in one's backyard would be irresponsible, even if it means harsh internal priorities and (unavoidable) displacement effects," and the real question is therefore "whether the science made possible by intense neutron beams is the right area to prioritize in Sweden" (Hallonsten 2013b: 54). The referral round was the first instance when the scientific communities of Sweden, and its various organizations, had a chance to investigate, discuss, and debate the prospects of establishing the ESS in Sweden on scientific grounds, but it can be concluded that they largely missed the opportunity. One key to this lies certainly in the design of the investigation and referral round process: Organizations were asked to respond to the investigation and its contents, and the Larsson investigation contained no appraisal of the scientific arguments for and against bringing the ESS to Sweden, or a Swedish investment in the ESS of several billion SEK (which could be foreseen in 2005 although no exact figures were available).

Some positive and negative responses deserve special attention. Lund University quite naturally stands out, not only because it is exuberantly positive, but because it points out specific favorable effects for the university itself, in an open and transparent expression of self-interest. But Lund University also advocates a PPP solution, because "it is important that the main financing is accomplished outside of the ordinary allocation of research funding, so that the flow of financing to faculties and research councils is not negatively affected" (Lund University 2005). At this point, it deserves to be noted, Lund University could not, of course, foresee either the dramatic increase in governmental appropriations for research (chapter 4), which eclipsed and thus contained the Swedish investments in the ESS (chapter 11), nor the money that it would itself have to pay to both the ESS and MAX IV, which led to some internal resistance and resentment (chapter 9).

The two academic organizations in Sweden which express outright negative opinions about the ESS in their responses are Karolinska Institute and Uppsala University. Before discussing these, the cautiously positive response from the Royal Academy of Sciences deserves some attention. The Academy would later, in 2008, come out as a strong critic of the ESS project and publicly question the Government's apparent prioritization of the ESS over MAX IV (see chapter 7), but in its response to Larsson's investigation, it approves of the ESS plans and notes that it would be "of benefit to Swedish science if a large international research facility would be established in Sweden." However cautious regarding priorities, the Academy also makes its support for the ESS conditional upon a realization of the PPP solution, that complementary investments are made so that Swedish science can benefit from the project both during construction and operation, and that a thorough investigation is made of the scientific pros and cons of the project. Moreover, the Royal Academy of Sciences complains that the investigation has had too much of a "top-down character," and would have benefited from greater involvement in the work by the Swedish research community (Royal Swedish Academy of Sciences 2005).

In its referral response letter, Karolinska Institute, a medical university, argues that "the relatively low interest in neutron scattering in biomedical research" makes it hard for it to recommend a Swedish ESS bid and approve the Larsson investigation's conclusions, and that it therefore instead recommends that Sweden "withdraws any offer to host the project in question" (Karolinska Institute 2005). Uppsala University's response questions the focus and scope of Larsson's investigation, and its relevance for evaluating the prospects of Sweden hosting the ESS, arguing that "scientific value should guide the discussions over an investment in research of this magnitude," and claiming that the "results presented in the report lack legitimacy in the scientific community" (Uppsala University 2005). Their conclusion is that Sweden should not prioritize the ESS. The letter of response also questions the PPP solution, and "makes another appraisal than the sketch in the report, of the interest of private industry in co-funding the ESS," partly on the basis of "knowledge about the conditions that rule at other neutron sources" (Uppsala University 2005).

An interesting letter of response is that from the Swedish Agency for Public Management (*Statskontoret*), a governmental agency with origins in the 17th century and with the key mission of investigating, evaluating, and following-up governmental operations and governmentally sponsored activities, and overall issues of the function and development of public administration in Sweden. The agency rejects the proposal that the Government should declare itself willing to host the ESS, not with reference to any evaluation of what type of research Sweden should invest in, which it states is not a concern of the agency, but because of the financial risks involved, more specifically that a future ESS might be much more costly in the longer run than initially expected (Swedish Agency for Public Management 2005). The argument is interesting because it might carry some relevance for the ESS and the Swedish commitment to the ESS going far beyond the time period covered in this book, and will be returned to briefly in the concluding chapter of this book.

There are several other points of criticism in the letters of response. Spontaneous responses point out the risk of bias in the investigation arising from Allan Larsson's role as chair of the board of Lund University, which is the organization in Sweden with the largest potential benefits from the building of the ESS in Lund (Jönsson 2005; People's Campaign Against Nuclear Energy and Nuclear Weapons 2005). The criticism is, of course, not without relevance, given the prehistory of the Larsson investigation, including the way that the ESS was brought to the attention of Allan Larsson by Lund University administrators and most likely presented in a very favorable manner right from the start, as well as the suspicion that he, in an informal sense, appointed himself rather than was appointed, and the fact that the ESS Scandinavia secretariat actively assisted Larsson in the investigation (see above). Quite predictably, the environmental aspects and Allan Larsson's neglect of these also came under fierce criticism. Although for the most part expressed in the spontaneously submitted responses, which make value judgments when arguing that there is a lack of a "serious environmental trial" of the project (Action Skåne Environment 2005) or claim that statements concerning land use and mercury are "deceptive" (Wennergren and Olsson 2005), these letters also point out that the Larsson investigation is selective and fails to acknowledge some important aspects of a future ESS in Lund. The mercury issue is one of the most serious aspects in this regard: At the time of the Larsson investigation, the Swedish Government and the European Union were both involved in preparatory work toward a complete ban on the use of mercury (Swedish Government 2005b), which entered into force in Sweden in 2009, and while Allan Larsson mentions this work he merely concludes that the 15,000 kg of mercury that would be used in the target station of the ESS would be "used in a closed system," referring to a study made by Studsvik AB on the matter.

Larsson admits to the neglect of environmental aspects, and defends himself by claiming that it was not part of his assignment (Larsson 2005a: 13; *Sydsvenskan* 2005d) and that a future ESS facility would quite obviously be subject to due process in Sweden, including a legal trial in a Swedish Land and Environmental Court (*Mark- och miljödomstolen*) (see chapter 8) (Larsson 2005a: 82–84). True, the assignment from the Government did not contain an investigation of environmental impacts or anything like that (Swedish Government 2004d), but Larsson nonetheless cites the positive arguments put forward by ESS Scandinavia in other fora, including the long-term benefits for the environment that the future research at the ESS would produce (see also chapter 10) (Larsson 2005a: 85), which creates a skewed impression.

ESS Scandinavia was, of course, keen to present the round of responses as very positive overall. In a general sense, and on an aggregated level, they were, although several things in the report and in the responses (see above) should perhaps have been subject to deeper debate locally, nationally, and, not least, in the scientific community and research policy. It is particularly interesting to note that while Larsson's entire investigation, and the conclusions it draws, rests on the premise that a PPP solution is the preferred (or only) means of financing the ESS, this idea evaporated completely after the referral round. People involved in the preparations for the Swedish Government's declaration of intent in 2007 (see chapter 7) discarded the PPP solution as unrealistic (Interviews: Honeth, Leijonborg), and Larsson himself avoided the issue of whether it was ever plausible (Interview: Larsson). The fear of displacement effects, which Larsson is "clearly at pains to remove, or at least substantially reduce" (Granberg 2012: 130), and which are very prominent in many of the responses, did surface at some points in the years to come but were mainly preempted by the extraordinary increase in the Government expenses on R&D, which probably calmed most such fears, although it did not completely do away with the risks of displacement effects (see the discussion in the concluding chapter).

Although the report by Allan Larsson caused some hope within ESS Scandinavia that some kind of decision by the Government was on its way (ESS Scandinavia 2005b), nothing happened other than Larsson's appointment as "special negotiator" being prolonged.

Allan Larsson remembers "bumping into the official responsible at the Ministry of Education in the late fall" of 2005, and asking him "how the continued work would be structured." The response was not what Larsson expected – the ESS was "not a matter of priority" – and Larsson decided it was "pointless to argue with him." The election campaign of 2006 drew closer, and the Government gave no further indications, "either for or against" (Larsson 2019: 107–108). Occasional requests for clarifications from members of parliament eager to see the Government declare their interest in bringing a European megaproject to Sweden were answered with diplomatic statements from Leif Pagrotsky that the Government "will take a stand on the issue as soon as it can responsibly do so" (Pagrotsky 2005a, 2005b, 2006).

Nonetheless, Larsson's investigation set the ball rolling. Several people bear witness to the enormous energy and gusto that Larsson brought to ESS Scandinavia, and having retained his title of "special negotiator" after the Government renewed his appointment by six months, from July 1, 2005 to December 31, 2005 (Swedish Government 2005c), Larsson could continue to travel around Europe and lobby for ESS Scandinavia. An interesting detail concerns a photo on the last page of the annual report for the European neutron facility ILL for 2005, where "A year in photos [at the ILL]" is depicted. The photo in question bears the caption "Sweden joins ILL (4 April)" (ILL 2006: 106). True, the Swedish Research Council had decided to join the ILL in the spring of 2005 to secure continued access to neutrons for Swedish researchers after the closing of the Studsvik research reactor (chapter 4), and council representative Johan Holmberg is in the picture, together with Patrik Carlsson, who was employed at the ESS Scandinavia consortium, and Allan Larsson. Judging from the picture, Larsson is the one who has sealed a deal at the ILL, as he is seen shaking hands with then ILL director (and later ESS Scandinavia director) Colin Carlile. In his memoirs, Larsson describes how the meeting with Carlile was "a meeting of minds" (Larsson 2019: 96).

Another episode that is telling in terms of Allan Larsson's lobbying abilities concerns a visit to the Royal Academy of Sciences to present the ESS and his own investigation, in late 2005. As discussed in the previous section, the Academy had been among the least positive in the referral round of the investigation, and would later speak out against the Governmental ESS campaign (see chapter 7). According to then secretary of the Royal Academy of Sciences and professor of plant physiology at Umeå University, Gunnar Öquist, the Academy was "not impressed" by the idea of a Swedish ESS, and also "not impressed" by the ESS design. But when Allan Larsson presented his case to the "class for physics" (one of the ten field-specific divisions or member categories of the Academy) in late 2005, his strengths as a political lobbyist became clear: the group was impressed and left the room thinking that "this is not such a bad idea" (Interview: Öquist).

But while Allan Larsson was a formidable addition to the lobbying group of ESS Scandinavia, the scientific credibility of the campaign was relatively low. Lund University, still in a sense the owner of the campaign by virtue of acting as employer of its staff and hosting its offices, understood the necessity of recruiting scientifically experienced and qualified people to Lund and ESS Scandinavia to strengthen the image of the campaign and its reputation abroad. Consequently, initiatives were taken for "VIP visits" to Lund for people viewed as strategically important, to give them "thorough information about what Lund had to offer" (Larsson 2019: 36). The first was Colin Carlile, "a person I felt we needed on our side" (Larsson 2019: 36). Carlile, a physicist by training, had been an active neutron user since the early 1970s and worked at Rutherford Appleton Laboratory and the ILL in Grenoble, as a scientist and instrument developer. At the ILL, he had risen first to the position of Associate Director and Head of Division Projects and Techniques, in 1999, and then to Director-

General, in 2001. When in 2006 his tenure as ILL Director ended, he was approached by Lund University for a guest professorship. Carlile's visit and tour of Lund took place on September 15, 2005 (ESS Scandinavia 2005d). In a "thank you letter" to Allan Larsson and Lars Börjesson after the visit, Carlile allegedly noted that "…the case to site ESS in Sweden is difficult to resist" (Larsson 2019: 105).

Half a year later, a guest professorship in "instrumentation for the ESS" at Lund University was created, and the work to convince Colin Carlile to accept the position on an initial three-year contract began (ESS Scandinavia 2006b). The professorship was funded jointly by Lund University and Region Skåne (ESS Scandinavia 2006a; Bexell 2006). On July 8, 2006, Colin Carlile accepted the appointment (Carlile 2006), and a year later, was appointed head of the ESS secretariat at Lund University (see chapter 7).

Carlile's personal motivations for coming to Lund are hard to assess. With the benefit of hindsight, his own view of things is that he was among those who saw the potential of the Lund ESS candidacy early on, and decided to act to help bring the ESS to Lund in spite of the slim chance of success, and that he also saw the chance to finally resolve the issue of a site for the ESS, after nearly two decades of discussion. "The one thing that I've got to do is to put my energy into provoking a site decision. And so I came to Lund with that determination, site decision!" (Interview: Carlile). "My goal is to help getting the funding together for the ESS, and get the facility built in Lund. […] For this, we need to build alliances and I will use the contacts I have to do so" (*Sydsvenskan* 2006b).

The recruitment was, of course, crucial for the campaign. Another recruitment that was less obviously significant but certainly extremely valuable in the years to come, when a scientific case was to be built for the ESS (see chapter 8), was the guest professorship for Christian Vettier, who had been science director both at the ILL during Carlile's tenure as director, and also at the European Synchrotron Radiation Facility (ESRF) in Grenoble. The first contact with Vettier was made in December 2006 (Berggren 2006), and an offer was made in January 2007, for a three-year guest professorship in "scientific applications of neutron scattering and synchrotron radiation" that would start in June the same year (Bexell 2007b). Vettier had been a member of the Science Advisory Committee to the ESS R&D Council that produced the 2002 four-volume ESS project reports (ESS R&D Council 2002a), and also participated in the meetings of the ESS Initiative that was founded by Peter Tindemans in 2003 (see chapter 5), occasionally together with Colin Carlile. Vettier's personal motivation for coming to Lund was that he wanted to continue working with the ESS, and saw Lund as the preferred site among the then-contenders; stating that he "could have gone elsewhere" and that the other site contenders had also approached him, he nonetheless believed Lund was the best site and made the decision to come to Lund (Interview: Vettier). Another distinguished neutron scientist, Hungarian Ferenc Mezei, who had worked at the ILL for many years and who had been the "task leader" for instrumentation in the ESS R&D Council and hence made significant contributions to the 2002 ESS project reports (ESS R&D Council 2002a), was offered a three-year guest professorship at Lund University in January 2007 (Bexell 2007a). A month later, Mezei accepted a 25% guest professorship starting April 1, 2007 (Mezei 2007), but soon left to join the Hungarian competitor to host the ESS (ESS Hungary 2008).

In November 2006, in a letter to Region Skåne asking for financial support, ESS Scandinavia outlined its planned activities for 2007, citing competence building, including recruitment of guest professors, workshops on accelerator technology and target station, interaction with Swedish research organizations, enterprise and governmental agencies, filing of applications for permits (environmental, radiation safety, building/construction), communication and PR (ESS Scandinavia 2006c). By then, signs of activity in Stockholm had already started to appear.

7. The Swedish ESS bid (2006–2009)

The Government launches its campaign

When the four party "Alliance for Sweden" ("*Allians för Sverige*") won the Swedish parliamentary elections on September 17, 2006, it was the first time in 24 years that a majority government could be formed in Sweden, and it also meant an end to a 12-year Social Democratic rule. The four parties that formed the alliance and the new government that took office on October 6, 2006, were *Moderaterna*, *Centerpartiet*, *Folkpartiet*, and *Kristdemokraterna*, and as noted in chapter 4, each had their own profile areas that to some extent, and with attention also paid to the relative strengths of the parties (counted in percentages of the popular vote, or seats in parliament), determined which specific ministries they would occupy, and which policy areas they would be responsible for, in the new government. As noted in chapter 4, *Folkpartiet* had claimed education and research as a profile area under chair Lars Leijonborg and vice chair Jan Björklund, and the two became obvious choices for Minister for Education and Research, and Minister for Primary and Secondary Education, with Leijonborg as head of the Ministry of Education, from October 6, 2006.

The recruitment of Peter Honeth, long-time director of administration at Lund University, as deputy (the highest civil servant in a Swedish ministry, called *statssekreterare* in Swedish, which translates to state secretary, the term henceforth used) to Leijonborg was decisive in many ways: Lars Leijonborg himself has highlighted how Peter Honeth "came to play an enormously important role in the reform of Swedish higher education and research policy" (Leijonborg 2018: 338). It would also prove to be the safest way to bring the ESS campaign to the attention of the new minister and the "inner cabinet" of the four party chairs of the Alliance for Sweden, of which Leijonborg was one, and through which all major policy decisions passed during the reign of the 2006–2014 four-party government. Mats Johnsson, who had worked in the ministry since 2000 and handled the ESS internally since 2002, remembers his first thought when he learned of the appointments of Leijonborg and Honeth as Minister and state secretary:

> What happens on the day the Government is appointed is that all new ministers gather in the Prime Minister's Office, in the canteen, and the administrative director of each ministry collects his or her minister and guides him or her to the ministry in question, to present him or her to the staff which has gathered. We stood there in our large hall, and then Lars Leijonborg emerges, with Peter Honeth right after him, and my first thought was "Now we will get the ESS." Because of course, if Peter is state secretary, he will obviously push the issue, and if he does, it will become reality. (Interview: Johnsson.)

Several people testify to a "sharp contrast" between the previous Social Democratic government and the new Alliance government when it came to the ESS (Interviews: Börjesson, Berggren, Wickberg). Karl-Fredrik Berggren remembers how Leijonborg "came in full of ideas" and had already revealed that the Government "was going to increase the research budget unprecedentedly" during the election campaign. Berggren was suddenly able to arrange a meeting with Leijonborg at very short notice, and met with him and Peter Honeth, together with Patrik Carlsson, on November 10, 2006. By then, the Government had already presented its budget bill for 2007 to parliament, in which it was noted that "preparations are underway within the Government offices concerning a possible Swedish hosting of the European Spallation Source (ESS)" (Swedish Government 2006b: 282). Berggren's recollection of the meeting is very positive, calling it "an experience" to meet Leijonborg, "a generally very positive person," and that the Minister had two critical questions. The first was why the ESS should not be located in Uppsala, "and Peter [Honeth] started to get uneasy but I said 'Sure, in principle it could be established in Stockholm or Uppsala, but the momentum, the enthusiasm and the people working on this are in Lund, so if you want results it should probably be Lund.' And he was happy with that answer." The second was about the costs, and Berggren and Carlsson explained what Sweden would be expected to pay, "and the numbers did not deter him" (Interview: Berggren).

Leijonborg himself remembers this meeting and the other preparatory work in the fall of 2006 in somewhat more modest terms. His first reaction, according to his own memory, was cautiously positive: "If there are objective reasons for a spallation facility in Europe, then why shouldn't it be built in Sweden?" (Interview: Leijonborg). Peter Honeth confirms, noting that his knowledge and opinion "played a role, to get Lars interested," but that the Minister "did not buy it immediately" (Interview: Honeth). Leijonborg remembers several "warning signs" around the ESS, including the risk of too much money and too many projects ending up in Lund, a risk that it was especially important to consider given the worries throughout the Swedish academic system that Peter Honeth's presence in the ministry would tilt the balance in favor of Lund. Another thing was, of course, the large investment of taxpayers' money, and the risk that new technology would emerge further down the road that would make this huge investment redundant, or the risk that the investment would have significant displacement effects. All in all, Leijonborg remembers that he tried to "gain as much knowledge about the issue as possible," and that he could not get behind it unless "other scientists" and not just "the devoted few" could provide him with good reasons to go ahead (Interview: Leijonborg). Allan Larsson reportedly courted the Minister with appeals for decisive action: "If you make this happen, you'll get the credit," he reportedly said (Leijonborg 2018: 340). In parliamentary deliberations, Leijonborg underscored that the matter needed to be thoroughly prepared before the Government could take any initiative (Swedish Parliament 2007a).

In this regard, one meeting in particular, on December 11, 2006, was pivotal. John Wood, director of the British CCLRC and chair of ESFRI, visited Stockholm and met with Leijonborg in the offices of the Government. According to Thomasson and Carlile (2017: 22), an invitation to John Wood to the Nobel Prize festivities had been "proffered at short notice"; in reality, Carlile had extended an invitation to the Nobel dinner to Wood without having any mandate or ability to do so, or guarantee that there would indeed be a seat for Wood at the festivities (Interview: Berggren), and the stories diverge on how this issue was resolved. According to some, Lund University's Chief of Protocol Carin Dahlgren saved the day (Interview: Wickberg), while others claim the ministry arranged through its official channels for the invitation to be made out to Wood (Interview: Honeth).

There were, allegedly, several reasons for inviting John Wood to attend the Nobel dinner, and meet with Lars Leijonborg the day after. Some claim the purpose was "to lobby him in order to balance negative sentiment in the UK against the ESS" (Thomasson and Carlile 2017: 22), while according to others, the main purpose was to obtain an outside opinion of some weight to convince Lars Leijonborg that the ESS was a project worth taking a gamble on. Leijonborg was reportedly "impressed" by this esteemed British scientist's high regard for the ESS and ESS Scandinavia (Interview: Berggren); the meeting with John Wood was thus decisive for him (Interviews: Honeth, Leijonborg).

The internal process at the Ministry of Education and in the Government is somewhat more difficult to trace. The incoming four-party government was ambitious and could rely on a majority in parliament, and Leijonborg himself intended to make a difference. "His opinion was that Sweden should have a European flagship project to put Sweden on the map" (Interview: Flodström). Other ministers and ministries appeared to agree. The restraint towards large and costly projects in the Ministry of Finance, and its near-veto role in the Government, is well-known (Interview: Johnsson), but Minister for Finance Anders Borg had several important connections with local and regional politicians in Skåne, who probably helped influence him in a positive direction (Interview: Kinhult). Foreign Minister Carl Bildt, a veteran of the *Moderaterna* party, was positive, which helped internally (Leijonborg 2018: 342) and would prove important later on, when the campaign was taken to bilateral and multilateral negotiations at European level and the Swedish embassies in a number of European capitals could be used for meetings. Given the previous controversies over the issues of mercury and the use of premium farm land for the facility, the Ministry of the Environment was also a key passage point. According to *Sydsvenskan*, the final piece in the puzzle fell into place the Friday before the announcement, when the Ministry of the Environment confirmed that the ESS could be built in accordance with Swedish environmental legislation (*Sydsvenskan* 2007a). Although the press release from February 26, 2007, stated that the Government had "decided" to seek to host the ESS in Lund (Swedish Government 2007b), it was not a formal government decision, and hence there is no trace of it in the official proceedings of the Government of January and February 2007. This also means that the strict procedures for how government decisions are taken did not apply, but instead the almost as strict procedures for joint preparations in the offices of the Government and between the ministries, where the coordination offices and ministries need to give a written "*vidi*" (meaning "go-ahead"), normally an email, for the matter to be approved for further processing (Interview: Honeth).

On February 26, 2007, when the Swedish Government made the announcement that it endorsed ESS Scandinavia and would start working actively to bring the ESS to Lund, it came as a surprise to many. Prior to the 2006 elections and the subsequent shift of power from a Social Democratic government to a center-right four-party coalition, the ESS project had received no official support from the Government, but by the time of the decision in February 2007, it had become a national science policy priority (Berggren and Hallonsten 2012: 27). In March 2007, Allan Larsson was appointed the Swedish Government's chief negotiator for the ESS for the period March 15 to December 21, 2007 (this was later renewed, in stages, until finally expiring on June 30, 2009). The Swedish Research Council was instructed to assist Larsson in scientific matters, and "the Swedish embassies concerned" were also given orders to lend their help (Swedish Government 2007d).

The memorandum that was issued by the Government in connection with the two press conferences in Stockholm and Lund on February 26, 2007, contained only general descriptions of the ESS and its purposes and use, of the costs (at the time estimated at 1.2 billion Euro over ten years), and of the strategy to move forward with the project, very rudimentarily described: "Mr Allan Larsson will lead Sweden's negotiations with other European countries," and the "ambition is for the facility to be completed by 2018" (Swedish Government 2007c). The Government's pledge, detailed in the memorandum and repeated in the following years as the fundament for any negotiations with other prospective partner countries, was to cover 30% of the construction costs and 10% of the operating costs, if the facility was built in Lund. While hence clearly in the same range as that estimated by Allan Larsson in his report two years earlier, the Government's declaration of intent contained no mention of a public-private partnership (PPP) solution; the Swedish bid was to be covered solely within the framework of the ordinary governmental R&D appropriations. The reasons for this departure from the Larsson model were never declared, but key people note that the PPP solution for funding the ESS had been an unrealistic idea all the way (Interviews: Leijonborg, Honeth) (see also chapter 6 and Hallonsten 2015a: 420). Regarding the details of the financial arrangement, the Government referred to its coming research bill, due to be presented to parliament in the fall of 2008.

The new government's role in the redirection of Swedish research policy, and the unprecedented increase in governmental spending on science under its direction, was discussed in detail in chapter 4, and has clear relevance to contextualize the Governmental initiative in favor of the ESS. Lars Leijonborg himself viewed the ESS as the "moonlander project" of the research policy offensive (Interview: Leijonborg), and Peter Honeth is eager to put the ESS in the perspective of the general increases of the Governmental expenditure on R&D in these years, in whose company it was not a very big investment (Interview: Honeth). Figures 1 and 2 in chapter 4 clearly show that the ESS was indeed funded outside of the Governmental science budget, and thus nominally produced no displacement effects, although the issue is more complicated than that and will be returned to in the concluding discussion in chapter 11. An interesting detail concerns how the unprecedented resource increase, which gave the annual governmental appropriations for R&D more than a 60% increase (inflation-adjusted) over one decade (2004–2014) was hammered out. Lars Leijonborg, who was not only Minister for Education in these years but also one of four party chairs in the coalition, and hence at the center of the negotiations that determined policy directions and priorities, confirms that these negotiations concerning resource increases to research never went into detail, and that the ambitious "SEK five billion" increase in resources to science over the years 2009–2012 (Swedish Government 2008c: 1), was decided in such a negotiation without any details of exactly how the money should be spent, and with a simple figure as the only outcome, in this case SEK five billion. This, he argues, in effect, means that it would probably also have been five billion without the ESS (Interview: Leijonborg), something that other sources confirm (Interviews: Johnsson, Honeth).

The SEK five billion were announced in the 2008 research bill entitled "A boost for science and innovation" ("*Ett lyft för forskning och innovation*") and delivered to parliament on October 20, 2008. It is reasonable to expect that the Swedish research communities, ignorant of the Government's plan on how to eventually fund its 30% pledge, should the ESS be built in Lund, waited in anticipation for the 2008 governmental research bill and what it would reveal in this regard. But the bill itself was almost just as vague as the 2007 declaration of intent – if anything, it gave hints that the ESS investment could indeed infringe on other important investments. The summary of the bill states that financing of the ESS "will be done within existing frameworks" (Swedish Government 2008c: 2), an expression habitually used in government bills and memos, ardently imposed by the Ministry of Finance as a way to keep a lid on funding streams by making clear that the ministry responsible for a project (in this case the Ministry of Education) will bear the costs within its appropriations, and that any increase of these appropriations is a separate matter (Interviews: Johnsson, Leijonborg).

In the section of the 2008 research bill specifically devoted to research infrastructure, the Government announced an increase in the appropriations to the Swedish Research Council for research infrastructure of SEK 150 million annually from 2009 and on, "to be used to partly fund the possible Swedish hosting of the European Spallation Source (ESS)," and that otherwise, the future Swedish funding of the ESS would be covered by "the increase of the R&D appropriations of SEK 400 million announced in 2007" (Swedish Government 2008c: 190). Here, it certainly looks as though the ESS would consume a large share of the general increase in funds to Swedish research that the Government had previously announced. The bill also elaborated on the very rudimentary motivation for Sweden hosting the ESS that was communicated in the memo of February 26, 2007, arguing that in order to remain "an area of strength," Swedish materials science needed "access to world-leading equipment," i.e. the ESS (Swedish Government 2008c: 192), but it neglected to discuss the larger issue of whether, and to what degree, the ESS facility could be properly absorbed by the Swedish science system and made use of by Swedish scientists on a scale that motivated the investment (see also chapters 6 and 11). The bill also noted that negotiations had begun with Denmark over "the possibility of a shared hosting of the facility," that Estonia, Latvia, Lithuania, and Poland had "announced their support to locating the ESS in Lund" (Swedish Government 2008c: 195).

The Government had created an ESS project secretariat at Lund University through a decision on June 20, 2007 (Swedish Government 2007e), intended to be time-limited and cease operations on December 31, 2008. The instruction from the Government specified that the secretariat should

> plan and analyze the research activities that are needed to construct the ESS; accomplish a buildup of competence and recruitment of scientists and technicians for the construction of the ESS, in collaboration with the university; prepare technical collaborations pertaining to the ESS with laboratories and universities in other countries; assist in the negotiations with other countries over Swedish hosting of the ESS; […] investigate the possibilities of constructing the ESS in accordance with Swedish environmental legislation and associated regulations; investigate the prerequisites for construction of the ESS in line with the Swedish strategy for mercury use; […] investigate the prerequisites for constructing the ESS on basis of planning and construction regulation and environmental legislation and associated regulations; partake in preparations and collaboration within the EU framework programmes within the neutron area, and collaborations with enterprise pertaining to the participation and use of the ESS by enterprise. (Swedish Government 2007e.)

On July 5, 2007, the Government appointed Colin Carlile as head of the secretariat, for an initial period from July 6, 2007 to December 31, 2008 (Swedish Government 2007f). The Vice-Chancellor of Lund University, Göran Bexell, made the secretariat a unit within the "special activities" ("*särskilda verksamheter*") of the University, and it began its work on September 6, 2007 (Lund University 2007). Three new positions were created and filled in the summer of 2007: information officer, manager, and unit administrator (ESS Scandinavia 2007a). The Government also appointed an advisory group to the secretariat, consisting of Göran Bexell (chair), Lars Börjesson, Henning Christophersen, Lena Gustafsson, Hasse Johansson, Pia Kinhult, and Barbro Åsman (Swedish Government 2007g).

The creation of the secretariat meant the dissolving of the ESS Scandinavia Consortium (ESS Scandinavia 2007b), and the resignation of Karl-Fredrik Berggren as project director. "It is time to change crew," Berggren noted, "proud to hand this over to someone with a fresh pair of eyes" (*Sydsvenskan* 2007d). The campaign was also entering a new stage: "I managed the national scene but Colin was better suited to handle the international" (Interview: Berggren). Carina Wickberg also left. By September 2007, the secretariat had taken shape, with seven employees, in addition to Colin Carlile, including Christian Vettier who was appointed guest professor at Lund University from October 1, 2007 and took the role of ESS Scandinavia Science Director (ESS Scandinavia 2007b). The budget of the secretariat was set by the Government decision in June, to SEK 13.5 million in 2007 and SEK 20 million in 2008 (Swedish Government 2007e). In early 2009, ESS Scandinavia had a staff of 30, a technical advisory group chaired by Kurt Clausen of the Paul Scherrer Institute in Zürich and consisting of 20 international experts, a scientific advisory group headed by Robert McGreevy of ISIS and consisting of ten scientists from all over Europe plus Christian Vettier, and a "stakeholders group" chaired by Henning Christophersen and representing local and regional interests in industry, government, and academia (ESS Scandinavia 2009a: 6, 29).

Not much national political debate accompanied the Government's efforts to reach a funding solution and establish the proper organizational structure for ESS Scandinavia. The four-party government rested on a firm parliamentary majority, historically unusual in Sweden and a stable ground for reform, and science was already a relatively uncontroversial policy area in Sweden. Parliament in general seems to have been positive to the ESS and eager to see results (Bernhardsson et al. 2007, 2008; Jönsson et al. 2007, 2008; Olander 2007; Granlund et al. 2008), with the predictable exception of the Green Party, who voiced moderately negative opinions about the ESS in a motion filed in response to the 2008 research and innovation bill (Wetterstrand et al. 2008). Outside of parliament, to the extent that the issue was ever openly debated, it mainly concerned the question of whether the MAX IV synchrotron radiation facility should be prioritized over the ESS (see below). The minor, and predominantly local, opposition to the ESS project, mainly directed at environmental concerns, followed the campaign from its launch in 2000–2002 and all the way to decision making in 2009–2014, as discussed in a previous chapter.

With the Government's decision to back the Lund candidature and the additional funding provided by the Government to the secretariat, the campaign was ready to go to a new level. A timely event, that had been planned several years previously, was the quadrennial European Conference on Neutron Scattering (ECNS) that took place in Lund on June 25–29, 2007. The conference attracted some 700 scientists from all over Europe and ESS Scandinavia had proposed Lund as the venue in 2003, which the local Lundian ESS lobby saw as a "sign of the good reputation that the ESS Scandinavia bid enjoys among European neutron users" (ESS Scandinavia 2003n). As it turned out, with the Scandinavian ESS campaign entering a new phase – the European phase – the ECNS conference in Lund "provided an ideal opportunity for Sweden and the ESS project team to market Lund to the rest of the neutron research community as the natural place to build the ESS and it was an opportunity that was seized upon," not least by having Lars Leijonborg give the opening speech (Thomasson and Carlile 2017: 23).

The European phase

The years following the 2002–2003 setbacks discussed in chapter 5 was a time of regrouping on the European stage, with new actors emerging and old ones dropping out. Allan Larsson had thoroughly investigated the competition to host the ESS as part of his work in 2004–2005, and his memoirs are a formidable source of information on what he did, and how, part of this work, which in reality was wider than merely investigatory, certainly involved negotiation and lobbying on European stage: In 2004–2005, the "overarching goal" of his work was already to "rally as many actors in as many countries as possible around ESS Scandinavia, so that it would be pointless or very hard for any of the competing projects to reach a similar critical mass" (Larsson 2019: 89).

The first effort was to mobilize a Nordic cooperation, without which "there would be no chance of success," and so Larsson's first trip in 2004 was to Copenhagen and meetings with senior Danish government officials. The reaction was unexpected: The ministry official reportedly began the meeting by asking "if the ESS was invented to enable Sweden to keep the nuclear reactor in Barsebäck running" (Larsson 2019: 89). The Barsebäck nuclear power plant, visible from Copenhagen, had caused resentment in nuclear energy-averse Denmark for several decades, and the late 1990s decision by the Swedish Government to close the two reactors had not completely eliminated the Danish distrust. But the grumble seems to have been largely tactical. "After this eruption," writes Larsson (2019: 89), "we could enter into a more serious discussion." The Danish were clear that the ESS was a complex political issue, given the size of the required investment, but that they were generally positive to a Scandinavian ESS. A similar sentiment was reportedly expressed by the Government representatives visited throughout Larsson's "road show" in the fall of 2004 to Oslo, Helsinki, Tallinn, Riga, Vilnius, Warsaw, and Berlin. The meeting with German officials yielded positive messages that would become "decisive, in many ways, for the continued work," namely that the German government advocated that the ESS be built, that Germany would however not stand as candidate, and that if the choice was between the UK and Scandinavia, Germany would opt for Scandinavia (Larsson 2019: 90–92). The message from the French government was slightly more disappointing; officials asserted that a deal had once been made between France, Germany, and the United Kingdom, in connection with the launch of the ILL in Grenoble, that the next major neutron facility in Europe would be built in the UK, and France was going to honor this agreement (Larsson 2019: 92–93).

Larsson's conclusions, as detailed in his 2005 report, were that the main competitors to a Scandinavian ESS were Germany, Hungary, and the UK. The Hungarian and UK governments had recently begun investigatory work not unlike his own, and the two German candidates (Jülich and Leipzig/Halle) had regional governmental support but no backing from the federal government (Larsson 2005a: 52–53). In addition, the Basque Region in Spain had proposed Bilbao as the future site for the ESS (ESS Initiative 2004a) and the Spanish government had indicated it might give its support (Larsson 2005a: 23).

Almost two years had passed between this assessment and Larsson's second appointment as negotiator for the Swedish Government, in March

2007, and the playing field had changed slightly. The UK had not given up completely, and the Yorkshire candidature officially remained a competitor until late 2007 (ESS Initiative 2007). But the most serious contenders, and the most active trio in the ESS Initiative, were Hungary, Spain, and Sweden (Interview: Tindemans). In early 2007, authorities in Spain and Hungary had made similar commitments and declarations of intent to the Swedish Government; both offered to pay 30% of the construction costs (ESS Site Review Group 2008: 15–16). In an interview in February 2007, the Swedish soon-to-be chief negotiator Allan Larsson estimated Sweden's chances of ending up as the host of the ESS at 50% (*Sydsvenskan* 2007a).

In March 2007, the work was set to start. A planning meeting was reportedly called with Allan Larsson, Lars Börjesson, Peter Honeth, Karl-Fredrik Berggren, and Colin Carlile, plus some high-level officials in various governmental agencies, including the director general of the Swedish Research Council, Pär Omling, and the director general of the Swedish Agency for Innovation Systems, Per Eriksson. The purpose was to draw up a strategy for the coming negotiations, and it was agreed "not to rush off and ask for statements in favor of Lund as a location for the ESS, or start discussions over financing" because it was important to "avoid gridlocks" and instead "make it possible for our collaborative partners to hold an open discussion without the interpretation that they took a specific position" (Larsson 2019: 112). All these insiders understood how hard it would be to simultaneously achieve a funding solution and a site decision, and therefore the "Swedish team" decided to go for a site decision first and a funding solution later; apparently a very "Swedish" approach (Thomasson and Carlile 2017: 24). The strategy was summarized by Larsson in his memoirs as "taking such small steps that nobody can say no, and taking so many steps that only yes remains" (Larsson 2019: 118). In practice, it meant negotiations in phases, with round table meetings as key events, every six months. The three phases would have different foci. The first would merely be to inform other countries about the Swedish offer and gather information about how to proceed further. The second would emphasize scientific and technological issues, and aim to build alliances with universities and institutes across Europe, where leading competence and capacity in accelerator technology, instrumentation, and the scientific use of neutrons was found. The third would be geared to obtaining the necessary political support for bringing the ESS to Lund, and begin structuring the joint work of European governments and scientific communities of updating the technological design and scientific case, and reaching a solution for the financing of the facility (ESS Scandinavia 2009a: 5). Bilateral meetings with researchers and politicians from European countries were held in Swedish embassies across Europe, with invitations to scientists and politicians extended by ambassadors, which lent a certain dignity to the whole process. The strategy would prove successful, and with the benefit of hindsight, both Lars Leijonborg and Allan Larsson credit the stepwise procedure for the favorable outcome. The prospective partner countries apparently needed the time to process the issue internally and position themselves in relation to each other and the Scandinavian proposal. In the earlier phases, the common response in dialogues was that scientists in most countries were in favor of the ESS project, but the politicians needed to think the issue through carefully, not least with regard to the question of funding this kind of project, and whether perhaps one of the other site candidacies were better (Leijonborg 2018: 342). "This was based on my experience of decision making, you can't just come in and push ahead and think you'll get a decision, you have to let it take time, but it is a way of building a coalition" (Interview: Larsson).

The first of several *round table meetings*, to sum up the phases of negotiation, was held in Copenhagen in October 2007, and was attended by representatives of no less than 23 countries (*Sydsvenskan* 2007e). The meeting was intended to demonstrate "right from the start that ESS Scandinavia was based on cooperation between Sweden, Denmark and Norway" (Larsson 2019: 118). It was followed by round tables in Delft/The Hague in February 2008, Riga in September 2008, Prague in April 2009, and Krakow in September 2009 (Larsson 2019: 122, 132, 156, 160). Between the round tables, Allan Larsson criss-crossed Europe, sometimes seconded by Colin Carlile, sometimes by Lars Börjesson, holding meetings in Swedish embassies. For the ministry, this was for the most part, but not entirely, a period of silence. Larsson "took care of it" (Interview: Johnsson), but "his way of doing it also caused some antagonism in some countries, not least with Denmark, where I had to step in and sort things out" (Interview: Honeth). In Larsson's memory, several European governments declared support for the Swedish candidature early on, including Denmark, Norway, Belgium, and the Netherlands, and also Germany, whose representative had allegedly said that "Lund is the site" at the Copenhagen round table meeting in October 2007 (Larsson 2019: 121).

But this was before the 2008 financial crisis. Originally aiming for a conclusion of the process in late 2008, under the French EU presidency, the Swedish ESS negotiators now had to switch gears and adapt. "With my experience of how governments function in times of crisis, I knew that the research ministers would not have access to finance ministers to convince them to pay money to foreign countries for a new scientific project. Forget it. So we rearranged our agenda and focused on only the site decision. And then we'll deal with the other issues later" (Interview: Larsson). A technical argument came to the support of this approach: a site decision would enable the drafting of a site-specific design, which would enable better cost estimates to base funding negotiations on (Larsson 2019: 130–131).

In December 2007, ESFRI had decided that something had to be done to relieve the gridlock regarding the location of the future ESS, and formed the ESFRI Working Group on ESS (EWESS), chaired by Paul Zinsli of Switzerland, which prepared a questionnaire for the three contenders, about the most important aspects relating to building the ESS on each of the sites. The Bilbao site questionnaire response outlined details of the planned technical design of the ESS, the applicable Spanish and regional legal frameworks, an outline of the cost and financing model, and environmental and socio-economic aspects. It also described the financial offer, citing a pledged contribution from the Basque and Spanish governments of "at least" 29.25% of the estimated construction costs and 15% of the future operations costs, plus land and "site preparation" (ESS Bilbao 2008: 15). The ESS Scandinavia site questionnaire response had a slightly stronger sales approach and presentation than the Bilbao report, with imagery and graphics, echoing the Expression of Interest from 2002 (chapter 6) and foreshadowing the massive advertising campaign of 2009 and forward (chapter 10). Otherwise, it covered roughly the same elements and declared that the Swedish Government was prepared to pay 30% of the construction costs and 10% of the operations costs for the ESS, should it be built in Lund (ESS Scandinavia 2008a: 20).

After the responses were collected, an independent Site Review Group (SRG) was set up "to assess the answers by the sites in order to enable ESFRI and the ministers to proceed with a decision on the preferred site for ESS" (ESS Site Review Group 2008: 7). The SRG was chaired by Catherine Cesarsky, director-general of the European Southern Observatory (ESO), and composed of Norbert Holtkamp, principal deputy director of the ITER (International Thermonuclear Experimental Reactor) organization, Thom Mason, director of the Oak Ridge National Laboratory in Tennessee, and Peter Tindemans as scientific secretary. The report, delivered on September 10, 2008, comprised 12 pages of careful review of all three sites. Eager not to take a clear stand, the SRG nonetheless made quite clear that the three sites were differently prepared for the task. "We tried very carefully to use exactly the same words for all the sites. But just make a slight change, so that it's clear for the very careful reader. So we were actually saying that Lund is the best, but Bilbao would be very good. And Debrecen would not be a choice" (Interview: Tindemans).

Discarding socio-economic benefits in their analysis of the pros and cons of the three sites, the SRG focused entirely on scientific and technological aspects, and features of the three proposals that would maximize "science and innovation for Europe as a whole" (ESS Site Review Group 2008: 7). In terms of organizational arrangements, the site review highlights no differences between the three sites, noting that all three envision an ESS ERIC (see chapters 3 and 11), and as a secondary option, a limited liability company (ESS Site Review Group 2008: 10). With regard to local costs "which are especially relevant for conventional facilities," the SRG noted a small advantage in Bilbao. All three sites were judged to be well suited with regard to the physical site and the respective organizations' preparatory work and risk assessment work, and all three had minor issues that the report highlighted for further investigation, such as the proximity to a public road in Lund, and the limited load bearing capability in Debrecen which implied some technical challenges (ESS Site Review Group 2008: 17). The local organizations built up to achieve a timely ramp-up of efforts to prepare for eventual start of construction, including available resources, show some differences. For example, the SRG is "impressed by the determination displayed by the Lund team and proceeding towards construction without delay will position the team well should the bid be successful, reducing the risk of start-up difficulties" (ESS Site Review Group 2008: 19). Lund had a slight edge with regard to local conditions for the successful operation of an international knowledge-intensive organization, with its availability of housing at reasonable prices, its accessibility and proximity to Malmö and Copenhagen, its international atmosphere, its vibrant regional innovation system including "a high number of R&D jobs" in university and industry, and the "business-oriented culture of cooperation and decision making in Lund to which public and private parties contribute alike" (ESS Site Review Group 2008: 22–23). Bilbao, similarly, is praised for its equally international air, knowledge-intensive economy, "business-oriented culture of cooperation and decision making," and good housing opportunities (ESS Site Review Group 2008: 23–24). The Debrecen site, in contrast, had some challenging issues, including limited accessibility given the distance to Budapest and its airport (two hours), comparably unsatisfactory "working and living conditions for foreigners" and "an obvious language barrier." "No doubt this will improve but catching up takes time" (ESS Site Review Group 2008: 22).

The SRG report specifically mentioned MAX IV as an advantage for the Lund site candidature, treating it as *fait accompli*, not only praising its key role in the "scientific and industrial environments" of Lund and the Øresund Region which "are very amenable to ESS," but also noting how its colocation with the ESS offers "the possibility of shared technical and support facilities with potential for operational cost savings" (ESS Site Review Group 2008: 8). The SRG had asked Allan Larsson specifically if MAX IV was going to be built, and was told that it would, "with 99% certainty" (Larsson 2019: 55, 133). This is a spectacularly bold statement back in 2008, given that the Minister for Education Lars Leijonborg made it clear, just months later, that the Government would not fund "two accelerator facilities in Lund" (see next section) (Royal Swedish Academy of Sciences 2008b) and the complete absence, at the time, of any other credible idea for funding MAX IV. Larsson would of course be proven right, but the comment was presumptuous and probably a good example of the type of gambles that skilled negotiators like Larsson made, and that contributed to the eventual success (see also the concluding discussion in chapter 11).

ESFRI endorsed the report in September 2008 (ESFRI 2009: 9), and while it wasn't immediately published, the contents were leaked. Most insiders appeared to agree that Lund was in the lead (Larsson 2019: 129; *Sydsvenskan* 2008e). The Hungarian and Spanish site contenders had already entered into a strategic partnership in late 2007, "to coordinate the search for international partners," and appointed a joint project director and joint international advisory board (ESS Initiative 2007). In the fall of 2008, apparently realizing that the Lund candidature now had an edge, Hungary and Spain strengthened their collaboration to enhance their competitiveness, and formed a consortium that would work to locate the facility either in Bilbao or Debrecen, which involved an agreement that the winning side would somehow compensate the loser financially, as well as agreements that parts of the future facility would be developed and built in both locations (*Sydsvenskan* 2008a).

Meanwhile, the Swedish Government intensified its efforts. A meeting with some 30 senior officials at the Foreign Ministry in September 2008 led to the launch of a "focal point" concerning the ESS, which meant that documentation was prepared that would allow ministers and deputies to bring up the issue "when appropriate," and spread information to other ministries, with the management of the Foreign Ministry stressing the importance of the project (Larsson 2019: 128). A "preliminary deadline" for the negotiations was set to December 2, which was when the Swedish Government hoped to have a site decision from the ministers of the prospective member countries at a meeting in Versailles (Larsson 2019: 138). But instead of a conclusion, this meeting, chaired by French Minister for Higher Education and Research, Valérie Pécresse, prolonged the process and produced no outcomes except that some positions were clarified; a news report held that the UK was close to declaring its support for Sweden, whereas France was leaning towards the Spanish candidature (*Sydsvenskan* 2008i).

A loose "ESS Community" was set up, consisting of the three site contenders and the prospective ESS member countries, and led first by Valérie Pécresse, during the French EU presidency of July to December 2008, and then by the Czech Minister for Higher Education and Research, Vlastimil Růžička, during the Czech EU presidency from January to June 2009 (Larsson 2019: 140–141). Meetings between the three competing countries "on neutral ground" and chaired by the ESFRI delegate from Switzerland "increased mutual understanding and laid down some principles but could not bring forward a site agreement between proposers," not even after "replacement projects, compensations or distribution of tasks" had been put on the table. Meetings in the winter of 2008–2009, without the three site contenders but with interested countries, chaired by Beatrix Vierkorn-Rudolph of the German Ministry of Education and Research, produced some "indications of preferences of interested countries but no conclusive results" (ESFRI 2009: 9). As part of these discussions, however, a timetable was decided: The three site contenders should update their applications and deliver these on April 30, 2009, at the latest, and then the fate of the ESS was to be sealed at two meetings in May 2009; the first by discussing the three applications and the second by bringing the site issue to an agreement (Larsson 2019: 140–141).

The Swedish efforts were now redirected to work towards heads of state and government, and their advisors, in "key countries" – a "counter-offensive" that built on the results of the previous year of negotiations. Prime Minister Fredrik Reinfeldt, Foreign Minister Carl Bildt, Finance Minister Anders Borg, and of course Minister for Education and Research Lars Leijonborg, "took a very active role" in "convincing their colleagues to support the Swedish-Danish alternative" (Larsson 2019: 145–146). The efforts immediately bore fruit: Four agreements were signed in 2008 and

2009, with Estonia, Latvia, Lithuania, and Poland, that secured the support of each of these four countries in designating Lund as the preferred site for the ESS and their contribution to the work to locate the ESS in Lund, including securing a funding agreement (Poland 2008; Estonia 2008; Lithuania 2009; Latvia 2009).

But it had also become clear, in the words of Lars Leijonborg, that the joint funding bid of Sweden and Denmark "was not as impressive as one might have hoped" (Leijonborg 2018: 342–343), and so work intensified to raise it. In early 2009, the Swedish Government raised its own bid from 30% to 35%, but the exact level of the Danish and Norwegian commitments were not yet settled. While Denmark had been represented in the ESS Scandinavia consortium all the way, as noted in chapter 6, Danish politicians and Danish scientists had been ambivalent from the start. The Swedish-Danish co-hosting that had seemed so obvious in the documentation for the bid presented at the Bonn conference in 2002 – "This is a joint Swedish, Danish and Norwegian initiative," as the preface to the Expression of Interest stated (ESS Scandinavia 2002a) – was not mentioned in the memo outlining the governmental declaration of intent on February 26, 2007 (Swedish Government 2007c), and the preceding Larsson report had indeed argued that "the declaration of intent to host the facility should come from the country where it would be built, i.e. Sweden," hence not counting on Danish support (Larsson 2005a: 58). Nonetheless, of course, the dynamic Øresund Region and the majestic Øresund Bridge that tied it together and established a crucial link between the future ESS and the world by connecting it to the Copenhagen international airport in Kastrup, were among the key selling points that made the ESFRI siting group favor Lund over Bilbao and Debrecen in its 2008 report. Further, the likelihood that Denmark would contribute a significant share of the construction costs undoubtedly improved the chances of the Scandinavian bid. As late as summer 2007, the negotiations with Denmark "still had a long way to go" and Danish politicians and civil servants apparently needed evidence that the ESS would benefit Danish science and enterprise (Larsson 2019: 115–116).

In April 2008, the Danish Government appointed a delegation to negotiate the Danish participation and the forms of cooperation in a joint Danish-Swedish hosting of the ESS in Lund (*Sydsvenskan* 2008b), consisting of Uffe Toudal Pedersen, the highest-ranking civil servant in the Ministry of Science, Technology and Innovation, Inge Maerkedahl, director-general of the Danish Research Agency, and Michael Christiansen, who had been the highest-ranking official in the Danish Ministry of Defense, Chairman of the Øresund Bridge Consortium, and Director of The Royal Theatre in Copenhagen, reportedly a "skilled and constructive negotiator" (Larsson 2019: 134). The first meeting between the Danish and Swedish negotiating teams, where the latter was reinforced with the Swedish ambassador to Denmark, Lars Grundberg, took place in May 2009. The Danish team immediately launched an idea that would prove crucial for the two countries to agree, namely to define the collaboration as "co-siting" rather than "co-hosting" (Larsson 2019: 135).

On October 13, 2008, a first "draft agreement" was signed by Danish and Swedish government officials, outlining the structure of the joint bid to host the ESS. It established that the preferred organizational form for the future ESS was ERIC, which had been proposed by the European Commission earlier the same year (chapter 3), and should this not be possible, a limited liability company under Swedish law was seen as the realistic alternative. The two governments also agreed to start, immediately after a joint European agreement regarding the site, a review of the 2002 reference design and develop a detailed design of the future ESS facility in collaboration with "a core group of countries, willing to participate as partners in ESS" (Denmark 2008). Most importantly, though, the draft agreement included the "ESS Data Management, Computing and Software Centre" in Copenhagen, "an important resource for the stewardship of all aspects of data curation and management generated by the ESS as regards developing, maintaining and running the software to handle data, to perform data analysis and to simulate instruments and experiments" (Denmark 2008). The idea of an ESS data center in Copenhagen had already been floated in the spring of 2008 (*Sydsvenskan* 2008d), and in the view of Allan Larsson it presented a "problem" because it wasn't part of the 2002 reference design, which meant that it would increase the costs and hence weaken the position of Lund in the competition with Bilbao and Debrecen (Larsson 2019: 138). But the negotiation skills of the Danish Government and its delegation (Larsson 2019: 134) were apparently enough to make the Data Management and Software Center (DMSC), as it would later be called, an essential part of the deal between the two countries. Denmark obviously needed some reason to invest a couple of hundred million Euro in the ESS on foreign soil, and to pledge to do so long before any other countries had declared their support; the reward came with the DMSC being located on the northern campus of the University of Copenhagen and employing at least 62 full-time equivalents (Denmark 2009). Several sources indicate that this agreement, and the outsourcing of a vital part of the ESS organization to Denmark, was "the political requirement for Denmark to join the project" (Thomasson and Carlile 2017: 18) and that "without it, Denmark wouldn't have joined" (Interview: Honeth). In the 2014 letter from the Danish Minister for Education and Research, Sofie Carsten Nielsen, to her Swedish colleague Jan Björklund, reaffirming the Danish commitment to funding 12.5% of the ESS, the DMSC is described as "absolutely essential for the Danish contribution to the construction of ESS" (Denmark 2014).

The Danish bid had been to fund 10% of the construction costs of the ESS, but Sweden expected 15%, and the compromise solution of 12.5% was agreed at ministerial level, and codified in the April 2009 agreement that followed up on the "draft agreement" of 2008 (Larsson 2019: 138). On April 3, 2009, Lars Leijonborg and Helge Sander signed the "Joint Statement" ("*Fælleserklæring*" in Danish) about the ESS, in which they state that their governments "are in agreement that they will jointly seek to host the European Spallation Source (ESS) and work to achieve a principal decision, within 2009, and a joint European agreement that encompasses organization, financing, and operation of the facility" (Denmark 2009).

Norway's participation with 2.5% of the ESS construction budget was also the subject of some negotiation. Apparently "shaken by the financial crisis and the fall of oil prices" which made the timing bad for sharp negotiations (Larsson 2019: 148), Norway initially refused to make any commitments but agreed to continue discussions. In early 2009, with the help of Uffe Toudal Pedersen of the Danish Government, an agreement was hammered out where Norway agreed to pay 2.5% of the ESS construction costs, predominantly through in-kind contributions, which in effect took the Scandinavian bid to the crucial 50% mark (Larsson 2019: 149).

But the key was, of course, to win the support of the three big players: France, Germany, and the United Kingdom. While both Germany and France had dropped out of the race to host the ESS at an early stage, the UK had stayed on, albeit seemingly half-heartedly. In 2007–2008, writes Leijonborg (2018: 342), "it seemed many in the United Kingdom were hoping for a gridlock situation to arise, so that they could make an offer, at just the right time, to locate the ESS there." John Wood, director of CCLRC and chair of ESFRI, had spoken well of ESS Scandinavia at the meeting in Stockholm in December 2006 (see previous section), but other influential British representatives had different opinions, including the director of ISIS, Andrew Taylor, who had perhaps not given up the idea of having the ESS brought to his lab (Larsson 2019: 119, 151–152). Among the three big countries, the UK was the only one that did not come out in favor of Lund during 2009. In fact, it took until 2011 before the United Kingdom joined the preparatory phase of the ESS (chapter 9). As noted above, back in 2004 Germany had softened the position it had adopted after the 2002 *Wissenschaftsrat* evaluation (chapter 5) and communicated to Allan Larsson that it would support the ESS but not seek to host it, and that given a choice between Scandinavia and the UK, it preferred Scandinavia. In January 2009, Lars Börjesson and Allan Larsson met with the director-general of the Federal Ministry of Education and Science in Bonn, Beatrix Vierkorn-Rudolph, to initiate deliberations about a bilateral German-Swedish agreement that reportedly would entail broad ambitions of a "scientific collaboration between Northern Germany and Southern Sweden / the Øresund Region," involving ESS and MAX IV in Lund, the synchrotron radiation facility PETRA III at DESY in Hamburg, and the European XFEL, also in Hamburg (Larsson 2019: 154–155) (see chapters 9 and 10).

France also demanded wide-ranging political dealmaking. The major international fusion energy research facility ITER (International Thermonuclear Experimental Reactor), had been located to Cadarache in Southern France in an international agreement between six entities (the European Union, India, Japan, China, Russia, South Korea, and the United States) in 2005 (McCray 2010: 298). In 2008, large cost increases plagued the project and made French science diplomats nervous and unwilling to risk causing antagonism towards any country that could contribute to filling the deficit. France would, therefore, probably wait to declare its position on the site of the ESS until other countries had done so. Bernard Bigot, Director General of the French Commission for Atomic Energy (*Commissariat à l'énergie atomique, CEA*), "one of the heavyweights when decisions on large research projects are to be taken," had advised the Swedish negotiating team to involve their Prime Minister in the campaign, because the issue would be resolved "at the highest political level" (Larsson 2019: 143–144). "All issues are resolved in the Elysée," Bigot reportedly told Allan Larsson, and the right way to proceed was to pay a visit to Bernard Belloc, science advisor to President Sarkozy (Larsson 2019: 152). The "hyperactive" President of France (Reinfeldt 2015: 345) was unpredictable and self-seeking, demanding quid pro quo through agreements on bilateral research collaborations, which he also got (see chapter 9). On March 20, Sarkozy reportedly promised Swedish Prime Minister Reinfeldt support for the

Danish-Swedish candidature (Leijonborg 2018: 345), but the deal was not yet settled.

Ahead of the ministerial meeting of the European Space Agency (ESA) in 2008, the Swedish Government had decided to withdraw from the *Ariane* space rocket program, based in France but run collaboratively by a number of European countries, through ESA. The decision had been preceded by investigatory work by the Swedish National Space Agency, and negotiations between the agency and ESA, that had concluded that Sweden would phase out its participation due to perceived cost inefficiency and lack of transparency. The procedure of Swedish withdrawal began after the ministerial meeting of ESA (Swedish National Audit Office 2013). While the decision had been taken completely independent of the ESS, and on the basis of a tight budget situation for the agency, which simply necessitated prioritization (Interview: Honeth), for the French, the withdrawal was important and threatened their commitment to the ESS in Lund. A domestic Swedish crisis came to the rescue. In early 2009, the car manufacturer Saab, based in Trollhättan in Western Sweden, had gone bankrupt, and thousands of jobs were on the line, but since the Swedish participation in Ariane consisted largely of the manufacturing of engine components at Volvo Aero, which was also based in Trollhättan, a continuation of the Swedish membership in Ariane would make sense in order not to lose additional jobs in an already troubled town. The matter was settled in collaboration between the Ministry of Education, the Ministry of Finance, and the Ministry of Enterprise. "Even the Ministry of Finance, which routinely says no to everything expensive, acted swiftly and positively, for once" (Leijonborg 2018: 347). "So, we found new money to retain our involvement in Ariane" (Interview: Honeth). The ESS was undoubtedly part of the equation, and while Peter Honeth argues that Sweden would probably have remained in Ariane even without the ESS, due to the pressing industrial policy situation in Trollhättan (Interview: Honeth), at the time he admitted to *Sydsvenskan* that the ESS played an important role for the decision (*Sydsvenskan* 2014e). Lars Leijonborg claims that this was one of very few instances where the scientific grounds for his decision making around ESS were complemented by matters of "political prestige" and "employment issues" (Interview: Leijonborg). On June 8, 2009, Sweden re-entered the Ariane project. But this was not to be the last crisis that involved France. A thoughtless remark by Swedish Foreign Minister Carl Bildt (where he compared French resistance to Turkish EU membership with the previous French resistance to EU membership for the UK, championed by de Gaulle) apparently annoyed

French President Sarkozy to the extent that he cancelled an official visit to Sweden in June 2009, and decided to withdraw support for a Swedish ESS. Apparently, Valérie Pécresse resolved the situation, managing to convince the President to reverse his decision (Leijonborg 2018: 351).

In the early spring of 2009, the window for a decision was beginning to close. France and the Czech Republic had agreed to take the informal role of leading the process by virtue of their successive EU presidencies (see above); Sweden would hold the EU presidency in the fall of 2009, and be followed by Spain in the spring of 2010, which was generally expected to bring the ESS process to a halt for a full year, unless a decision was made before (ESFRI 2009: 9). Meanwhile, Sweden's chances were slowly but surely improving. Lars Leijonborg concludes that "Germany was on our side" but that "competitors had understood that we were in the lead and tried to block a decision on a date when the issue was to be decided, and they succeeded with this." His overall assessment at the time, seen in retrospect, was that if it would come to a decision, "we estimated that our chances were very good" (Leijonborg 2018: 347). In February 2009, Allan Larsson gave his guess in an interview, noting that the ESS issue had been lifted from the level of scientists to the level of politicians, and that a decision on the site could come in May (*Sydsvenskan* 2009d).

MAX IV

But before a conclusion on the European stage, an important matter of domestic Swedish research policy needed to be resolved, namely the funding of the MAX IV synchrotron radiation facility. MAX IV had become a key bargaining chip for Sweden in the ESS negotiations, and was highlighted by the ESFRI Site Review Group as a great advantage for Lund. In fact, it was treated as *fait accompli* in the SRG report, in no small part because of Allan Larsson's claim that it was a 99% certainty that it was to be realized (see above), and it was pushed by the Swedish negotiating team as one of the real advantages of the Lund site (Larsson 2019: 133, 170).

MAX IV was inaugurated on June 21, 2016 and is, at the time of writing, underway with the operation of several of its first 16 instruments. It can therefore be concluded that the fears that the ESS bid would disable a commitment to MAX IV, expressed not least by the Swedish Royal Academy of Sciences (*Kungliga Vetenskapsakademien*, *KVA*) (see below) were unnecessary. But the story certainly does not end there, because there are also signs that MAX IV suffers from structural shortcomings in its funding and organization model (Lund University 2012; Swedish National Audit Office 2012; Swedish Research Council 2013; *Sydsvenskan* 2018a; Swedish Research Council 2018b) whose origins can be partly traced back to the campaign to locate the ESS to Lund, and how this campaign knocked MAX IV down on the list of priorities of research policy and also made MAX IV into a marketing piece and a campaign asset rather than a project in its own right, worthy of attention and focus for strategic priority on the part of the Government.

MAX IV is the latest incarnation, and by far the greatest expansion yet, of the synchrotron radiation facility MAX-lab that was originally built up in Lund in the late 1970s and early 1980s. Starting as a small-scale university project, MAX-lab grew over two to three decades to become a Swedish national research facility and international user facility, with the total number of annual users peaking at over 1,000 in 2014, with close to half coming from outside Sweden (Hallonsten and Christensson 2017a: 62). By then, MAX IV was already under construction, having been first proposed in the early 2000s as the natural next step for MAX-lab and for the growing synchrotron radiation user community in Sweden after the previous three MAX accelerators, inaugurated in 1986, 1995, and 2007, respectively. While the MAX IV history requires its own book to do it justice, it is clear that the idea for a larger and more advanced synchrotron radiation facility in Lund developed increasingly in the shadow of the Swedish ESS bid and with a significantly more scientific approach than ESS Scandinavia. MAX IV was built bottom-up, conceived as an R&D project by the accelerator physics group of MAX-lab in Lund in the late 1990s. The ESS also had its origin in the grassroots of the European neutron communities and was built on its world-leading expertise in accelerator technology and neutron scattering technology, but from a Swedish perspective, the ESS was almost entirely politically motivated – the clearest proof for this was the complete lack of scientific evaluation in the 2005 Larsson report (chapter 6). MAX IV, on the other hand, was almost completely lacking in politics, seeking the support of international evaluation panels (as instructed by the Swedish Research Council) rather than the Government. Benner (2012: 165) describes MAX IV as "a relatively clear-cut example of how a professional approach can produce aggregate results," where "the political system had been an addressee rather than a patron" in the efforts to mobilize support for the project.

In October 2008, appeals to build MAX IV seemed to be "coming from everywhere" (*Sydsvenskan* 2008g) – from senior representatives from industry, prominent researchers across the whole of Sweden, and from Region Skåne, which also declared its active support for MAX IV and its readiness to contribute financially to make it reality (*Sydsvenskan* 2008h). The Government research and innovation bill of October 2008 made no funding pledges but merely noted that "it is, in the Government's assessment, urgent that MAX IV becomes reality" (Swedish Government 2008c: 194). The Government seems to advocate a PPP-like funding solution for MAX IV – "since the facility is of importance for industry and enterprise, a large share of its funding should come from these and from other national and international stakeholders" (Swedish Government 2008c: 194) – which is strange in a way, since the Minister (Leijonborg) and state secretary (Honeth) had so clearly deemed a PPP solution for the funding of ESS unrealistic, at an early stage (see above). Perhaps the suggestion was just a way for the Government to smooth over the message in the bill that there would be no direct governmental financing of MAX IV. On December 4, 2008, the Government appointed Anders Flodström, professor of materials physics and former Vice-Chancellor of Linköping University and KTH Royal Institute of Technology in Stockholm (and early MAX-lab user), to investigate possible funding models for MAX IV and prepare a decision for its funding in negotiation between different organizations (Swedish Government 2008e). Though it was not prepared to put up any money of its own, the Government simply needed to put a solution for MAX IV in place before it could take the ESS issue further toward a decision at European level. A month into the work, Flodström noted in an interview that it was "politically sensitive" to try to locate both the ESS and MAX IV to Lund, but seen in a European perspective, MAX IV was a strength for the Scandinavian ESS candidacy, and could very well help in bringing the ESS to Lund (*Sydsvenskan* 2009a).

Anders Flodström and Allan Larsson toured Scandinavia and Northern Europe together in the winter of 2008–2009, "with different agendas [but] unquestionably loyal to each other" (Interview: Flodström). Delivering his report in early April 2009, Flodström demonstrated a truly realist view on the very limited opportunities for MAX IV in national governmental research policy, and focused his efforts on finding a compromise solution among those actors in the system that had declared support and that seemed willing and capable to contribute. Flodström attributed part of the difficulties to find a funding solution for MAX IV to the ESS, a governmental priority in the same area that was competing for the same foreign investments, at least on a Nordic level (Interview: Flodström), and concluded this comparison of the two projects by noting the very different treatment of the two projects in governmental policy, asking the rhetorical question: "Why does ESS have 'political' funding and not MAX IV?" (Flodström 2009: 8–9).

The answer is both simple and complicated. The slip-of-the-tongue remark by Lars Leijonborg at the presentation of the Governmental research bill on October 23, 2008, that "there won't be two accelerator facilities in Lund" (see below), was apparently not entirely inaccurate; the Government was not prepared to promise any funding for MAX IV, but continued to praise it in its research bill and consecutive budget bills, and had its priorities clear. "Of course, our judgment was that it wasn't possible, in the internal governmental work, to propose another big project for Lund" (Interview: Honeth). Nonetheless, as has already been noted, MAX IV had shown itself to be a formidable bargaining chip in the negotiations and campaign for the ESS at European level, and could very well become the *unique selling point* that finally tilted the balance over to Sweden in the competition with Hungary and Spain. While not prepared to pay, the Government nonetheless needed a decision regarding MAX IV in place ahead of the final ESS battle on the European stage. Through intense bi- and multilateral discussions in the winter of 2008–2009, a group of possible actors crystallized: The Swedish Research Council and Lund University would take key responsibility to secure funding for MAX IV, and be complemented by Region Skåne and Vinnova (Interview: Holmberg). These four organizations were, consequently, named by Flodström as the "key players" and requested to found a MAX IV company and jointly fund the accelerator complex, totaling SEK 900 million, together with a "consortium of Swedish and Nordic universities." A separate company should build and own the buildings and lease them to Lund University, thus bearing the investment cost of SEK 1 billion; and as for experimental equipment, a Nordic university consortium plus the Swedish Research Council and Vinnova should form a group "to draw up a basis for applications for experimental equipment for MAX IV" to be submitted to private research foundations in Sweden and Denmark, especially the Knut and Alice Wallenberg Foundation (KAW). Further, funding for the "Strategic Research Areas" launched in the 2008 governmental research bill (chapter 4) could be reallocated to pay for instrumentation. Future operating costs should be paid by Lund University, Nordic and Baltic contributors, and "Nordic industry" (Flodström 2009: 9–11).

On April 28, 2009, the Swedish Research Council, Vinnova, Lund University, and Region Skåne – later colloquially named the "Gang of

Four" – announced at a joint press conference that they had reached a four-party agreement and signed a declaration of intent to "form a consortium to initiate the process for establishing" MAX IV and work to achieve a quick go-ahead for the project, with start of construction in 2010 at the latest (Lund University 2009). The Flodström report's plan for financing MAX IV was largely reaffirmed in the declaration, which furthermore stated that a project secretariat located at Lund University would carry the process forward. An oversight committee was given the task of "preparing an agreement between the stakeholders and funders, including steering, organization, and funding"; ensuring that the construction proceeds as intended; and seeking additional funding (Lund University 2009). Erna Möller, executive of the Wallenberg Foundation, commented in June 2009 on the expectation, from Flodström and others, that the Knut and Alice Wallenberg Foundation (KAW) would make substantial contributions (of several hundred million SEK) to instruments at MAX IV. Möller said the Foundation was positive, but would use the same internal process as always. "There are many who think we should contribute, and we might, but there needs to be an application for funding first. [...] They are counting on us, and probably they should, but we won't be asking them if they need money from us" (*Sydsvenskan* 2009i).

As noted above, MAX IV was used as a bargaining chip by ESS proponents both in negotiation and marketing, and for the Swedish ESS campaign, the April 27, 2009, decision by the Gang of Four to set MAX IV on track towards realization was, therefore, timely. Quite clearly, the MAX IV decision was an asset for Sweden in the run-up to an ESS decision on European level – "a strong argument" (Interview: Honeth), that was "actively used" (Interview: Leijonborg). "Without MAX, I'm not sure we would have been as attractive" (Interview: Larsson). But the MAX IV decision also put to rest the domestic debate over ESS crowding out MAX IV, which in the views of many was a wiser investment. And while evidence suggests that the Swedish ESS investment did not infringe on other projects, the funding solution for MAX IV, and the additional funding arrangements for MAX IV that have since emerged (including the SEK 400 million from KAW, the SEK 160.25 million from 13 Swedish universities, and the rental contract for the MAX IV building which essentially postponed part of the construction costs) amount to circumstantial evidence for crowding-out effects, a matter that will be returned to in later chapters.

What is important in this context is the radical difference between the ESS and MAX IV seen in the perspective of Swedish science and science policy. While first promoted by a small lobby of devoted scientists, the ESS was made into the Government's "moonlander" project for its research policy offensive (Interview: Leijonborg), and it lacked the anchoring in Swedish science that MAX IV clearly had, which was proven several times in the process that led up to the decision in the spring of 2009 to go ahead and build the facility (Swedish Research Council 2006a, 2006b; Interview: Flodström). Initially, there was no competition between the two projects; MAX-lab was an active part in the forming of the ESS Scandinavia initiative and consortium in 2000–2002, and co-host of the meeting in Lund where ESS Scandinavia was inaugurated, in October 2000 (chapter 6). MAX-lab had strong backing for its MAX IV plans across Sweden, and the evaluations of the technical and scientific cases for MAX IV that clearly endorsed the plans and strongly recommended the Swedish Government to go ahead and fund MAX IV (Swedish Research Council 2006a, 2006b) probably raised the hopes among MAX-lab staff and users that the Government would take an initiative to make the project reality. The rather impressive track record of MAX-lab (Hallonsten and Christensson 2017a), the strength of the MAX IV proposal, and the confirmation of both in evaluations by panels of internationally renowned scientists, would perhaps normally have taken MAX IV far in national research policy, but the decision of the Government to back the ESS proposal and pledge some SEK 3–4 billion for its construction (a sum of money in the same range as the cost estimations for a fully built-out MAX IV, in 2006–2007) shattered these hopes, although of course a scenario where Sweden lost the ESS bid could have turned the tables again, in favor of MAX IV.

It is against this background that the brief but clear wave of resistance against the Government's ESS bid, in 2008 and on, should be understood. This was conducted predominantly by the Royal Swedish Academy of Sciences (*Kungliga Vetenskapsakademien, KVA*) whose concerns grew when the Minister for Education and Science, according to tradition, presented the new governmental research bill to the Academy in the fall of 2008. Shortly after the presentation, the Royal Academy wrote a letter to the Swedish parliamentary subcommittee on education and research, voicing an interpretation of the message by the Minister at his presentation, that the Government's stake on the ESS precluded governmental investment in MAX IV. The Academy quoted the Minister as saying, during the presentation, that "there won't be two accelerator facilities in Lund," which the Academy interpreted as a flat-out rejection of the MAX IV project in favor of the ESS. This, argued the letter, was highly undesirable – in the

Academy's view, MAX IV should "doubtlessly" be given priority over ESS, given MAX-lab's strong track record and broad, and strong, user base in Sweden and the Nordic countries (Royal Swedish Academy of Sciences 2008b). The Academy reiterated the call it had made in its 2005 response to the Larsson report, that the Government should make a final decision whether or not to pursue the building of the ESS in Lund only after a thorough evaluation of the scientific and economic advantages and risks of the project, involving the whole breadth of the Swedish scientific community, had been undertaken (chapter 6). Moreover, the Academy points out, the ESS should be funded according to the plan outlined in the Larsson report, and not as currently, through the ordinary and "already strained" Government R&D budget (the Academy apparently chose to ignore the aforementioned major funding increases launched in the 2008 research bill) (Royal Swedish Academy of Sciences 2008b). The Academy urged the Government to "reconsider this and put MAX IV first" (Royal Swedish Academy of Sciences 2008a). According to Gunnar Öquist, then secretary of the Academy, "It is wrong to say that the Royal Academy of Sciences was against the ESS," only that the apparent priorities of the Government were wrong, and that a Scandinavian ESS would have difficulties recruiting the proper expertise to design and build the ESS, and the Academy simply believed Sweden did not have the capacity domestically (Interview: Öquist).

The site decision

The "ESS Community" formed in late 2008 under the leadership of French Minister for Higher Education and Research Valérie Pécresse, was a way to put some kind of process in place for how to reach a deal, and when on January 1, 2009, the Czech Republic took over the EU presidency from France, the torch was passed to Czech Minister for Higher Education and Research Vlastimil Růžička, and the Minister for Education, Youth and Sports, Miroslava Kopicová (Leijonborg 2018: 344–345). Valérie Pécresse, who had declared that France would bring the ESS site decision to a closure within 2008, but who had failed to do so, reportedly "voiced rather too audibly that if France could not solve this problem then the Czechs, who were next in line for the EU presidency, certainly would not be able to" – and this reportedly offended the Czech Government and made Růžička determined to succeed (Thomasson and Carlile 2017: 27). A timetable was laid out, where the proposals from the three site contenders would be discussed at a meeting in Prague on May 4, and the final issue of location would be settled at an extraordinary and informal meeting with the site contenders and the future ESS member countries in connection with the meeting of the EU Competitiveness Council on May 28, 2009, in Brussels. The Competitiveness Council itself is a configuration of the Council of the European Union, and therefore an official European Union decision making body, composed of ministers of science, education, enterprise, innovation, and so on, from all EU member states, and has regular meetings at least four times a year. The Czech Government had tried to put the ESS site issue on the agenda of the regular Competitiveness Council meeting, but this was reportedly "vetoed by the Spanish and Hungarian site contenders," who probably used the fact that the ESS was not really a matter for an official EU body as an excuse to try to block a decision. Invitations were instead extended to an "informal dinner" after the regular meeting on May 28, and at a separate and neutral venue, to "representatives of governments 'interested in participating in ESS'" (Thomasson and Carlile 2017: 27).

The meeting in Prague, on May 4, 2009, gathered representatives of 15 countries at the Imperial Hotel close to the Old Town, and the three site contenders presented their proposals. Peter Honeth spoke of a "positive meeting," his Danish colleague Uffe Toudal Petersen reported on "a good discussion," and Czech minister Vlastimil Růžička called the meeting "productive." It was confirmed that a final decision would be made in Brussels on May 28 (*Sydsvenskan* 2009e). As the Brussels meeting approached, Allan Larsson expressed some pessimism, noting that the Czech Government appeared to want the financing of the ESS in place before any site decision. Larsson regarded this as unrealistic, arguing that a better option would be "to only concentrate on finding the best place for the facility" (*Sydsvenskan* 2009f).

The Brussels meeting on May 28, 2009, was not held in an official EU building but in Hotel Leopold, just two blocks from the offices of the European Parliament. Those present were ministers for science and similar from "a number of countries, a sufficient number, represented by authorized delegates, either ministers or others, so that a decision could be taken" (Interview: Honeth). The exact details of what happened at the meeting are not known, as only fragmented excerpts are available. The meeting was "chaotic" (Interview: Leijonborg) – there was no formal agenda, no predefined voting procedure, and really no rules. Spain, "which had likely reached the conclusion that their chances of winning a vote were slim," questioned the legitimacy of the meeting, and declared that whatever the outcome, they would continue to argue that Bilbao was the best location for the ESS. After the three candidates had presented their bids, they were asked to leave the room and a vote was cast. On re-entering the room, Lars Leijonborg learned that "a clear majority of the assembled countries, led by Germany and France, had declared that the ESS should be built in Sweden" (Leijonborg 2018: 351–353). Hard feelings were expressed towards the end of the meeting, when the Spanish delegation realized their support had evaporated, and they reportedly started "protesting wildly and arguing that formal errors had been made" (*Sydsvenskan* 2009g). According to other news reports, a stunning majority voted in favor of Lund – Denmark, Estonia, France, Germany, Italy, Latvia, Norway, Poland, and Switzerland. Only Portugal voted for Bilbao (*Science* 2009). The Lund candidacy had certainly appeared to be the strongest, and with the benefit of hindsight, it can be concluded that the crucial support offered by France and Germany a few days ahead of the Brussels meeting (Interview: Honeth), should have been enough to make the Swedish negotiators confident. But it was nonetheless a surprising victory, judging from the media reports the day after, and the testimonies of those present. Lars Leijonborg emerged from the room at 23:15, saying "Well… hrm, we'll have to conclude that the meeting has had a, from a Swedish viewpoint, very pleasing result" (*Sydsvenskan* 2009g).

The Spanish and Hungarian delegates were, naturally, both annoyed and displeased. Hungary would remain in opposition to the outcome of the Brussels meeting for some months (see below), but Spain apparently took steps towards a concession the same evening (Interview: Honeth). The following day, Spain declared that they would align and support Lund, and suggested an agreement where key components of the ESS be built in Bilbao, to make use of some of the funds already pledged for the project by the Spanish Government, and to smooth over some of the local disappointment. Leijonborg (2018: 353–354) argued that the solution was not optimal, but "acceptable" from the point of view of the future design and construction of the ESS; however, it was certainly a good idea in a political perspective, given that it would put an end to the competition from Spain. Leijonborg and Honeth traveled to Madrid a couple of weeks later to seal the deal, and established a "trustful" relationship with their Spanish colleagues (Interview: Honeth). An agreement, signed on June 10, 2009 in Madrid, stated that the Swedish and Spanish governments had agreed that the ESS was to be built in Lund, with the data management facility in Copenhagen, and "an important infrastructure site, an ESS Laboratory Test Facility and Accelerator Components Factory, will be located in Bilbao in Spain" (Spain 2009). This way, the Spanish contribution to the ESS construction costs – "expected to fall in the range of 150 M Euro" which corresponded to approximately 10% of the total, with the cost estimates at the time – would cover the Bilbao site and therefore stay in the Spanish economy and be "considered as an in-kind contribution from Spain" (Spain 2009).

Hungary was less cooperative, and took some time to come around. In a news report in *Nature*, the project director for ESS Hungary, Lázló Rosta, called the May 28 vote in Brussels a mere "opinion," acknowledging that there is "sort of a Swedish advantage, but there is not a final decision" (*Nature* 2009). An official statement from ESS Hungary in July 2009 argued that the procedure of voting at the May 28, 2009 meeting had been "unfair" and called for a more "transparent" process, "in line with values and ethics of the EU" and "European rules" (ESS Hungary 2009). There are no clues in the statement as to what the representatives of ESS Hungary meant by "European rules," other than a general appeal to rational criteria for choosing the best site, but given what available historical research says on the matter (chapter 3), the criticism rings hollow. A period of silence ensued, as ESS Scandinavia began the transition from campaign organization to project organization, and the Swedish and Danish governments began the work to achieve legally binding funding commitments to follow the support expressed on May 28, based on a full design update and new cost estimates, including technical design, scientific case, and all other aspects of building the ESS on a greenfield site outside Lund (chapters 8 and 9). On December 18, 2009, ESS Hungary conceded and the Hungarian Government declared its recognition of the site in Lund, which made Hungary the 14th partner member of the newly created Steering Committee for the ESS (ESS 2010a: 14). In December 2009, Allan Larsson's mandate expired (Interview: Larsson).

One important theme remains to be discussed before concluding this chapter and turning to the period after the site decision. Clearly, many people in and around ESS Scandinavia were very keen on presenting the outcome of the Brussels meeting as decisive in favor of Lund as the site for ESS, and the billboard of local Southern Swedish newspaper *Kvällsposten* on May 29, 2009 read: "EXTRA News just in: Lund wins fight for ESS" (Hallonsten 2012: 10). But nothing much was really decided, except that it was more or less clear after May 28, 2009 that if the ESS was to be built at all, it would be built in Lund. None of the letters of intent, memoranda of understanding, or joint declarations signed by the Swedish Government and the governments of Denmark, Estonia, France, Germany, Italy, Latvia, Lithuania, Poland, and Spain, in 2008–2009, were legally binding. "It was not a formal decision" (Interview: Honeth); the support for ESS Scandinavia offered by these countries "could be reconsidered if something judged to be more important would come up" (Leijonborg 2018: 354), because all they had done was to decide "which site they would work with in the continued planning process," and most importantly, "the financing was still a completely open question" (Interview: Honeth). Importantly, the first legally binding agreement that fifteen European countries would jointly fund and build the ESS in Lund was not reached until August 2015, when the statutes of the European Spallation Source ERIC came into force. Several countries made commitments, with detailed sums of money, during 2014, and some of these would probably count as legally binding if their status would have been tried in a legal procedure; and in July 2014, a funding solution could be presented by the Swedish Government that was based on commitments reliable enough for the ESS Steering Committee, the board of the ESS company, and the Danish and Swedish governments who were the owners at the time, to decide to proceed to start of construction (chapter 9). Before that, a Memorandum of Understanding between sixteen countries was signed, on February 3, 2011, and was "the first multinational agreement on the ESS" (ESS 2011a: 8), but this document also lacks the status as a legally binding intergovernmental agreement.

Therefore, nothing was decided in Hotel Leopold in Brussels on May 28, 2009, other than that the joint work of a number of European countries to design an ESS facility and eventually, probably, build it in Lund, would continue. The decisive events whereby Lund emerged as the default site were the concessions by Spain and Hungary, since these meant that no competition remained. But there is still nothing to suggest that if, for example, one of the large countries such as France, Germany, or the UK would have changed its mind, the ESS project could have collapsed. In other words, even after May 28, 2009, and the site decision in favor of Lund, the project had a long way to go.

8. Planning the ESS (2009–2014)

A growing organization

Before the site decision of May 28, 2009, each of the three candidate sites had their own organization with its own legal status, and there were different loosely knit organizations in place to drive the project forward at European level, such as the *ESS Initiative* chaired by Peter Tindemans (chapter 5), which was dissolved at some point in 2007–2008, and the *ESS Community* driven from late 2008 by French and Czech ministers, leading up to the Brussels meeting (chapter 7). The ESS Scandinavia consortium had been dismantled in 2007 and replaced by an ESS Scandinavia *project secretariat* funded by the Swedish Government and administratively located in Lund University. In the spring of 2009, when the secretariat's mandate was prolonged, an interim board was created "to bridge the transition period" before a limited liability company could be formed to take over responsibility (ESS 2010a: 28).

The two years that followed would bring an "extreme" change for the ESS (Interview: Börjesson). The campaigning and negotiating would certainly continue (chapters 9 and 10), but now the organization in Lund suddenly "had to be serious" and show partners locally, nationally, and abroad that it could get the task done – "preparing a budget, preparing a project scheme, time plan, and so forth" – and it was "difficult to adjust, to move from these campaigning things onto a real planning project" (Interview: Vettier).

When nuclear physicist and accelerator designer Mats Lindroos joined the ESS (see below), in the spring of 2009, it was a small organization – "we sat around one table, Patrik, Colin, me, and some others" (Interview: Lindroos) – and had its offices in a corridor in the medieval city center of Lund, on Stora Algatan. At the end of 2009, the number of employees was slightly more than 40 (ESS 2010a: 24), and some months later, it was about

75 (ESS 2011a: 8). The real expansion came after the *Technical Design Report (TDR)* had been published and approved by the ESS Steering Committee in the spring of 2013 (see a later section), and especially after breaking ground in 2014: the number of employees grew from around 140 in mid-2012 (ESS 2012b: 5) to 194 at the end of 2013 (ESS 2014a: 7) and 289 at the end of 2014 (ESS 2015a: 7). The ESS Activity Reports from these years give a good impression of the growth and transformation; while they have roughly the same appearance, they change considerably in character over this time, shifting focus from the marketing campaign and highlighting the future wonders that the ESS will bring, to detailed descriptions of the work processes and management structures of the ESS company.

Although all major comparable projects around the world are resource-intensive and in need of many key recruitments at various points in their buildup phases, most of them have been built within preexisting structures of national laboratories or similar; the key exceptions being the ILL and ESRF in Grenoble. The UK neutron source ISIS was always part of the Rutherford Appleton Laboratory (RAL) in Harwell, and the Swiss neutron source SINQ is located at the Swiss governmental research center in Villigen that is nowadays called the Paul Scherrer Institute (PSI). The same is true for most other large-scale scientific facilities around the world; almost without exception, if not built within existing government institutes and Big Science centers, they have been developed at universities (Hallonsten and Heinze 2015; Rush 2015). The ESS was a *greenfield project*, without such preexisting organizational structures to rely on, and its Activity Report 2011 consequently notes that this presents "perhaps one of the greatest risks" for the project as a whole (ESS 2012c: 12).

The first and most palpable reminder of this was the need to recruit the right people. This proved to be "enormously difficult" (Interview: Argyriou), for many reasons, foremost because key people were hesitant to take a chance on the ESS as a new employer, and Lund as new place to live and work, until there was more financial, political, and organizational certainty around it (Interview: Möller). "An expert that had a reputation and a world standing didn't want to come to ESS to perhaps find themselves out of a job in six months or two years" (Interview: Argyriou). Prominent exceptions existed; as noted in chapter 6 and will be discussed below, reputable scientists Colin Carlile, Christian Vettier, and Mats Lindroos came in early and were soon joined by Ferenc Mezei and Dimitri Argyriou. What became known as the "Saab gang," with Kjell Möller at the helm, was also successfully recruited in early 2011 to take charge of project management, and when Sven Landelius was appointed chair of the board of ESS AB, he came from a fourteen-year tenure as CEO of the Danish-Swedish consortium that had built and operated the Øresund Bridge with undisputable success.

Some claim that the ESS "hired too many people, too early" and that this might have been part of a plan to create a *fait accompli* that would be difficult or impossible to later back away from, both for the Swedish and Danish governments, and the other prospective member countries (Interview: Johnsson). One major risk of early recruitments was that the recruited competence was not optimal for the design update and pre-construction phases, and to some extent, ESS CEO Colin Carlile seems to have made some faulty choices there, as he "started to build an organization like it would look during operations, and then he had to build the facility and then he had the wrong organization"; more specifically, an organization filled with highly qualified scientists who "didn't know anything about building instruments or building a facility" but "would be fantastic at operating the instruments eventually" (Interview: Matic). An all too academic working atmosphere, not particularly suitable for big projects, with scientists who "work three months, and then disappear to do research somewhere for six weeks," characterized the ESS in 2010–2011 (Interview: Möller), but informants agree that from a political point of view, the strategy to recruit scientists and establish an academic organizational culture was probably apt because it secured interest and support in scientific communities. Another challenge of recruitment was that the typical profile of people seeking employment at the ESS caused some human resource management challenges in the long term, because they all came with their own personal image of what it was that they would contribute to achieving – "everyone has their version of the shiny castle on the hill" – but eventually there would be only one ESS and "that means some dreams die [...], and I can tell you, all dreams have still not died, and that makes decision-making processes harder, it drives up cost and it builds inefficiencies" (Interview: Argyriou).

In the fall of 2009, at the fifth European round table discussion held in Krakow in September 2009 (Larsson 2019: 160), Sweden and Denmark invited other countries to a first meeting of an ESS Steering Committee (Interview: Börjesson), which would become the highest governing entity of the ESS and transform into the ESS Council after the organizational transition in 2015. It held its constituent meeting in Copenhagen on 22–23 October 2009, consisting in this first iteration of 28 delegates from 14 countries (Denmark, Estonia, France, Germany, Hungary, Iceland, Latvia, Lithuania, Italy, Norway, Poland, Spain, Switzerland, and Sweden) and four representatives from the ESS organization (ESS 2010a: 27–28, 42). Different countries were at very different stages in the process of committing financially to the ESS project and joining as members, which showed in the composition of the committee in the years 2009–2012. The UK, for example, was not represented at all, and would not be until 2011. Sweden was represented by Peter Honeth (who was also the chair until early 2010), Lars Börjesson of the Swedish Research Council, and Mats Johnsson of the Ministry of Education and Research. Denmark, close to financial commitment and determined to soon enter into partnership with Sweden as co-host and co-owner of the ESS limited liability company (see below), was likewise represented by a ministry official, and also an academic representative. Several countries, including Germany, Hungary, Poland, and Switzerland, had similar arrangements, with one representative of a ministry or a national research funding agency, and one academic/scientific representative. Spain was represented by its ministry and by the ESS Bilbao organization. Some countries, including France, were represented only by one or two ministry representatives, and some, including Italy, only by members of the scientific community and domestic institutes (ESS 2010a: 42; ESS 2011a: 30; ESS 2012b: 50). Countries that sent representatives from universities, institutes, and facilities signaled most of all "a specific technical and scientific interest, and an interest in participating in construction" with the future roles of their domestic institutes front and center (Interview: Börjesson), and many of these would also later become heavily involved in the construction of the ESS, and make big in-kind contributions.

As stipulated both in the articles of association for the limited liability company ESS AB and in the Memorandum of Understanding between sixteen countries signed in February 2011 (see below), the Steering Committee was the highest decision-making body of the ESS organization, although the formal responsibility lay with the board of the ESS company, which was obliged to follow the "recommendations" of the Steering Committee as long as these were not "contradictory to the Articles of Association of the company or Swedish law" (Swedish Government 2010d). This meant that as early as late 2009, the Steering Committee had a central role in the governance of the ESS organization, and a key say in all the major decisions regarding design, technical and scientific choices, and funding and governance.

The company *ESS AB* ("AB" means "*aktiebolag*" in Swedish, i.e. joint stock company) was established through a series of decisions and actions by the Swedish Government in 2009–2010. The Governmental budget bill for 2010 asked parliament to authorize the Government to "form a company whose activities shall be to plan, project, fund, build, own, administer, run and maintain the European Spallation Source (ESS) and sell parts of the Government's shares to other parties" (Swedish Government 2009a: 19–20). On November 17, 2009, the Swedish Government decided to purchase all stocks of the company *Goldcup 5129 AB* (Swedish Government 2009b). At an extraordinary meeting of shareholders of *Goldcup 5129*, on April 14, 2010, the name of the company was changed to *European Spallation Source ESS AB*, and five people were elected to its board: Per Eriksson (Vice-Chancellor of Lund University), Katarina Bjelke (of the Swedish Ministry of Education), Lars Börjesson (of the Swedish Research Council and also chair of the ESS Steering Committee from 2010 and on), Lena Gustafsson (Vice-Chancellor of Umeå University), and Sven Landelius, with Landelius as chairman. The new articles of association adopted by the extraordinary meeting of the shareholders specified that "[t]he object of the Company's activities is to plan, design, finance, construct, own, manage, operate and maintain the European Spallation Source (ESS) together with any other activities compatible therewith" (ESS 2010d). The company began operations on July 1, 2010, with a first transaction consisting of the transfer of operations and assets of the ESS Secretariat from Lund University, on behalf of the Swedish Government, which constituted a shareholder's contribution of SEK 5.7 million (ESS 2010b: 7). At the same time, Colin Carlile was appointed CEO (Interview: Landelius).

Sven Landelius, an engineer by training, had been the CEO of the Danish-Swedish Øresund Bridge Consortium in charge of building and operating the fixed link between Malmö and Copenhagen, and decided to leave the consortium in 2006 to start a consultancy business and serve as a member of various boards. In early 2010, he was called up by Peter Honeth, who asked if he would be interested in working with the ESS. Landelius had done some investigatory work for Allan Larsson and Lund University concerning the exploitation of the land area between ESS and MAX IV, what would later be called Science Village Scandinavia (see chapter 10), and was well acquainted with the ESS project (Interview: Landelius). As he recalls, "it was a matter of course" to form a limited liability company for ESS operations during the time of buildup of the organization and the pre-construction phase (Interview: Landelius). What was not self-evident, however, was the future ownership of the company. Denmark was on its way in, and the ESS AB board admitted Danish representatives as guests even before the Danish stock purchase, but the main discussion early on concerned whether other countries would be allowed to purchase shares. According to the Government's budget bill for 2011, the original idea was for other countries to successively become shareholders as they joined the project formally (Swedish Government 2010b: 147), but in the fall of 2010, it was decided to keep ownership of the ESS company restricted to Denmark and Sweden, and wait to spread ownership among other countries until a new organization was in place (the ERIC, see below) (Interview: Landelius). In May 2010, the Danish parliament decided to purchase stocks in the ESS company (Swedish Government 2010b: 147), and in late December 2010, the Swedish Government sold 26.316% of the shares in the ESS company to the Danish Government (ESS 2010c: 10). This arrangement would remain in place until the transition to the ERIC some five years later (see chapter 11).

The creation of a state-owned company, and the joint ownership by Sweden and Denmark, "was symbolically important, in order to create confidence and seriousness in the process," sending a powerful message to the other prospective partner countries that the ESS project was being taken seriously by its two co-hosts and given a firm organizational structure, and it created a framework for accomplishing the task, including making recruitments and establishing collaborations with the necessary partners across Europe (Interview: Börjesson). The organization charts in the ESS Activity Reports from the years 2009–2012 give somewhat confusing and mixed messages concerning, in particular, the top-level governance structure, where several entities are crowded: the ESS AB board and CEO, the Steering Committee, and the three advisory committees the Technical Advisory Committee (TAC), Scientific Advisory Committee (SAC), and Administration & Finance Advisory Subcommittee. Peculiar arrows illustrating channels for input from the fourteen partner countries not only to the Steering Committee and advisory committees, but also directly into the organization hierarchy, come from the ESS organization's eagerness to (over)emphasize the collaborative nature of the project, and the commitment of the partner countries. The 2009–10 chart presents the ESS AB company (and, presumably, its board) as superordinate to the Steering Committee and the rest of the organization. The Science Advisory Committee (SAC) and the Technical Advisory Committee (TAC) are represented as advisory mainly (or only) to the Director General, whereas the Administration & Finance Advisory Subcommittee has an unclear role as subcommittee to the Steering Committee, or its equal, or yet another body giving "recommendations" to the company. Absent from the chart is the Stakeholders Group, formed out of the remnants of the ESS Scandinavia Consortium (chapter 7) and mentioned in the 2009–2010 ESS Activity Report as "an effective and necessary interface to local interests" (ESS 2010a: 28) but abolished later in 2010.

But it is the relationship between the ESS AB board, the Steering Committee, and the Director General that is most important from a governance perspective. In this regard, too, the organization charts published in the activity reports from these years give a confused impression and change the arrangement of boxes and arrows from year to year (ESS 2010a: 42; ESS 2011a: 30; ESS 2012b: 50). This is probably, to some degree, due to the fact that the ESS was a greenfield operation that expanded vastly and developed technology, science, and, importantly, also collaborative, administrative, and organizational routines and practices at the cutting edge. Interestingly, however, testimonies suggest that the confusion conveyed by these organization charts need not have been there; in fact, it seems the real situation, behind the charts, was fairly simple and clear. This would be an inversion of the common situation where reality is complex and changeable, and organization charts convey an oversimplified image (e.g. Pugh 1973; Mintzberg 1983). The highest decision making levels in the ESS organization in 2009–2012 seem to have had a far simpler command structure and division of authority than the organization charts convey: "The discussion was more around how to represent this graphically than who would do what. It was really clear who did what." There were certainly discussions about the very important function of the advisory committees, but this was reportedly settled at an early stage, so that that they would report not to the board but to the Steering Committee (Interview: Börjesson). The division of labor between board and Steering Committee was likewise very clear, according to key people involved: The board was simply responsible for the activities of the limited liability company, in accordance with the Swedish Companies Act (*Aktiebolagslagen*) (Interview: Landelius). Indeed, the "relationship between the European Spallation Source Steering Committee and European Spallation Source ESS AB (ESS AB)" had been described in detail by the Government in the spring of 2010: The Steering Committee, consisting of "representatives of the countries participating in the construction of the spallation source and its future ownership," is in charge of the "direction of research," the "scientific and technological management," "contributions in kind," and the "costs of different parts of the facility in relation to the technology and performance chosen for the different parts of the facility taking into account given budgetary frameworks." The ESS AB company "shall, concerning the actual construction of the spallation source, take into account the recommendations" of the Steering Committee provided that these do not contradict the articles of association of the company or Swedish law (Swedish Government 2010d). The Memorandum of Understanding signed by sixteen partner countries in February 2011, which formed the basis of the work of the international collaboration around the ESS, essentially reaffirmed these instructions, in addition to declaring ESS AB a temporary construct; "the stated objective of the owners, Sweden and Denmark" is that the company will "in due course" transform into "an international entity under the auspices of the ESS Partner Countries" (ESS 2011e).

In other words, the duties of the board of ESS AB was to oversee the operations of the limited liability company that was formally and legally responsible for the ESS organization in Lund, including all aspects of top-level governance of a corporation such as employer relations, taxes, and accounting/financing. The latter in particular was its top priority, since the operations were by definition loss-making and the company needed continuous capital injections from its owners (chapter 9). A lot of proactive effort, especially from the board, was required to make this work; both toward the Swedish Government (and later the Danish), who provided the funding, and inside the ESS organization, which had to prepare the budgets and activity reports to support the requests (Interview: Landelius).

The work of the ESS AB board was heavy, reports its chairman Sven Landelius, whose experience as chair of several companies was that the work typically represents 20–25% of a full-time position, but in the case of the ESS AB, it was closer to 50%, and "I have never been on a board with such strong participation" (Interview: Landelius). The original members remained on the board until the liquidation of the company in 2015, and from November 2010, the Danish Ministry of Science, Technology, and Innovation was represented by Lars Kolte and Inge Mærkedahl, who were replaced in 2012 by Hans Müller Pedersen and Bo Smith. In addition, from November 2010 Lars Goldschmidt (of the Confederation of Danish Industries) sat on the board (ESS 2010a: 42; 2011a: 30; 2012b: 50; 2013a: 41; 2014a: 38; 2015a: 42).

With the ESS AB board formally and legally in charge of the ESS organization and its operation, the Steering Committee had the key role of driving forward the crucial processes that would take the facility through the design update, all the necessary preparations, and toward a funding solution and start of construction. This meant that the center of power, overall, lay with the Steering Committee, and the mandate to push ahead or put the brake on the process. It was in the Steering Committee that "all the stakeholders, both the governments and the institutes that are interested in construction" were represented, which meant that the committee was "the driving force that made the whole process go forward." "All the big things in the design were decided in the Steering Committee," such as the choice to go for tungsten instead of mercury in the target station (see below) or the number of beam ports (Interview: Börjesson).

The February 2011 MoU affirmed the composition of the Steering Committee, specifying that it "consists of two representatives from each of the ESS Partner Countries to this Memorandum of Understanding" (ESS 2011e). The 2012 ESS Activity Report calls the Steering Committee "a more traditional governing body [...] populated by scientists with experience of large scientific infrastructures together with ministerial or research council personnel who are close to the funding authorities from their individual countries." It also notes that the Steering Committee "is aided in its job" by the Administrative and Finance Committee (AFC), "formally a subcommittee" of the Steering Committee and composed of delegates from the partner countries (ESS 2013b: 5). The central role of the AFC was "overseeing all administrative and financial functions" and especially "preparing the future legal framework and organisational scheme" for the ESS. A specific task, carried out by a working group on charge by the AFC, was "to explore which legal form is best suited to building and running ESS" (ESS 2011a: 9). The 2010–2011 iteration of the AFC had representatives of 12 countries (Denmark, Estonia, France, Germany, Italy, Latvia, the Netherlands, Norway, Poland, Spain, Sweden, and Switzerland), mainly from ministries and research councils but also occasionally a facility or institute representative (ESS 2011a: 30), and was complemented by a UK representative in 2011 (ESS 2012b: 50). The 2012 ESS Activity Report also notes that the SAC and TAC are "directly related to the ESS organisation itself" but "the source of independent advice" and "populated by international scientific and technical experts and are nominated by the ESS," not necessarily corresponding evenly to the seventeen partner countries counted at the time (ESS 2013b: 5). In their first years of existence, the SAC and TAC were composed of scientists from a variety of organizations in Europe, Japan, and the United States, including in particular universities and laboratories with particular strength in neutron instrumentation and its use, such as the ILL, the PSI in Switzerland, J-Parc in Japan, and the Jülich Research Centre in Germany (the SAC), and ISIS, CERN, Oak Ridge National Laboratory in Tennessee, and Istituto Nazionale di Fisica Nucleare (INFN) in Italy (the TAC) (ESS 2010a: 43; 2011a: 31; 2012b: 51).

The division of labor between Steering Committee, board and management was discussed at some length in the first "light review" (Interview: Landelius) that was undertaken in the spring of 2012. The review recommended that more "clearly defined roles, responsibilities, authorities, and accountabilities" were developed between "the ESS board, the International Steering Committee (STC), and the new CEO," especially after the design update phase, when the organization transferred into "project execution," and urged the board of ESS AB "to delegate powers to the CEO so he/she can deliver the very ambitious programme without constantly seeking board approval" (ESS 2012e: 3). More specifically, the review identified a "lack of mutual confidence" between the board and the management, having to do with "confusion" regarding division of responsibilities (ESS 2012e: 5). This seems to some extent to have been a question of personal chemistry. "It was no secret" that Colin Carlile and Sven Landelius "did not get along very well" and had "very different views on things" (Interview: Lindroos). Whereas Sven Landelius was completely geared to running the project in a controlled process and as a proper limited liability company, with quarterly interim reports, budget systems, and controlling, Carlile was partly stuck in the campaign organization that had been so successful in bringing the ESS to Lund, and partly a scientist with considerable management experience but few or no project leader or project management skills. "We were not always on speaking terms. He thought I went too far and was too rigid and all that. [...] I think it is a personality issue. He is not a project person" (Interview: Landelius). When Colin Carlile was replaced by James Yeck in 2013 (see below), the animosity between the board and the CEO seems to have vanished.

The ESS organization, from the CEO and down through the divisions and directorates, was built up gradually in 2009 and on, with the science directorate, machine directorate, and project directorate created back in 2009, headed by Christian Vettier (science director), Mats Lindroos (machine director), and Patrik Carlsson (project director). In April 2010, internationally renowned neutron scientist and ESS Hungary champion

Ferenc Mezei was recruited as Head of the Target Division (ESS 2010a: 24). In spring of 2011, the Programme Office was established, with the responsibility of driving the design and cost update processes (ESS 2012c: 8–9). The recruitment of a program director and a team with experience of project management was crucially important, since prior to this, the ESS "did not have a project organization to build the whole thing," but was still "more like a campaign organization" (Interview: Matic). On came what was popularly called "the Saab gang," with former employees of Saab Aerospace in Linköping as a backbone. "There was no project structure. They came from marketing, had launched the ESS, and then 'OK, what do we do now?'" (Interview: Möller). The Saab gang created "the first embryo of an organization that would be able to execute the project technically" and "did a great job setting it up as an industry project" (Interview: Lindroos). "They came in and added all such competence" (Interview: Matic). "It was really important, internally and externally, that professional project management knowledge came in. [...] And it enabled a move from science to orderliness in the project" (Interview: Börjesson).

Kjell Möller, an engineer by training and with long experience from project management, marketing, and strategy in the Saab Aerospace Organization, including periods of tenure as Executive Vice-President, Manager of Industrial Cooperation, Marketing Manager, Marketing Director, and Program Director, had worked with several large projects, among them the *JAS Gripen* military aircraft which was developed in collaboration between Saab Aerospace and the Swedish Government. Asked by ESS management in 2010 to review the job description of program director that they had drafted, he realized "there's no one on earth that can fulfill all the demands they had specified [...], everything technical, everything in project management, everything in financing, everything in management – this is a team, not one person" (Interview: Möller). Consequently, Möller was given the task of putting together a team that could match the job specification, "and that could do the job for a period of three to four months" (Interview: Möller). Möller himself became project manager, with Johan Leander, who had been the project manager of the Gripen aircraft project, as deputy, and a couple of other people "working one day per week or so, implementing basic principles of running projects like this." The work was mainly geared to establishing routines and control mechanisms (Interview: Möller), with the original plan being to make a shorter-term contribution of implementing project management standards and routines, and then leave. But the role was made permanent, and at a meeting of the Steering Committee in

February 2011, Möller was offered the job as ESS Program Director (ESS 2011a: 31). Gradually, all kinds of issues connected to construction and procurement became the responsibility of the program directorate and Möller – permit process, in kind, negotiations. “So my organization grew quite large” (Interview: Möller). With the Saab gang in place, the ESS organization was equipped with the competence necessary to develop the design of the facility and prepare all parts of the construction project.

With the ESS organization in Lund expanding and recruiting domestically and all over Europe, in order to be able to take on the complex and demanding set of tasks that the design update and all the other preparatory work demanded, another part of the ESS was also being set up. A key part of the deal with Denmark in the run up to the site decision in 2008–09 had been the Data Management and Software Center (DMSC) that was located in Copenhagen and secured realization of the “co-siting” idea of Denmark which allowed large parts of the Danish investment to stay in the Danish economy (see chapter 7). The DMSC was founded in 2010 and initially hosted by the Niels Bohr Institute at the University of Copenhagen, formally under the Science Directorate of the ESS organization, and with professor of computer science Stig Skelboe as project leader (ESS 2011a: 18). Although the key function of the DMSC would be to take care of data management and the development of data processing techniques in the operations phase of the future ESS, the DMSC also had a role to play in the design update phase of 2010–2012, namely to develop computer software simulations of various parts of the operation of the ESS, from accelerator, through target, and to instruments, in order to optimize the design of these components. The longer-term work of preparing for data management in the operations phase was also initiated early on. The *Technical Design Report (TDR)* for the ESS, published in early 2013 (see below), provided detailed descriptions of the functions of the DMSC, making highbrow claims that the data center would manifest “a new approach for software and data management [...] that intuitively integrates control of the neutron instrument and its sample environment” by “delivering a 24/7 e-science service programme to cover the complete research cycle from idea to publication” (ESS 2013b: 132). In more concrete terms, the DMSC would “host the ESS staff and IT equipment” and be responsible for “High Performance Data Storage,” “Web & Cloud Services,” and “Infrastructure for Data Analysis & Scientific Computing,” with “Instrument Control,” “Live Data Monitoring,” and “Data Reduction” remaining the responsibility of the main ESS site (ESS 2013b: 133). After the publication of the TDR, when the

ESS moved from the design update phase to preconstruction phase, the DMSC was formally established by the ESS AB board as a "permanent establishment" ("*fast driftställe*") of the ESS organization (ESS 2013c: 5), and soon thereafter, physicist and experienced neutron user Mark Hagen was appointed its director (ESS 2014a: 8).

Preparing the site

The ESS Scandinavia campaign, described in great detail in chapter 6, had worked purposefully to "win" the competition at European level and managed to win the support of the Swedish Government and key people who could bring the process to a successful outcome in the spring of 2009 (chapter 7), but when faced with this *fait accompli*, a number of crucial processes had to be set in motion in order for the ESS to become a reality beyond the handshakes and colorful pamphlets. A major element in this was, of course, the buildup of an ESS organization (previous section) that could take care of the design update and gradually proceed toward start of construction (later sections). While difficult and challenging in a myriad of ways, this work was under the control of the ESS community of scientists, technologists, and science administrators, and with the inclined help of politicians in Sweden and Denmark, and the involvement of project management professionals like the Saab gang and Sven Landelius, it achieved the milestones it had to in order to keep momentum and approach its goals.

But some things were outside the control of the ESS organization, its friends and collaborators abroad, and its patrons in government. In particular, physical planning, including the process to make the necessary land area available, and the permit process, whereby the ESS would pass all the controls of the Swedish public administration's various agencies and courts, were processes of crucial importance that led their own lives.

Since the conclusion of the very early mini-site competition process in Skåne in 2000–2001, a land area of some 1.2 square kilometers in the *Brunnshög* area northeast of Lund had been the designated location for the ESS. Situated between the E22 highway connecting Lund with towns immediately to its north and northeast, and the Odarslövsvägen road that ends up in the tiny village of Östra Odarslöv some three kilometers northeast of Lund, the land area was primarily used for farming, and owned by one of Sweden's few remaining Entailed Estates ("*fideikommiss*" in Swedish), namely the *Björnstorp and Svenstorp Estate*. Entailed Estates in

Sweden are regulated by a special will that dictates that the estate, in full, is inherited by the first-born heir, and cannot be split or sold in parts – the term "*fideikommiss*" stems from the Latin *fideicommissum*, which means "entrusted estate." Key to this ancient and blatantly outdated legal framework is that the estate cannot diminish its land holdings, which means that land cannot simply be sold but has to be exchanged for a holding of at least the same size or value.

In the spring of 2006, Lund University had entered into a declaration of intent concerning future leasing of land ("*Viljeförklaring beträffande upplåtelse av mark*") with the Björnstorp and Svenstorp Estate. In the document, Lund University declared itself willing to lease the land on behalf of ESS Scandinavia, and the Estate agreed to keep the land available for lease to Lund University, "or the representative of ESS Scandinavia that might replace the University," until the end of 2009. Any decision "that implies that the ESS is located to Lund" would prompt further negotiations of the terms of the lease (Lund University 2006). As a site decision drew nearer in 2008–09, however, it became increasingly clear that building the ESS facility on leasehold land was unrealistic, and the 2008 ESS Scandinavia response to the ESFRI Working Group on the ESS site questionnaire (chapter 7) consequently made clear that the designated land on the Brunnshög area was to be purchased (ESS Scandinavia 2008a: 28). The task of brokering a deal, and enabling the transfer of ownership of the designated land area from the Björnstorp and Svenstorp Estate, fell on the chief financial officer of Region Skåne, Harald Lindström, commissioned by the chair of the regional executive committee Jerker Swanstein in November 2009. A farm property on the northern edge of the City of Lund was suddenly available for sale, but with the vendor's condition that the sale had to be "considered done" before the end of the year, "for some tax reason" (Interview: Lindström). The opportunity that had opened was to exchange this land, which was more than double the area needed for the ESS, for the land under ownership by the Björnstorp and Svenstorp Estate, the designated ESS site. Negotiations ensued, and an agreement was eventually reached where Region Skåne would purchase the land – formally called "Lund Hoby 20:2 and parts of Lund Hoby 20:1" (the *Hoby Land Area* hereafter) for SEK 74 million, and then exchange it for the land area at Brunnshög, formally called "parts of Lund Västra Odarslöv 3:1, 5:1, s:2, Lund Östra Odarslöv 12:1 and 13:1" (the *Brunnshög Land Area* hereafter) with a difference of SEK 6 million paid by the Estate (Interview: Lindström; Region Skåne 2010a).

On December 15, 2009, the Regional Executive Committee decided to recommend that the Regional Council buy the Hoby Land Area for SEK 74 million, and give the Executive Committee authority to seal this deal and the task of starting negotiations with the ESS concerning access to the Brunnshög Land Area (Region Skåne 2009c). In a memorandum to the Regional Council that outlines the background to the matter, it is noted that "the ESS secretariat lacks the capacity to purchase land" and that "no other actor, at the moment, can take the responsibility of purchasing land for the ESS facility" except Region Skåne (Region Skåne 2010a). "Nobody else had the money or the ability to accomplish this but Region Skåne" (Interview: Lindström). The paperwork that was needed to "consider the land sold" was accomplished within December 2009 (Region Skåne 2010c), and on March 2, 2010, the Regional Council made the formal decision to buy the Hoby Land Area, and to trade this land for the Brunnshög Land Area, and to thereby also receive compensation of SEK 6 million (Region Skåne 2010b). The lease agreement between Region Skåne and the ESS was later formalized and made available the Brunnshög Land Area for a yearly fee of SEK 1,000, and giving the ESS the option to purchase the land, which it later did, after its transition from corporation to ERIC (see a later section), "so that the region got its money back and the tax payers were held without harm" (Interview: Lindström).

In a letter to local newspaper *Sydsvenskan* in March 2011, journalist Björn af Kleen and local Left Party politician Mats Olsson argued that the whole transaction was a "golden deal" for the Björnstorp and Svenstorp Estate, and that the region's tax payers were the losers (Olsson and af Kleen 2011). The claim is not entirely true; while it was probably a good deal for the Estate, if any "losers" is to be found in this story it must be the ESS, which eventually bought the land from Region Skåne "at the price it had paid, plus indexation" (Interview: Lindström). Meanwhile, it is hard to make a comprehensive assessment of the whole deal since the Brunnshög Land Area was part of the package that brought the ESS to Lund, and thus simply had to be made available, at almost any cost: "Representatives of the Swedish Government had negotiated in Europe and came home from Brussels [in May 2009] with the result that the ESS is now going to be built in Lund. The negotiating position I had didn't give me a lot of latitude" (Interview: Lindström).

The ESS Activity Report 2009–2010 notes that "the land earmarked for the ESS site is now in the hands of Region Skåne" and "will be used as the location of the ESS" (ESS 2010a: 30). But it would take until the go-ahead decision by the ESS Steering Committee in the spring of 2013, and the transition from the Design Update Phase to the Pre-construction Phase that it brought, before real preparations for buildings and construction could start, including contracting for civil construction. In the meantime, a key process to be set in motion and brought to a successful result was the permit and environmental trial procedure.

Lund University, in its capacity as principal of ESS Scandinavia, had filed applications to local authorities for trials of the site for the facility, including radiation safety, environmental planning, and construction planning, back in June 2005, prudently aware that such trials typically take several years (ESS Scandinavia 2005a). The environmental trial and the radiation safety trial were particularly important in this regard. The latter was especially delicate, given the Swedish decision to dismantle its nuclear power plants and the controversies around the ESS, in that part of Denmark as well as locally in Lund and in the Swedish Parliament. A prestudy made in 2008 by Studsvik Nuclear AB had highlighted that in comparison with a reactor, the safety aspects are simpler because no fissile material is present on site, and no fission products are produced in the facility, which is a great benefit from a safety and waste management point of view (Eriksson 2008). Nonetheless, when, in August 2010, the ESS formally notified the Swedish Radiation Safety Authority (*Strålsäkerhetsmyndigheten*, *SSM* hereafter) that it intended to apply in late 2011 for the necessary permits to start facility construction, the SSM replied that although the ESS should indeed be regarded a non-nuclear facility, it would possibly have to use the same standards that it would in the case of nuclear facilities, "because of its uniqueness as a neutron spallation source" (ESS 2011a: 24). In practice, this meant that the ESS had to submit a *Preliminary Safety Analysis Report (PSAR)* to SSM, which it did in March 2012, containing information on the site and the design of the accelerator and target station, and all aspects of radiation safety pertaining to the procedure for constructing the facility. The PSAR application was to be followed by similar documentation regarding operation, decommissioning, and a specific strategy for managing radioactive waste (ESS 2012b: 20).

Sydsvenskan reported recurrently on the progress of the radiation safety permit process in 2012–2013. In July 2012, the ESS was requested to complement their previous application, which only concerned the target station, and provide assessments of all other parts of the facility. Only after this could SSM begin the trial process. "When we regard the application as complete, we can enter into details and scrutinize different safety aspects,"

Peter Frisk, project leader for the permit trial of ESS at SSM explained (*Sydsvenskan* 2012b). The chief safety manager of the ESS, Peter Jacobson, was "not surprised" and viewed the requests for complementary information as a "normal and continuous process" (*Sydsvenskan* 2012c). Not surprisingly, the customary referral round of the application submitted by the ESS to SSM yielded negative comments from the local branch of the Swedish Society for Nature Conservation (*Naturskyddsföreningen*), which had been one of the more vocal opponents of the ESS in earlier phases of the campaign (chapter 6), and who argued that "permission to build the ESS in Lund should not be granted" (*Sydsvenskan* 2012a).

In July 2014, over two years after the radiation permit process had been initiated and a mere two months before start of construction (chapter 11), the SSM approved the ESS plans to conduct operations with ionizing radiation and decided to grant the necessary permits for start of construction. Importantly, permits for the start of operations and for the transition into steady state operation, remained and will be granted on the basis of ordinary future trials (*Sydsvenskan* 2014m).

In parallel with the radiation permit application of March 2012, a license application was also submitted to the Land and Environment Court, including "preliminary safety and environmental impact assessment" that had been worked out partly on the basis of "initial consultations" with representatives of the Court and SSM. An ESS AB interim report from the first quarter of 2012 claims that the ESS "has set voluntary safety and environmental objectives, which are more stringent than the applicable laws" (ESS 2012d: 5). The environmental impact assessment appended to the license application included environmental impact factors for both the construction and operation phases, and assessments of the impacts on landscape, geology, water supply, and effects on local communities, including the results of geological investigations of the site that yielded crucial information on groundwater conditions and soil composition (ESS 2012b: 20). In December 2012, the Land and Environment Court in Växjö decided to postpone the environmental trial of the ESS until the European Commission had made a decision regarding environmental risks of the handling of radioactive waste. The ESS appealed the decision, claiming that it would jeopardize the time plan, and in March 2013, the court reversed its previous ruling, stating that it would be unrealistic to await the decision of the European Commission since such a decision required documentation on a level of detail that could, in practice, only be achieved after start of construction of the ESS (*Sydsvenskan* 2013a).

The proceedings of the Land and Environment Court took place in April 2014, in a temporary location at the *Dammsgård* conference center close to the designated ESS site. A delegation of some 20 people "with different expertise" represented the ESS and made the case strongly that "no other location than Brunnshög is more suitable, and that the radiation from the operations of the facility will be very limited." The delegation also appealed to the Court for a speedy process in order to avoid costly delays (*Sydsvenskan* 2014f). On June 19, 2014, the Land and Environment Court in Växjö approved the ESS facility, noting in its opinion that it was "improbable" that another site with better conditions from an environmental point of view could be found without "unreasonable extra costs" (*Sydsvenskan* 2014i).

The trial in the Land and Environmental Court naturally reawakened some of the debate that had been ongoing in Lund, and to some extent spread to national level, during the ESS Scandinavia campaign in 2000–2006 (chapter 6), but that had died out during the European phase of the campaign and the site decision in 2009. It was not only the local branch of the Swedish Society for Nature Conservation (*Naturskyddsföreningen*) that restated their previous criticism; they were joined by Action Skåne Environment (*Aktion Skåne-Miljö*) and the Federation of Swedish Farmers (*Lantbrukarnas Riksförbund, LRF*). Citing ionizing radiation, too close a proximity to existing buildings, and the "most premium farm land of Northern Europe," the three organizations proposed another site several kilometers east (*Sydsvenskan* 2014f). Just as in 2000–2007, the opposition was on a relatively small scale and never reached national political level (see chapters 6 and 10, and Stenborg and Klintman 2012), but the ESS organization took the resistance seriously and held a series of "stakeholder dialogues" from 2010 and on, especially with local residents (ESS 2010a: 33; ESS 2012c: 27).

A main theme in the ESS work to preempt criticism and resistance on environmental grounds was its "energy policy," which took some rather inane expressions as part of the local marketing campaign (chapter 10) but which also entailed ambitious work, intertwined with several of the processes of the Design Update, to achieve an energy plan that would make credible claims to sustainability and a minimized carbon footprint. The 2012 ESS Project Review – called the "light review" by some (see below) – noted that "in addition to the planning around site function, flow, and layout there has been a significant focus on sustainability and energy use, which is a noteworthy feature of ESS planning" (ESS 2012e: 10). The ESS

Annual Report 2011 and the ambitious 197-page *Energy Design Report* published by the ESS in January 2013 detailed the "ESS energy concept," which was summarized as "Responsible, Renewable, Recyclable," implying that the ESS would be as energy-efficient as possible, only use electricity from renewable sources, and also reuse "as much as possible" of the surplus heat; all of which meant that the "ESS will become the first major research facility with a sustainable operation" (ESS 2012c: 27). Among other things, the Energy Design Report noted that if built conventionally, the mere energy use of the ESS facility would cause emissions of 165,000 tons of carbon dioxide annually, but the "ESS energy concept" would lower this emission volume by 190,000 tons in total, meaning that the eventual ESS facility would go beyond carbon neutral and become a net contributor to a more sustainable society (ESS 2013d: 9, 194).

The design update

A key rhetorical point had been made by proponents of the ESS project in the early 2000s, even before the failure of the Bonn meeting (chapter 5), that Europe is "not optimally organized" (Tindemans 2000) and would see its world-leading position in neutron instrumentation and use overtaken by the United States and Japan in just a few years' time, when the Spallation Neutron Source (SNS) at Oak Ridge National Laboratory in Tennessee and the Japan Proton Accelerator Research Complex (J-Parc) in Tokai some 100 km northeast of Tokyo, would come on track in 2007 and 2009, unless European governments made an effort to remedy the political inability that had been allowed to delay the realization of the ESS. There was probably some truth to this claim, especially seen in the perspective that the ESS, when the claim was made, had been on the drawing board for almost a decade and that both the Japanese and US spallation facilities were rapidly approaching start of construction, using key parts of the 1997 ESS reference design (ESS Council 1997; see also chapter 5) and thus capitalizing heavily on the work of Europeans in the 1990s. But the tables would eventually be turned: The fact that the Japanese and US neutron sources were completed before there was even a site decision for the ESS created an advantage in the Design Update Phase of the ESS in Lund, because it could make use of the experiences and achievements of SNS and J-Parc in its design work. This is a known pattern in the history of instrument and facility development, and although the scales are somewhat different, the most well-documented similar cases are the developments in the US system of National Laboratories in the 1960s and on, when several labs reoriented their activities and designed and built new experimental facilities for materials science, including neutron and synchrotron radiation sources, in part based on experiences and findings of previous projects, all inside what has been called the Big Science "ecosystem" (Westfall 2010), where competition and collaboration always coexist and where new projects always build on the experiences of previous ones, staff move between them, and technical components are reused, retooled and recombined. There was probably some frustration in the European neutron community when it saw the solid 1997 ESS reference design implemented by competitors in the USA and Japan, while Europe was stuck in negotiations over funding and location (chapter 5), but history would correct this injustice, at least in part, when the ESS organization embarked on the design update in 2009 and forward and could make use of all the advancements made at Oak Ridge and in Tokai.

At its first meeting in October 2009, the ESS Steering Committee launched the three-year Pre-Construction Phase, which entailed "the building-up of the international organisation that will construct and operate the ESS, the concluding of the licensing and planning processes, and the finalising of financial and in-kind contributions," and the Design Update Phase, which would engage "a large number of scientists and engineers from leading laboratories" in the work to "optimise and update the ESS technical design" (ESS 2010a: 15). A year later, in the ESS AB interim report for the fourth quarter of 2010, the Preconstruction Phase was defined as stretching "from 2010 to early 2013" and consisting of "three lines of activities": (1) "The Design Update of the whole facility, taking into account the major technological advances since the design was tabled in 2003" and also using the experience of the US and Japanese spallation sources, to which "close active links" have been developed; (2) "The Preparation to Build work programme," which entailed "prototyping key components of ESS" as well as the licensing process and "other activities such as the surveying and monitoring of the site" and the development of the energy policy; and (3) "The Integration Activities" which were more vaguely defined but deemed "essential to ensure that all stakeholders are well-informed of our activities" and which have had "significant emphasis [...] in the past six months," probably referring to the marketing campaign that is the topic of chapter 10 (ESS 2010c: 5).

Work to prepare for the Design Update had already begun in 2008–2009, before the site decision, as a way to strengthen the ESS Scandinavia team and its technical competence and build a credible technical case, and to respond to the criticism that had been voiced by the ESS Site Review Group (SRG) that ESS Scandinavia had paid more attention to the political negotiating process than to technical aspects (ESS Site Review Group 2008: 19). In late 2008, Patrik Carlsson and Colin Carlile visited CERN to build collaborations, and among other people they met with Mats Lindroos, originally a nuclear physicist with a doctorate from Chalmers University of Technology in Gothenburg, and with a sixteen-year tenure at CERN in various capacities including accelerator operation and development. Lindroos was certainly a well-reputed accelerator physicist, at least within the CERN organization, and had recently taken over the role of technical coordinator, and managed a substantial upgrade, of the ISOLDE (Isotope Separator On Line DEtector) which fires protons on a stationary target to create radioactive isotopes, a process not unlike spallation (see chapter 3) (Borge 2016).

"I noticed they were eager, and they wondered if I would be prepared to possibly come to Lund and help them," Lindroos remembers. "I think they were realistic and realized that they wouldn't be able to recruit a whole lot of people with Swedish university salaries, compared to permanent positions at CERN" (Interview: Lindroos).

Coming to the ESS as its first Head of Accelerator Division meant building up the division, recruiting, and starting the design update work from scratch. "Twelve people" worked at ESS when he arrived, "some physicists and the rest were PR people," and with respect to the technical design of the future ESS facility "they were mainly waffling about the same things that had been said ten years earlier" (Interview: Lindroos). So Lindroos simply began with pen and paper, to draw up the basic principles of the ESS proton accelerator, making use of his experience from ISOLDE, and also from the Superconducting Proton Linac (SPL) project at CERN that he had been involved in, his prior participation in the design of the *Eurisol* radioactive ion beam facility, and his extensive network across Europe. "They said 'We'll support you. We're behind you.' [...] I knew I had the ability to put together the collaborative project that would accomplish it. I could find the specialists. I knew I couldn't do it on my own" (Interview: Lindroos). Thus, a first "rudimentary" collaboration was set up that met regularly and drew up a first design draft. The ESS Accelerator Design Update (ADU) project, as it was formally called, had its first meeting in March 2010, and gathered experts from six institutes and universities in five countries (Denmark, France, Italy, Spain, and Sweden), plus the ESS organization in Lund, and the ESS Bilbao (ESS 2011a: 10). The first thing they did was to "throw away" the design work done by the ESS R&D Council and at the Jülich Research Centre prior to 2003. "There were flaws in that design, very unusual ideas that hadn't been tested and were quite risky, so we reverted to what we thought to be a very safe solution, but which made use of the latest technology" (Interview: Lindroos).

But the time around and after the site decision in 2009 was, nonetheless, politically sensitive and characterized by positioning by the prospective member countries and their institutes and universities, who wanted to come out as favorably as possible in the competition for contracts and deals on in-kind contributions to the facility. A general call for interest in participating in the design of the accelerator was blocked by the newly created Steering Committee, where countries protected their interests, and instead, the network contacts were used to "create a bottom-up pressure" and form a coalition of the willing. The bilateral agreement between Sweden and France, discussed in chapters 7 and 9, also became crucial in facilitating the collaborations between the ESS and French accelerator physicists (Interview: Lindroos). In 2011, an agreement was reached with Uppsala University to create a test facility for accelerator technology, making use of competence and experience at a reasonable distance from Lund (*Sydsvenskan* 2011b), and in the same year, a project funded by the European Regional Development Fund was launched under the name CATE – Cluster for Accelerator Technology, to train companies in the Øresund Region to enable them to deliver accelerator components to the ESS, through educational activities and conferences. It was unsuccessful in the sense that it did not lead to any procurement contracts in the region, but it contributed to building up competence and establishing connections between companies and other facilities abroad, including CERN (*Sydsvenskan* 2013b; Interview: Lindroos).

The collaborative framework established in the early years built in part on the ESS Preparatory Phase, a project funded in part within the European Union's Seventh Framework Programme for Research and Technological Development (FP7), that ran between April 2008 and March 2010. It was a collaboration between eleven institutes, universities, and research councils in nine countries (France, Germany, Hungary, Italy, Latvia, Spain, Sweden, Switzerland, and the UK) and was coordinated by the Paul Scherrer Institute (PSI) in Switzerland. The Swedish partner was Lund University, and the total budget was 6.6 million Euro, of which 5 million came from the EU. The overall goal of the project was "to pave the way to and to facilitate a site decision and a final decision to construct and to operate the ESS." This included some "specific investigation" on technical and scientific aspects, but the key function of the project was to bring together interested partners across the European continent in collaborations that could pave the way for future, deeper cooperation around the design and eventual construction and operation of the ESS (European Commission 2017). When the project was concluded on March 31, 2010, it was already almost a year after the site decision, and while the issue of contributions from institutes and universities across Europe was politically sensitive, Lund was the designated site and the ESS organization was based in Lund, focused on building the ESS in Lund, with the Steering Committee gathering the prospective member countries. Therefore, it was only a matter of time before an ESS community, geared to contributing actively towards realizing the facility in Lund, was formed and could start to expand.

The details on collaborations available in the ESS Activity Reports make little or no difference between accelerator, target, and instrumentation, but list impressive numbers of collaborations: As of summer 2011, "more than 40 laboratories in 20 countries" were part of "an enthusiastic and growing ESS community" (ESS 2011a: 8). A series of activities also took place in the years 2010–2014 that sought to involve industrial companies in the Preconstruction Phase for ESS, to prepare them for future contributions (ESS 2010a: 11; ESS 2012b: 46).

But more interesting, from a general point of view and as part of a deeper analysis of the process of making the ESS a reality in the design and preparatory phase between site decision and groundbreaking, is the much-discussed phenomenon of in-kind contributions. The model has been around in international scientific collaborations, and in international collaborative technology projects, since at least the 1950s, and CERN is in no small part built with in-kind contributions, which simply means that instead of paying cash contributions to a facility or project, countries can contribute with technology, personnel, and knowhow. There are drawbacks, most clearly risks of suboptimal procurement, meaning that competitive alternatives might be precluded in the process, and that contracts might create technological lock-ins. The administrative burden is severe and the negotiating process necessary to harmonize the ambitions and capabilities of member countries to contribute to the facility budget with the ambitions and capabilities of universities, institutes, and companies in these countries, is a tough job (Interviews: Weeks, Yeck, Börjesson).

The TDR of 2013 estimates the extra costs incurred by the whole in-kind process at 100 million Euro, and identifies the in-kind process as the "key issue" in keeping time schedules and budget (ESS 2013b: 5).

But the advantages are at least as significant: Countries can let their money stay in their domestic economies, instead of seeing it go to the boosting of the local and regional economy around a facility construction project in another country, which already reaps the site benefits (see chapter 3). In the political situation for the ESS, where it took five years, from the site decision in May 2009 to the summer of 2014, to put the funding solution in place (chapter 9), the use of in-kind contributions was a key means to sweetening the deal for many countries. Universities, institutes, and companies in the respective partner countries were thus given advanced scientific and technological development projects, and in some cases very lucrative high-tech contracts. Most of all, however, extensive use of in-kind contributions is a way – perhaps the only way – of soliciting the foreign competence and knowledge that is absolutely necessary for a pan-European project like the ESS to even come close to realizing its ambitions and potential, technically and scientifically, and to tie the community together and make all stakeholders happy. The ESS, built on greenfield land in a small country with comparably minuscule competence and knowledge in the areas concerned, was absolutely dependent on in-kind contributions from all over Europe, especially for bringing the knowledge and knowhow in. If these labs and institutes across Europe would have provided blueprints of instruments and technical components, "that wouldn't have worked," Mats Lindroos explains. "They have developed these technologies for 20 years. If we would have started from scratch here, and hired engineers, we would have lost 20 years." With the in-kind arrangement, "they get the credit for building these things and they get to put their flag on it, and so they contribute with know-how for free" (Interview: Lindroos). Lars Börjesson concurs, stating that any other *modus operandi* would be "a waste of resources, in a European perspective" and that with the in-kind system "everyone's involved" (Interview: Börjesson). "The people with the money wanted to see their communities engaged by any means. And we engaged their communities" (Interview: Argyriou).

On the accelerator side, the work to engage labs across Europe began very early, in 2009, and the ESS was able to build on the enthusiasm and inspiration that the design of the facility created. "It was easy for us to sell all the exciting stuff" (Interview: Lindroos). But when the first meeting of the Steering Committee decided that 70% of the accelerator should be built using in-kind, pressure built up. "We took the map and looked 'where are the accelerator labs?' and we went there, talked to them, and tried to get them on board. That way, we got up to 50%" (Interview: Lindroos). Complaints surely surfaced, for example from Germany in whose view "the whole accelerator is French and Italian," but who didn't act when the opportunity opened: Lindroos and his team had tried hard to get German accelerator labs involved, "but they were over head and ears in their own projects and didn't want to participate" (Interview: Lindroos).

On the science side, the constant tug-of-war between the scientific motivations for an instrument, and the ambitions of specific institutes that wanted to make specific contributions, made in-kind contributions extra complicated. Institutes in Germany, for example, "had their own list of instruments that they wanted to build" (Interview: Argyriou) which did not necessarily correspond to the overall, moderated ambitions of the existing and future European neutron user community. Therefore, some more formalization was required, compared to the accelerator side, and several Science and Technical Advisory Panels (STAPs) were set up, one for each instrument or instrument group, consisting of independent groups of scientists and instrumentation specialists that advised the instrument teams, whether they were internal to the ESS organization or in a partner lab elsewhere, on how to build their science case and how to construct their instrument project, and to solve their technical problems. The work built mainly on the reports from the science symposia (see below), and translated these into technical and scientific requirements, like flux of neutrons, or resolution, and then to performance parameters, which in turn were translated to technical parameters. This was the only way to guarantee that the instruments that were built would perform to a level that made them interesting to a broader European science community, and not just to the small group people at the institute(s) that designed and built them. In other words – in the tug-of-war between the scientific case for the ESS and what the in-kind contributors would want, science won (Interview: Argyriou).

In September 2011, the first IKON – in kind instrumentation meeting was held in Lund, gathering some 110 scientists from the ESS partner countries in a joint planning of future instruments for the ESS. The IKON meetings, organized twice a year, welcomed all "current and future ESS Partner Countries" to participate in the work to develop the ESS instrument suite through in-kind work packages (ESS 2011f). The ESS Activity Report 2011–2012 calls the IKON meetings "one of the main vehicles for exchanging ideas and discussing progress on instrumentation and neutron technology" that "bring together a large community of neutron scientists, stimulating the cross-fertilisation of ideas and new collaborations that will benefit neutron science in general" (ESS 2012b: 31). In addition to the IKON meetings, an Instrument Design School was organized in 2011 "aimed at training the next generation of instrument scientists" (ESS 2012b: 31). In November 2012, the first meeting of the ESS in-kind review committee was held, and the first five in-kind agreements (with Denmark and Spain) were approved and forwarded to the Steering Committee for approval at their first meeting in 2013 (ESS 2012f: 5). As soon as the Design Update Phase had been concluded and the ESS moved into the Preconstruction Phase, in the spring of 2013 (see below), a formal Call for Expressions of Interest (EoI) was issued, announcing the move of the ESS project into the construction phase, and inviting in-kind contributions from institutes, universities, laboratories, companies, or combinations of these, from current or prospective ESS partner countries, in the form of "technical and other components and/or material for the accelerator complex, target, and experimental instruments of the facility; personnel for a range of specific tasks during the construction phase; or other material or services that may be of significant value to the construction of the facility." The call was not for "fully developed proposals" and "not a call for procurement" but a step in the process of delivering in-kind to the facility (ESS 2013f). Jim Yeck, who by this time had succeeded Colin Carlile as CEO of the ESS, identified the Call for EoI as "an important step in defining concretely who will contribute to the project and how ESS will be built" (ESS 2013g).

Estimates of the ratio of in-kind contributions to cash contributions in the construction phase have varied over the years. In the fall of 2009, Colin Carlile told *Sydsvenskan* that he hoped that some 50% of the total ESS construction would eventually be financed in kind, not more, "because we need cash as well," and that the most realistic outcome was that Denmark and Sweden would pay most of the cash contributions (*Sydsvenskan* 2009p). The ESS Activity Report 2010–2011 noted that the signatories of the 2011 MoU had "agreed to an in-kind contribution of 55 per cent and a cash proportion at 45 per cent" (ESS 2011a: 22), but it is unclear if this referred to commitments made by the member countries themselves, or an overall model for how the ESS facility in full was to be funded. The ESS Activity Report 2011–2012 estimated the total in-kind contributions from partner countries to be "of the order of 50% of the total construction costs"

(ESS 2012b: 9), whereas the *Technical Design Report (TDR)*, published only months later, stated that during the work of the Design Update Phase, the ESS organization "worked under the expectation that 75% of the capital cost will be covered by cash contributions and that 25% will be in-kind" (ESS 2013b: 5). The estimation in early 2018 was that at least 30% of the eventual total construction of the ESS would be covered by in-kind contributions (ESS 2018a: 25). During the Preconstruction Phase, 83 discrete in-kind contributions, from collaborating organizations in eight countries (the Czech Republic, Denmark, Germany, Italy, the Netherlands, Norway, Spain, and Switzerland), were made, to a total value of 44.6 million Euro. German institutes, universities, and companies accounted for 33 of these contracts, and almost half of the sum (20.5 million Euro), thus already accounting for 10% of their total pledged contribution to ESS construction (ESS 2015b: 30).

The building of a scientific case for the ESS had been underway for as long as there had been an ESS project of any form, and the 1997 and 2002 ESS designs of course contained extensive yet summarizing accounts on the scientific opportunities the future ESS facility would open. ESS Scandinavia had worked with symposia and conferences, especially from 2008 and on, to begin the process of updating the scientific case and develop instrument concepts. In October 2008, an instrumentation workshop had been held on the island of Ven in Øresund, "to propose imaginative instruments," and this was followed up by similar meetings in Frascati in August 2009 and again on Ven in May 2010 (ESS 2010a: 19). A large number of workshops, meetings, conferences and seminars were organized by ESS Scandinavia and its successor ESS organization, or with its support, during these years, the majority of which had to do with science at the ESS, and it is difficult to compile comprehensive lists due to the fragmented information in the ESS Activity Reports. The following are, therefore, probably only a selection and should be read as an illustration of the extensive collaborative, networking, and outreach work of the ESS organization and its community as part of the development of the scientific case for the ESS in 2009–2014.

In 2009, a series of symposia called "New Science at ESS & MAX IV" was launched in collaboration with the Lund University Faculty of Science. The purpose was to "raise awareness" among scientists. To all of these, "renowned experts" from all over Europe and North America were invited "to present their research findings and views on the future of research at neutron and synchrotron x-ray facilities" (ESS 2010a: 18). A total of fifteen

"New Science" symposia were held between February 2009 and March 2011 (ESS 2010a: 18–19; ESS 2011a: 27). Major conferences, with several hundred attendees and with a broader scientific focus, were held in Prague in July 2011 and in Berlin in April 2012, under the headline "Science & Scientists," aiming to reach out to the scientific communities in the full breadth of potential applications of neutrons at the ESS, encouraging them to "get involved" in formulating the science that would be done at the ESS, primarily by participating in the proposing of new ideas for instruments (ESS 2012b: 26).

Other significant events in this context were the *Science Symposia*, sponsored by the ESS and held across Europe, in 2011 and on. The idea was to have prominent scientists organize the symposia in order to involve the grassroots, and the ESS provided a grant of 12,500 Euro to all workshops that were approved by the ESS Science Directorate. Proposals for a science symposium had to be accompanied by a guest list, and the organizers also had to make sure that high-level ESS representatives would attend and present the relevant technical specifications so as to take the discussion further toward new insights on how the ESS could be used in various fields of research. Each symposium resulted in a report whose contents were used to build the science case for the CDR and TDR. The response was considerable – the ESS Science Directorate was "flooded by requests" and had a stiff task selecting which to support (Interview: Argyriou). Eighteen science symposia were held during the Design Update Phase and the Preconstruction Phase, before start of construction in September 2014, in a number of locations all across Europe and concerning a wide range of scientific areas with direct interest in applications of neutron scattering. Another two were held after the groundbreaking, in 2014 and 2015.

In parallel with the science symposia and all the other meetings and gatherings, extensive work was undertaken to develop a set of instruments that would correspond to the ambitions and interests of the European scientific communities, and make the expected dramatically improved experimental conditions that would be offered by the ESS a reality. A key challenge was the necessity to look into the future, as explained in the fourth quarter interim report of ESS AB, where it was noted that the design of instruments would have to begin as early as nine years before they would first be used for scientific experiments (ESS 2010c: 6). Key to the process was to enthuse scientists, to bring the European neutron user communities on board, in order to have as good answers as possible to the very demanding questions of what the future would bring. Ideas for instruments, from the future ESS users, were necessary to start developing the science case for the future facility, to help partners propose in-kind contributions in the area of instruments, and to engage the scientific community (Interview: Argyriou). The science symposia and the other meetings and conferences discussed above obviously fulfilled much of this need, but there was also a need to translate the work accomplished in interaction with the scientific communities into technical specifications and ultimately instrument designs. One way of describing the work on the science case is to identify "two axes": scientific ambitions and instrument development (Interview: Argyriou). The work on the instrument axis began in earnest with what is internally called the "Vaals Meeting" since it took place in November 2010 in the small village of Vaals in the Netherlands, a mere kilometer west of Aachen and the Dutch-German border. Co-hosted by the Jülich Research Centre which is only 15 km to the north-east, the meeting gathered "just about 30 experts," "some of the best minds in neutron instrumentation" (Interview: Argyriou).

> We called up our friends. We talked to other friends, they made suggestions. [...] And we came up with a set of what types of instruments would work, and would be groundbreaking at ESS. [...] And what came out of Vaals was really interesting, because it was a sketch. Seven instrument categories. I call these "the magnificent seven." And when we actually started doing the detailed work, that sketch pretty much worked. It was absolutely frightening how good you can make things just by getting a whole bunch of people in a room and working. (Interview: Argyriou.)

But the real work to design specific instruments started in 2012. The first round of proposal and selection of instruments took place in 2012–13 and yielded four concepts that were processed internally and by the SAC, and three instruments were proposed to the Steering Committee in October 2013, which adopted all three and sent them from conceptual stage to construction phase (ESS 2013e: 1). In the second round, of 2013–2014, no less than 16 instrument concept proposals were received from around Europe, and nine of these were recommended for approval by the Steering Committee. A third round, opening in the fall of 2014, was the final round to propose and gain approval for instruments for the initial suite of 16 that is included in the current ESS design (ESS 2014f). The instrument proposal and selection rounds followed the precedent set by the Vaals meeting, making use of a thoroughly open and transparent, peer review-based process that was designed to avoid all suspicions of backroom deals and nepotism, and to secure, as far as possible, that the eventual selection of instruments for the ESS corresponded to the ambitions and needs of the European neutron user communities (ESS 2012a: 21). The Scientific and Technical Advisory Panels (STAPs) played an important role, as did of course the SAC, both of which had high standings in the communities. With the STAPs engaged hands-on in providing advice and feedback to those working on instrument proposals, the role of the SAC was mainly to take a "birds-eye view of the entire instrument suite," and make recommendations based on this view, and the final decision on which instruments to approve lay with the Steering Committee (ESS 2014f).

A special case, that can be used to illustrate the power of the open, peer review-based process of developing and selecting instruments for the ESS, is the aforementioned proposal by German institutes to have a number of instruments built for the ESS that would essentially fit their capacities and competences, but that might not have corresponded to what the European neutron user community really needed or wanted. Germany was "under heavy pressure to deliver in-kind," because parts of the Research Center Jülich's budget in particular had been conditioned upon it going into instrument development for the ESS, so that the German Federal Government could cover parts of its contribution to the ESS through the existing budget for the Helmholz Centres of which Jülich is part, "which makes it really important that there are instruments at the ESS that Jülich has the competence to build" (Interview: Matic). Here lies at least part of the reason for Germany's unilateral offer to construct six instruments as in-kind contributions. The ESS organization had only one option, namely to deal with the request "in an open and transparent way" and include the instruments proposed by Germany in the regular process, putting them through the peer review of the STAPs and the SAC. "So I made all these in-kind contributors part of an open process that was judged by the entire European community. They didn't like that. I got a lot of push-back on this. But we pulled it off. And in the end they got heavily engaged. [...] And the Germans won their instruments honestly, on their merits" (Interview: Argyriou).

A major milestone in the Design Update Phase, and a significant step toward the finalization of the *Technical Design Report (TDR)*, the cost reports, and all the other documentation necessary to move toward start of construction, was the publication of the *Conceptual Design Report (CDR)* in February 2012. Edited by Steve Peggs, Deputy Head of the Accelerator Division, the 246-page CDR was authored by 72 ESS employees and advisors, and outlined the ESS design in 11 chapters (Introduction, Neutron Science, Target Station, Accelerator, Control Systems, Data Management,

Conventional Facilities, Safety, Integration, Energy Systems, and Conclusions) and two appendices (Commissioning and Operations, Comparative Target System Concept). The Introduction began with a sobering but forward-looking paragraph, pointing at the major step toward realization of the ESS that the CDR, and the work that lies behind it, represents:

> The road to achieving a high power spallation source for Europe has been long and winding, with many twists and turns, but always with a determination to succeed. The production of this Conceptual Design Report is a concrete demonstration of the positive manner in which this project, long in gestation, is now maturing. Europe is working together in order to build, in southern Scandinavia, a slow neutron source of unparalleled power and scientific performance. (ESS 2012a: 5.)

It does not fall within the scope of this book to report in any further detail on the technical contents of the CDR (or the *Technical Design Report, TDR*, that followed a year later, see next section), but some things are worth noting. One is that as part of the work that led up to the publication of the CDR, the designated target material to be used in the ESS target station where protons from the linear accelerator knock out neutrons in the process that is known as spallation (chapter 3), was changed from mercury to tungsten. On October 19, 2011, *Sydsvenskan* reported that the ESS had decided to go for this solution, based on recent "scientific achievements" that have made tungsten (a chemical element with the symbol W and atomic number 74) a technically and scientifically better option, and also cheaper, than mercury (*Sydsvenskan* 2011d). The environmental concerns regarding the use of mercury were not the only, or even most pressing, reason for the switch, but it certainly helped that tungsten would be a significantly simpler solution in terms of future waste disposal, the ensuing permit process, and the PR campaign (chapter 10). Experiences from the use of liquid mercury in the targets of the SNS and J-Parc facilities in the United States and Japan also fed into the decision – "it didn't work as well as they had thought" – and the major reason was undoubtedly the technical and scientific advantages (Interview: Lindroos). In one sense, the decision and the switch was monumental, since it disarmed a major potential issue of controversy that could have heavily delayed the whole ESS construction project and caused significant negative PR for the project.

Other important things to mention relate to downscaling in the face of budgetary limitations. When the CDR was published, the ESS organization still planned for 22 instruments to be included in the ESS design, which it continued to do when drafting the TDR and the cost estimates (next section), only to lower its ambitions in 2014–15 and decrease the number of instruments to 16. A major decision communicated in the CDR was the discarding of the previous idea of designing the ESS to allow for the addition of a second target station to allow for more instruments, in favor of an upgrade possibility that settles on the possibility of adding more beam ports to the existing target station, "a more efficient use of the accelerator and the neutrons produced" (ESS 2012a: 71).

Transition

In the almost nine years that were covered in chapters 6–7 of this book, although its scale and the stage of its operations changed dramatically, ESS Scandinavia remained a campaign organization. As discussed earlier in this chapter, the transition necessary after the 2009 site decision was a difficult process. Colin Carlile had been a formidable recruitment back in 2006 (chapter 6), instrumental in the campaign and in the mobilization of support in scientific communities across Europe, and was the obvious choice for CEO of ESS AB in 2010. But this appointment had been made with the presumption that he would step down in February 2013, which is when he turned 67, the statutory upper retirement age limit in Sweden (Interview: Landelius).

Many people bear witness to Carlile's self-righteous leadership style and his manner of focusing on campaigning and advertising, and making decisions without anchoring them in the ESS organization or among stakeholders like the Swedish Government or Lund University (Interviews: Berggren, Matic, Johnsson, Kinhult, Lindroos, Yeck). The relationship between Carlile and the chairman of the ESS AB board, Sven Landelius, was particularly tense at times and their differences seem to have lain deep in culture and personality, as discussed in a previous section. The search for a replacement for Carlile was therefore strongly focused on finding a different kind of leader, who could build on the results of the design phase and move on to prepare for, and go ahead with, construction. A search committee was put together, comprising Landelius, board member Lars Kolte, and four representatives of the Steering Committee. A short list could be relatively easily be put together, based on recommendations from the Steering Committee, and among the "ideal candidates" was Jim Yeck (Interview: Landelius).

James "Jim" Yeck, an engineer by training, had over fifteen years of experience of project management of large scientific facilities in a US context, having shepherded several major projects through pre-construction and construction, including accelerator-based research facilities on the scale of the ESS. In this sense, he was the ideal person to take over as CEO of the ESS, and although the search committee met with several people, and there were many possible alternatives who had a less distinct project management profile and a somewhat stronger scientific resumé (Interview: Börjesson), Yeck quickly emerged as the first choice. "We found each other really quickly. And his presentations for the search committee and the interviews we had gave a fantastically stable image of someone anchored in science but not a scientist himself. [...] So it was quite swift once the discussion started" (Interview: Landelius).

The ESS announced the appointment of Yeck on October 4, 2012 (ESS 2012g), and his employment at ESS began January 1, 2013, with the transition to the CEO post occurring two months later, on March 1 (ESS 2012f: 6). The shift from Carlile to Yeck constituted a fundamental change in leadership and organization culture, giving the whole operation a new "momentum" (Interview: Lindroos) and leaving behind "the last traces of the campaign organization" (Interview: Börjesson). The timing was apparently perfect – "a little bit earlier, we still needed the visionary," and a little bit later, "we would have lost credibility" (Interview: Argyriou). Apart from leadership style, where the contrast between Carlile and Yeck was sharp, the change also brought a new culture of transparency to the whole organization, in its internal work and especially towards partner countries and collaborators, which in retrospect seems to have made a crucial difference. Jim Yeck, trained completely in the US system, was curious about Europe and its complicated ways of organizing collaborative Big Science projects. When he first heard about the ESS, around the turn of the millennium, "the working assumption [...] was that the facility in Europe wouldn't be built" due to the inability of European countries to come to agreement over funding and organization. "The US is not known for that creativity when it comes to these large facilities, they have come up with a very standard cookbook and they adhere strongly to the cookbook, they follow this process, and in Europe there really isn't a process but you're sort of making it up" (Interview: Yeck).

The "standard cookbook" refers to a model used by the US Federal Department of Energy (DOE) in its work to renew and develop the several national laboratories where Big Science facilities, including neutron sources, are built and operated, and the bearing element is something called *Lehman reviews*, a robust system of review procedures, benchmarks, and project management plans named after Daniel Lehman, the director of the DOE Office of Management since 1991. In practice, for any major project to proceed from idea stage to conceptual design stage, and further to approval, financing, and construction, it has to pass such a Lehman review at each stage (Hallonsten 2016a: 64–65). Jim Yeck, the project manager of a handful of Big Science facilities in the United States since the late 1990s, was well-acquainted with the system and its advantages in terms of building credibility and support for a project via careful and transparent review processes. In September 2011, Daniel Lehman had been invited by the ESS project management team (the "Saab gang") to give a presentation on "Insights and Lessons Learned: Large Facility Construction" (ESS 2012b: 53), and to discuss "what to do and what not to do, and how things can be accomplished differently" (Interview: Möller). Yeck remembers talking to Lehman either in connection with this visit or somewhat later, as part of his deliberations on taking the job as ESS CEO, and that "he was not positive" but "said 'they're not going to build it'," although perhaps he was too biased in favor of his own project management procedures (Interview: Yeck). Nonetheless, upon arrival in Lund in the spring of 2013, Yeck was determined to implement a strict system of recurrent reviews for the continued work on the ESS. The idea clearly broke with European tradition, where the highest governing body of a project (in this case the Steering Committee) and its advisory bodies (the SAC and TAC) typically run the show, and ad hoc reviews can be called in at specific points in time. "But he wanted to have periodic reviews, and I think that was really good" (Interview: Börjesson). The idea was not unanimously approved at once, but had to be gradually established in a change of the culture of the organization and its stakeholder relations. Swedish Government official Peter Honeth was reportedly worried that major challenges identified by transparent reviews might derail negotiations, foreclosing the financing, but was later convinced of the benefits of the reviews (Interview: Yeck). It is natural, some argue, that people are reluctant to put their work up for scrutiny, especially when unfinished, but the ESS organization learned quickly to make use of the reviews and the feedback they gave (Interview: Börjesson). "The effects here were so tangible, so quick, and so positive, so it was no big deal" (Interview: Landelius).

A "light review" had already been undertaken after the publication of the CDR in the spring of 2012 by a panel chaired by John Wood, former ESFRI chair who had been instrumental in pushing the internal Swedish debate over the ESS toward a governmental initiative (chapter 7), and with

Thom Mason, Director of Oak Ridge National Laboratory in Tennessee that hosts the US Spallation Neutron Source, and is also one of the members of ESFRI's Site Review Group of 2008 (chapter 7), and Wolfgang Meissner, who had experience of large construction projects in the private sector, and with Swedish Research Council official David Edvardsson as secretary (ESS 2012e: 24). The review was based on documentation, a two-day meeting in Lund, and "discussions with other people externally" (ESS 2012e: 20–21). The review called the work so far "impressive" and saw "no significant reasons for the project not to proceed to construction" (ESS 2012e: 2, 12). It also directly addressed several challenges, primarily recruitment, the envisaged extensive use of in-kind contributions, timing of procurement and design of critical components, integration of the organization to avoid silos, a swift publication of the TDR to allow for revised cost estimates, and some clarification of the roles, responsibilities and interfaces between the CEO, the board, and the Steering Committee (ESS 2012e: 12–13).

The major milestone that would conclude the Design Update Phase was the delivery, in December 2012, of the *Technical Design Report (TDR)*, the cost estimates, and all the other documentation that had been in the works since the "Saab gang" began their work of project management. This time plan was "sacred" and could not be changed (Interview: Möller), and the organization delivered. The TDR was the central document, a comprehensive report of 690 pages and twelve chapters, including introduction and conclusions, plus an executive overview. The "Neutron Science and Instruments" chapter is the most extensive, spanning 141 pages, with the "Accelerator" chapter (125 pages) and "Target Station" chapter (119 pages) close behind. The list of authors contains 459 individuals from 38 organizations (18 universities, one company, and 25 public sector research institutes, apart from the ESS itself and the ESS Bilbao organization) in 16 countries, among which all 17 prospective ESS partner countries except four (Iceland, Latvia, Lithuania, Poland) were represented, plus Slovenia, Slovakia, Russia, and the United States. 107 of the listed contributors have a noted ESS affiliation (some have double or multiple affiliations). At the end of the introductory chapter, which provides the reader with historical context and an overview of the work done so far, is found a statement of the purpose of the TDR:

> This Technical Design Report is the fruit of design activities that took place during the pre-construction period. The task of the design phase is to complete an exercise in conception, design, computation and, critically, project organisation. The goal

> is to deliver a design for the whole facility, rather than a ready-for-procurement design. Detailed design will be an ongoing activity using resources as and when necessary to ensure on-time procurement, manufacture and delivery. A central part of this process will be an updated time plan and an updated costing report indexed to 2013 values. (ESS 2013b: 6.)

The TDR and cost report were delivered to the Steering Committee at its meeting in Trieste in December 2012 (ESS 2012f: 4). The new cost estimate for the construction phase of the ESS, that remained in place and formed the basis of the funding solution laid down in the ERIC statutes of 2015 (chapter 11), was 1.843 billion Euro (ESS 2013c: 3). This meant a significant increase from the previous estimate from 2008, which had been communicated in ESS Scandinavia's response to ESFRI's Site Review Group, of 1.478 billion Euro (ESS Scandinavia 2008a: 16). Some claim costs had been deliberately underestimated – "Colin [Carlile] was also a politician. He wanted to sell this. [...] And I contributed to that because I put a figure on it and said it couldn't go higher, and he solved that by lowering some costs unrealistically" (Interview: Honeth). But even with the increased price tag, everyone was not convinced that cost estimates were realistic. Sven Landelius, who claims that the ESS AB board had not been provided with the detailed cost estimates ahead of their presentation to the Steering Committee in Trieste, immediately concluded that "it didn't add up" and was "more wishful thinking than realistic thinking." "The board was overruled. And I was quite upset" (Interview: Landelius). There are some signs at the time of writing (2019), that Landelius was right in his worries. The discussion will be returned to in the concluding chapter, but here, it suffices to note that the start of operations has been delayed a couple of years, and the 22 instruments outlined in the TDR (ESS 2013b: 49) have become 16, two natural consequences of setting a figure in stone – 1.843 billion Euro – and being forced to stick to it due to legally binding international agreements.

Regardless of whether the figure of 1.843 billion Euro was the result of "wishful thinking" or "realistic thinking," it was swiftly adopted by the Steering Committee together with other documentation, including the TDR, which meant a major milestone for the project. ESS Science Director Dimitri Argyriou calls the meeting "fascinating," remembering that the Steering Committee simply bought it: "We told them 'here's the project plan, here's the budget, and here is 1.843 billion Euro'. And there was silence in the room. It was just like nobody cared. [...] They approved the cost" (Interview: Argyriou). "We were home, if you like," says Kjell

Möller. "In my memory, that was a critical point in time, when we crossed a boundary" (Interview: Möller). The fourth quarter interim report of ESS AB of 2012 notes that the reports "were warmly received" at the meeting in Trieste "and will now undergo review before the partner countries can make a decision to proceed to construction" (ESS 2012f: 4).

Jim Yeck, who arrived in Lund just weeks after the Steering Committee meeting and was soon to take over the helm, has a slightly different impression, claiming that "there was nobody ready to buy it," which prompted an external cost review by the assurance services branch of Ernst & Young (EY), commissioned by the Swedish Government, which "didn't find anything, so there was nothing to argue about, we hadn't hidden anything" (Interview: Möller), and therefore helped build credibility (Interviews: Yeck, Argyriou). But it was clear that for the prospective partner countries and their representatives in the Steering Committee to be convinced, real reviews of the "mini-Lehman" type that Yeck had in mind were actually necessary. "I wanted to do a review where we bring in 30 or more experts [...] Openness, transparency" (Interview: Yeck). On November 12–14, 2013, the first "annual review" of the ESS project was undertaken by a group of 33 experts, organized in seven subcommittees, and seven "observants" (ESS 2013e: 1). Ahead of the three-day meeting, the ESS organization had prepared and provided "detailed project execution plans for ensuring the expected technical performance within the approved cost objectives" (ESS 2014a: 4–5). The conclusion, issued by the committee at the end of the meeting, was that the ESS "will be ready for start of construction in mid-2014," provided that the permit process and the funding negotiations were then successfully concluded, and that the recommendations of the review committee had been implemented (ESS 2013e: 1). The review report, containing "over 80 specific recommendations" was only one of several important outcomes of the review. The "open and transparent manner in which the review was conducted" was itself considered "of great value," and a major consequence of the review was that "all of the ESS stakeholders are properly informed about progress, plans, issues, and plans for addressing challenges and opportunities" and that there was now in place "a common baseline of understanding that is needed before we secure financial commitments to construction and launch into construction" (ESS 2014a: 4–5). In May 2014, a follow-up review meeting was held with nine of the committee members, who concluded that no major "showstoppers" remained and that the recommendations from the November 2013 review had been followed by the ESS (ESS 2014c: 2).

The transition from the Design Update Phase to the Pre-construction Phase was thereby initiated. A restructuring of the ESS organization, "to match the responsibilities of the construction phase" and the change in leadership from Carlile to Yeck (see above), was carried out in April 2013. The ESS Annual Report 2013 summarized the shift by noting that "ESS has gone from being an organisation with the task of preparing plans to an organisation focused on implementing them" (ESS 2014a: 7). But the funding for the 1.843 billion Euro construction project was still to be found.

9. Funding the ESS (2009–2014)

The domestic scene

From the perspective of the Swedish Government and its constituency, the Swedish taxpayers, the ESS campaign in the period of 2009–2014 was dominated by the work to secure funding for the ESS in order to approach start of construction and thus the start of the realization of promises made locally, regionally, nationally, and globally. Sweden, Denmark, and Norway had jointly pledged to pay 50% of the ESS construction costs, making these promises *in blanco*, meaning independent of the eventual final sum in Euro, which was not settled until late 2012. Other countries, however, were not prepared to make similar pledges. Although some indicated percentages in their letters of intent of 2008 and 2009 (see next section), these were indeed mere *letters of intent* and did not typically "prejudice the final decision about the financing of ESS after the redesign," as a document describing the Swedish-German agreement of 2009 put it (Germany 2009b). Most of the countries that had thrown their support behind Lund in May 2009 did not mention percentages and sums of money in these letters of intent or in other documents that followed them up. These did not emerge until the funding solution was presented in the summer of 2014. Clearly, the finalization of a funding solution hinged upon the completion of the complex process described in the previous chapter, of building up a credible organization and developing a credible scientific case and technical design for the ESS, including cost books and risk analyses, and several other crucial details. Therefore, the negotiation process on European level could not enter into any critical stage until the end of 2012, at the earliest. The bilateral *Memoranda of Understanding* and *Cooperation Agreements* signed in 2008–2010 between Sweden and Denmark, Estonia, France, Germany, Italy, Latvia, Lithuania, Poland, and Spain signaled no binding commitments. The February 2011 multilateral Memorandum of Understanding, signed by representatives of sixteen countries, laid down a basic framework for the Design Update Phase and the work toward start of construction, but stated clearly that it "implies no legal commitment for the construction and operation" of the ESS; and that the sixteen countries instead, by signing, "signal their best intentions to pursue these goals" (ESS 2011e; see next section).

This does not mean that 2010–2012 was a period of complete silence in the international negotiations over the ESS, only that they proceeded at a slower pace and in anticipation of the delivery of the TDR, the cost reports, and other crucial documentation at the end of 2012 (chapter 8). But the work of the Swedish Government to secure the funding for the ESS was not only an international affair, but also a domestic concern of some weight and with a complexity of its own. The several hundred million Euro that the Swedish Government had pledged from the very start had to be accounted for in governmental appropriations, and large parts of this amount had to be paid out to the ESS organization as early as 2010–2013, to keep its operations running. This exercise included a need to dampen the worries that the Government was favoring Lund and its new megaproject(s) at the expense of other academic environments and other research areas across the country, and a need to follow up on deals apparently made back in 2002–2003, between ESS Scandinavia and the regional authorities in Skåne, and with Lund University, regarding substantial financial contributions to the ESS, with new negotiations, new agreements, and in one case also new legislation.

The exact size of the Swedish commitment was itself subject to change. When, in February 2007, the Swedish Government announced its ambition to seek to host the ESS, the envisaged Swedish commitment was 30% of the construction costs, which at the time were estimated at SEK 11 billion, or approximately 1.189 billion Euro (Swedish Government 2007b) (conversion rates and inflation rates used here are from the Bank of Sweden, www.riksbank.se). The 2008 response of ESS Scandinavia to ESFRI's Site Review Group had estimated the total construction costs at 1.478 billion Euro (which included 101 million Euro for "project preparation") (ESS Scandinavia 2008a: 16), presumably in 2008 prices, and Sweden was still, at this time, pledging a contribution of 30%, which would consequently mean 443.4 million Euro, or 86.7 million Euro more than first estimated (not adjusting for inflation). In 2009, the Swedish commitment increased to 35% due to the need to match the Spanish and Hungarian bids that both promised 50% (chapter 7), which meant 517.3 million Euro or, in effect, another increase of 73.9 million Euro (also not adjusting for inflation).

In late 2012, the new cost estimate of 1.843 billion Euro was adopted by the ESS Steering Committee (chapter 8), and Sweden's pledged commitment was consequently adjusted upwards to 645 million Euro (in 2013 prices). Thus, in the years from the Swedish Government's announcement in February 2007, to the summer of 2014, when the final funding solution was agreed, the total estimated Swedish contribution to ESS construction rose by 80%, or 288 million Euro, which, for comparison, is more than any other member country eventually contributed in total.

The funding streams from the Swedish Government to the ESS Secretariat at Lund University and to ESS AB in the years 2008–2014 are complicated to trace and to make good sense of. The first line-item funding for the ESS in governmental appropriations is in the budget bill for 2009, where an extra allocation of SEK 120 million (approximately 13.3 million Euro) is introduced to Lund University "as partial funding of the costs" of the ESS in Lund (Swedish Government 2008d: 143). The increase was permanent, and although the message was clear – Lund University should use the money to make financial contributions to the ESS – the matter became controversial only some years later, which will be returned to shortly.

The Government's budget bill for 2010, presented to parliament on September 21, 2009, announced that the Swedish Government would fund the preparatory and pre-construction phases of ESS, and that other countries would "participate with technical and scientific expertise in natura" (Swedish Government 2009a: 70–71). The bill introduced a SEK 150 million increase of the annual appropriation to the Swedish Research Council, with the explicit instruction that it be used to finance the ESS (Swedish Government 2009a: 213), followed by a similar increase of the annual appropriation in 2013, of SEK 75 million (Swedish Government 2012b: 99), and a temporary extra allocation of SEK 120 million to the Council to fund its contributions to the ESS, in 2014 (Swedish Government 2013b: 293–294). These appropriations were the primary channel for the Swedish Government's funding of the ESS in these years. After specific decisions – capital injections to state-owned companies need to be approved by parliament – the Council, in turn, made payments to ESS AB. The reason for channeling the money through the Council (and through Lund University, see below) was probably practical – the Swedish Research Council had routines in place for allocating large sums of money on behalf of the Ministry of Education (Interview: Holmberg). The arrangement means that while it is in principle possible to trace the money through the annual government bills, the annual government instructions to the Council, and the accounts of capital injections in the ESS AB annual reports and quarterly interim reports, it is also difficult to make complete sense of the payments. The total allocations to the Council for funding the ESS, in 2010–2014, should amount to SEK 1.02 billion (in accordance with the allocations detailed in governmental budgets as discussed above), and the Government's instructions to the Council have, in the same period, authorized payments of "unconditional capital injections" to ESS AB of SEK 1.67 billion (Swedish Government 2010e, 2011b, 2012c, 2012d, 2013d, 2014d). It is unclear if the difference has been borne by the Council, over its regular budget. ESS AB notes, in its annual reports for the same years, total shareholders' contributions of nearly SEK 2.2 billion, but some of this money has come from Denmark (ESS 2010a, 2011d, 2012c, 2013a, 2015a). A peculiar detail in the Government's annual instruction to the Swedish Research Council is a line-item of SEK 39 million that shows up every year and, from 2010 and on, is intended to fund the ESS. The expenditure post is a remainder from when the Government, through the Council, funded the operations of the Studsvik nuclear reactor and neutron facility (chapter 4), which was dismantled after a decision in 2004. Previous commitments necessitated a continued use of this money to support access to neutrons for Swedish scientists, and so when money was needed for the ESS, the expenditure post came in handy (Interviews: Holmberg, Johnsson).

A previous attempt to make sense of the funding streams and evaluate the "unpreparedness and risk" of Sweden, as a small country with no experience of hosting or taking the main funding responsibility for an international research facility like the ESS, ended in the conclusion that the complexity and opacity of the funding arrangement for the ESS thus far "signals the lack of a long-term plan and preparation for unforeseen events and cost increases" (Hallonsten 2015a: 424). Allan Larsson claims the opposite, arguing that "once go-ahead has been given to Lars Leijonborg" in 2007, "the decision was in reality taken, and then the budget division of the Ministry of Finance takes it into account, and they are rigorous" (Interview: Larsson), which would mean that the ministry and the Government had full control over the Swedish commitment to ESS funding. While it is, of course, reasonable to expect that the Ministry of Finance keeps track of future commitments, there is no way that it would be able to know, in 2007, how much it would have to spend on the ESS, since the eventual size of the Swedish commitment, namely 645 million Euro (35% of 1.843 billion Euro, in 2013 prices), was only known after December 2012. Furthermore, the Government's budget bill for 2010 shows that the Ministry of Finance expected a Swedish contribution of a maximum of SEK 300 million during the preparatory and pre-construction phases (Swedish Government 2009a: 70–71). Strangely enough, the bill still estimated the total construction costs for the ESS at SEK 11 billion, or approximately 1.189 billion Euro, in line with the February 2007 estimation, although ESS Scandinavia had published revised cost estimates of 1.478 billion Euro in 2008 – possibly a sign of "wishful thinking" on the part of the Government (cf. the discussion at the end of the previous chapter), but certainly an underestimation.

Another matter that can be taken as a sign of unpreparedness on the part of the Swedish Government and Swedish scientific communities is the discussion on displacement effects; whether the ESS is a good investment for Sweden; and if the location of the ESS in Lund, together with MAX IV, would cause resentment or protest in the academic system or the scientific communities. It is clear that the Swedish scientific community did not unanimously laud the location of the ESS to Lund, but it is likewise clear that the resistance stayed mainly under the surface and only emerged in broad daylight a couple of times. One of these was when the Royal Academy of Science criticized the Government's prioritization of the ESS over MAX IV (chapter 7). Another was in the spring of 2014, when the Vice-Chancellor and Deputy Vice-Chancellor of Stockholm University, Astrid Söderbergh Widding and Anders Karlhede, publicly argued that the Swedish investment in the ESS lacked proper anchoring in Swedish science and thus made no sense in terms of science policy priorities, and that the funding for the project was very uncertain, implying a significant risk for Sweden, which had already invested hundreds of millions. "Start of construction is planned for this summer," writes Söderbergh Widding and Karlhede in a debate letter to *Sydsvenskan*, and "taxpayers' money has been casually invested" although the funding "and the benefit for [Swedish] research" are uncertain. Also criticizing the many offset agreements made in the negotiations (see next section), which mean that Sweden invests in foreign research projects in exchange for its financing of the ESS, they call the procedure an "irresponsible" way to make decisions on research projects (Söderbergh Widding and Karlhede 2014).

The bearing element of the criticism from Söderbergh Widding and Karlhede is that the ESS investment was never properly investigated from a scientific point of view, that the scientific communities of Sweden "were never asked by politicians" about the ESS, and that there "has been surprisingly little discussion" in public fora about "the financial risks" and the fact that the project was located to Sweden due to "political rather than scientific" reasons (Söderbergh Widding and Karlhede 2014). In a follow-up interview in *Sydsvenskan* (2014c), Söderbergh Widding and Karlhede argue that MAX IV answers to great demand in Swedish science, whereas the ESS, from a Swedish perspective, has "a narrow user base." In the article, the interviewer, *Sydsvenskan* journalist Cecilia Nebel, counters the argument by referring to ESFRI's judgment that "the ESS is one of the most prioritized research infrastructures," thus mixing up Swedish and European science policy. Representatives of the ESS and of local and national enterprise also question the criticism in interviews and counter-letters, citing the broad and general benefit of the ESS to science but ignoring the main point, which is the alleged lack of benefit, specifically for Swedish science (*Sydsvenskan* 2014c; Müchler and Rankka 2014). Minister for Education and Science Jan Björklund simply defends his government's policy by claiming that geographical envy lies behind the criticism: "If the ESS had been located to Stockholm they wouldn't have written the piece" (*Sydsvenskan* 2014d).

Although probably misplaced as a response to Söderbergh Widding and Karlhede, who mainly take a national Swedish perspective and claim, for example, that MAX IV would have been a better investment, the Minister's invoking of the Uppsala/Stockholm rivalry with Lund and Skåne is not irrelevant. The remarkably good dividend for Lund University in the broad increase in resources to Swedish R&D in 2007 and on (chapter 4) had been subject to some debate: Although mainly allocated in competition and on the basis of quality appraisals, there were also accusations that some of the funding programs had been designed to fit the profile of research at Lund University. Thus, the major investments in ESS and MAX IV in Lund clearly caused discontent in other parts of the Swedish academic system and especially Stockholm-Uppsala in the years 2009–2014, although it hardly ever surfaced. The opinion letter by Astrid Söderbergh Widding and Anders Karlhede was a rare exception, and there are signs that a lot of criticism was voiced behind the scenes: "I don't know if it ever exploded, but it was surely there" (Interview: Leijonborg). The 2010 launch of the Government investment in the Stockholm Science for Life Laboratory (SciLifeLab), "a laboratory for large-scale genome analysis and biology" and a "national research infrastructure for scientific collaboration" at Stockholm University, Karolinska Institute, and the KTH Royal Institute of Technology in Stockholm (Swedish Government 2009a: 71) can clearly be viewed as a compensatory move. The funding in 2010–2012 amounted to some SEK 200 million (Swedish Government 2009a: 170), and in 2011, Uppsala University was added as a partner in SciLifeLab (Swedish Government 2010b: 79).

Peter Honeth says he "never viewed" SciLifeLab as a compensation – it came about "on its own merits" and "fitted with the priorities of the Government at the time" – although "of course, it's there in the background, that it was good to do something in Stockholm" (Interview: Honeth). Lars Leijonborg adds that SciLifeLab "lived on its own merits" but "felt like a good thing also on an argument of fairness" (Interview: Leijonborg).

One peculiar detail in the outline of the plans for funding the construction of the ESS in the Government's budget bill for 2010 is the note that "part of the Swedish funding is intended to come from Lund University and Region Skåne," and that an agreement between these two partners has specified this funding commitment at 2.5% of the total construction costs (Swedish Government 2009a: 70, 130). In 2010, when the Government still estimated the total construction costs at SEK 11 billion, or approximately 1.189 billion Euro, this 2.5% corresponded to a mere 30 million Euro in total. But once the TDR and the cost estimates had been published and agreed to by the Steering Committee (chapter 8), the sum was suddenly 46 million Euro. It was going to increase further, way beyond the 2.5%.

Lund University had been a key member of the ESS Scandinavia Consortium almost from the start, and very active in the campaign (chapter 6). There is an argument to be made that Lund University, if playing its cards wisely, would reap significant benefits scientifically from an ESS facility in Lund, and perhaps it is therefore only reasonable to expect that Lund University should fund part of the ESS construction. Rumor has it that the then Vice-Chancellor of Lund University Göran Bexell made promises in 2003 or 2004 that Lund University would co-fund the ESS construction costs, should the facility be located in Lund, and that those promises were reiterated when the campaign entered its most intensive European phase in 2008 (Interview: Honeth). In 2013, *Sydsvenskan* cited an agreement "made in the beginning" that "Lund University would fund the construction of ESS with SEK 55 million annually for ten years" without giving any details on when this agreement was made, or between whom (*Sydsvenskan* 2013c).

Lund University was one of the greatest beneficiaries of the unprecedented increase in Government R&D funding in Sweden in 2007 and on (chapter 4), seeing its annual governmental block grant allocation for research increase by a total of SEK 700 million between 2006 and 2014. A major hike came in 2009, when Lund University received a total increase of SEK 145 million compared to 2008 (Swedish Government 2008d: 143). This accounted for almost half of the total increase of the block grant allocation to universities and university colleges in 2009, which was SEK 302 million, with the five other large, full breadth universities in Sweden (Uppsala, Gothenburg, Stockholm, Linköping, and Umeå) receiving an average increase in their block grant allocation of SEK 20.6 million (Swedish Government 2007a: 143–167; 2008d: 141–166). The bill is clear: 120 of the SEK 145 million were earmarked for ESS, but it is clearly supposed to be a one-time line item. The same year, the Government had decided to sell the shares that Lund University had held in the local science park *Ideon* since 1993, with the revenue apparently amounting to SEK 120 million, and in the fall of 2008 "it was decided that the money should go to the ESS" (*Sydsvenskan* 2008f).

This is apparently also what happened, with the budget bill for 2009, but the increase of SEK 120 million was made permanent, as no deduction of any sum close to SEK 120 million, or any mention of the sum, is found in any later budget bill. Peter Honeth, the highest-ranking civil servant in the Ministry of Education at the time, confirms it was supposed to be nothing but a one-time contribution to the ESS, funded by the revenue from the sale of the Ideon shares (Interview: Honeth). What is most interesting, however, is that the later budget bills (from 2010 and on) do not say that Lund University should fund the ESS with SEK 120 million annually, although the University kept the extra allocation, and other increases in the years 2010–2014 were very detailed, such as the extra funding as part of the Strategic Research Areas grants (chapter 4) and the reallocation based on a bibliometric evaluation. The Government's instruction to Lund University states clearly that money can only be paid out to the ESS "after specific decision by the Government" (Swedish Government 2010c: 4), and the only such decisions conveyed in the instructions in the years concerned here are SEK 10 million in 2009, SEK 38 million in 2010, and SEK 82 million in 2014 (Swedish Government 2010c; 2010f; 2014e), adding up to SEK 130 million in total, compared with the SEK 720 million of extra allocations in the same years (six times 120 million). The Annual Report of Lund University for 2012 notes that "according to the information available to the University, the Government counts on a contribution to the financing of the ESS with a total of SEK 550 million over ten years," starting in 2013 (Lund University 2013: 63), and the following year's Annual Report notes that "according to an agreement, the University is to contribute a total of SEK 550 million to the ESS" (Lund University 2014: 86).

The controversy over Lund University's expected contribution to the ESS, which arose in 2013, is therefore ironic. In a letter to the editor, Eva Wiberg, Deputy Vice-Chancellor of Lund University, claims that there is a risk that the money for the University's co-funding will have to be taken from other research, and that Lund University "is forced to take money from its basic allocation and future increases to fund a large part of the ESS," which might lead to cutbacks in other areas, which she believes is not right – the ESS is a "national concern and should not impact the other allocations to Lund University" (Wiberg 2013). The argument is, of course, correct in principle but rings hollow when following the money, especially perhaps in comparison with the very similar argument made in connection with the contribution to ESS construction by Region Skåne.

The regional authority had, like Lund University, been an early champion of ESS Scandinavia, and its elected officials had placed great hopes in what an ESS in Lund could bring to the region in terms of a boost to the knowledge-based economy and innovation-based economic growth. A promise had reportedly been made back in 2002, by then chairman of the Regional Executive Committee Carl Sonesson (in office 1998–2002), of a contribution of "at most a few hundred million SEK" to a future ESS, which his successor Uno Aldegren (in office 2002–2006) "confirmed and raised" (Interview: Kinhult). It was, however, never intended to be a contribution to the construction costs, since such a contribution by a Swedish regional authority would be illegal (see below), but instead an annual contribution to the operating costs. In 2005 or 2006, the agreed amount was reportedly a total of SEK 300 million (*Sydsvenskan* 2016b), but in the fall of 2014, when the ESS was nearing construction, the sum of money apparently settled in agreement between the Government and Region Skåne was suddenly SEK 800 million (*Sydsvenskan* 2014n), which was itself over 4% of the total construction costs. There are no written agreements to back up any of this, only stories of negotiations between key people, and phone calls "during the last trembling minutes of the negotiations" when the Swedish bid had to be raised and the Ministry of Finance applied the brakes (*Svenska Dagbladet* 2016; *Sydsvenskan* 2016b). In a rare moment of watchdog reporting and journalistic muckraking, *Sydsvenskan* revealed the "secret power games" behind Region Skåne's investment in the ESS, identifying a connection between the final deal, sealed in the spring of 2016, and the sudden decision of the Government to give Malmö University College its long-desired university status. When announced, this came as a total surprise – the Government has repeatedly turned down such applications, and the Prime Minister had made it clear as late as in the fall of 2015 that no change of its policy was imminent – but the announcement was made just days before the controversial decision by the Regional Executive

Committee and Regional Council to make the payment of SEK 800 million to the ESS (*Sydsvenskan* 2016b).

The secrecy of the whole process was likely also due to the fact that the Government expected contributions from Region Skåne toward the construction costs of the ESS, which at the time was illegal. "No decisions could be made. The law was clear. This wasn't something that the region had legal right to do," Harald Lindström explains (*Sydsvenskan* 2016b). According to the Swedish Local Government Act (*Kommunallagen*), Region Skåne was not allowed to involve itself organizationally or financially in matters that were the responsibility of the national government (Swedish Government 2004a: 9), and it was the Swedish Government and no one else that had guaranteed 35% of the construction costs for the ESS in international agreements, which meant that Region Skåne simply could not make contributions to its financing (Interview: Lindström). The only solution was a legislative change, pushed by the Ministry of Finance so that it could include money from Region Skåne in its funding solution for the ESS, and executed in 2014 (Interviews: Kinhult, Lindström). This consisted of an amendment to a 2009 law that allowed county councils to co-fund national infrastructure projects such as railroads and highways (Swedish Parliament 2009), so that the same exception also pertained to research infrastructure with the legal status of an ERIC (Swedish Government 2014a: 4). In addition, and by necessity, a change was made to the 1997 law on municipal accounting procedures. With Region Skåne compelled to take a loan to cover its SEK 800 million investment in the ESS, it was necessary to enable a 25-year write-off, which is allowed and commonplace for other investments but prohibited by the law which required municipal deficits to be restored in three years. The change made investments in infrastructures with ERIC status exempt from this balance requirement and thus enabled the ESS investment to be written off over 25 years (Swedish Government 2014a: 6), in other words by SEK 32 million annually instead of a giant (with Region Skåne measures) one-time expenditure post. A peculiar side-effect is that the investment in ESS is posted as an asset on Region Skåne's balance sheet, although the region has little or no influence over the ESS and its operations, which therefore cannot in any reasonable way be viewed as an asset for Region Skåne.

More problematic, however, was the question of whether this investment of SEK 800 million of Skåne's taxpayers' money – although perfectly legal after the cited legislative changes – was appropriate. The fact that the region took a loan to fund the investment, with a 25-year repayment period, likely increased the total cost far beyond 800 million, considering interest payments. But the manner of the final decision made by the Regional Council to make the payment is also a strange episode. A news story in *Sydsvenskan* in December 2016 claims that an "express decision" was made by "the top politicians of Skåne" to give the ESS SEK 800 million, on June 20, 2016, and that the Regional Executive Committee and Regional Council both pushed the decision through the same day with "no opposition from any politician" and with the decision "immediately checked and approved." The money was transferred on June 23. Lars Melin, a private citizen who appealed against the decision in the Administrative Court in Malmö (see below), argued that "the region tried to sneak this out the week before midsummer," and that the benefit of the investment for the tax payers in Skåne is non-existent (*Sydsvenskan* 2016a). On the latter point, opinions naturally differ, and the question whether co-funding of the ESS is within the core business of Region Skåne is also debatable. Henrik Fritzon, chair of the Regional Executive Committee 2014–2018, claims that Region Skåne's responsibility for "a positive development of Skåne [...] in many ways," including the labor market and research, makes the investment reasonable (*Sydsvenskan* 2016a).

A lawsuit filed in June 2016 by Lars Melin, questioning the legality of Region Skåne's SEK 800 million co-financing of the ESS (Melin 2016) was dismissed in May 2017, when the Administrative Court in Malmö ruled in favor of Region Skåne, citing the changed legislation concerning co-financing of research infrastructure and municipal accounting procedures (above), and noting that the financial contribution of Region Skåne to ESS is in violation of no law. The court made no judgment concerning the allegations by Melin that the region had been coerced by the Government to make the payment, or that deals had been made in complete absence of democratic transparency (Administrative Court in Malmö 2017).

The European scene

The several bilateral agreements made between Sweden and other European countries in 2008 and 2009, and the vote of May 28, 2009, meant assurances that the prospective partner countries of the ESS favored Lund over Bilbao and Debrecen, but nothing more. In chronological order, Memoranda of Understanding had been signed with Poland (May 16/23, 2008), Estonia (July 10, 2008), Lithuania (January 15/22, 2009), Latvia (January 27, 2009), Italy (March 25, 2009), Spain (June 10, 2009), Germany (July 14,

2009), and France (October 16, 2009). All of these made general statements that Lund was the preferred site for the ESS, that collaboration was the preferred way forward for the realization of the ESS, and that the countries signing the bilateral memoranda intended to participate in the ESS. None mentioned any estimated financial commitments (except for the Danish-Swedish agreement of April 2009, which has been analyzed in chapter 7, and the Spanish-Swedish agreement of June 2009, see below). Some of these MoUs, like the one with Italy, did not even mention the ESS but contained only general language about "scientific and technological cooperation" between two countries, in neutrons and other areas (Italy 2009).

In the aftermath of the Brussels meeting of May 2009, the relationships were especially complicated when it came to three countries: The two site contenders, Spain and Hungary, and the UK, which had remained ambivalent and still had not shown its cards completely when it came to its own ambitions to host the ESS or an upgraded ISIS that could be realized instead of the ESS, and whether or not it would be prepared to support and join the ESS in Lund. Allan Larsson noted that the position of the UK was still unclear in the fall of 2009 (Larsson 2019: 162). It would take until 2011 before the UK finally decided to join the ESS preparatory phase as a partner country. Matters were also complicated with Hungary, largely due to their refusal to withdraw from the competition (see chapter 7); At the round table meeting in Krakow, the Hungarian delegation reportedly claimed that it would continue to offer to host the ESS, and wanted Debrecen to be designated the "reserve location" if the Lund candidature would fail to meet its commitments, which Larsson and team "diplomatically declined" (Larsson 2019: 162–163). It took another couple of months before Hungary yielded and declared that they "accepted" Lund as the site for the ESS (*Sydsvenskan* 2009r). Spain was apparently a more honorable loser; on June 10, less than two weeks after the Brussels meeting, an MoU was signed that made Spain a prospective member state in the ESS in Lund, and a collaborator in the Design Update Phase, and also located an "ESS Laboratory Test Facility and Accelerator Components Factory" to Bilbao, which also enabled the MoU to specify a sum for the Spanish investment in the ESS, namely "150 M Euro corresponding to 10% of the capital costs of ESS" (Spain 2009).

Two other countries with whom agreements were especially crucial were France and Germany. While the role of the United Kingdom in European politics and the mainstream European integration process through the EC and EU has always been fickle and unreliable, the initiative of France and Germany, and their collaboration, has been called the "Motor of Europe" and repeatedly proven to be the crucial decision-making point of EC/EU politics, both formally and, especially, informally, at least since the early 1970s (Middlemas 1995; Judt 2005: 484ff), including European scientific collaboration (see chapter 3). In addition, France and Germany are European powerhouses in neutron instrumentation and neutron use, and significant contributions from their national laboratories, institutes and university research environments were certainly required in the Design Update Phase, Pre-Construction Phase, and the eventual construction of the ESS (see chapter 8). In the context of this chapter, however, emphasis is on the political importance of France and Germany. It is fair to say that representatives of these two countries still had much of the fate of the ESS in their hands, although the facility was to be located in Lund.

In the memories of the key people involved, Germany had supported ESS Scandinavia since the winter of 2008/09, and the German delegation was leading the demonstration of support at the May 28, 2009 meeting in Brussels (Interviews: Honeth, Leijonborg; Larsson 2019: 155–156; Leijonborg 2018: 347). The first documented step towards a German participation in the ESS is a Letter of Intent, signed in April 2009, between the ESS and the German federal research laboratory *DESY* (*Deutsches Elektronen-Synchrotron*) in Hamburg, home to a state-of-the art synchrotron radiation facility (*PETRA III*) and a free electron laser (*FLASH*), and also the German host institute of the *European XFEL*. The LoI launched a "collaboration on matters related to large-scale scientific infrastructures" between the two organizations and the "foundation of a Science Corridor stretching from Oslo in the north to Hamburg in the south and beyond to Berlin" (DESY/ESS 2009) (see chapter 11).

On June 15, 2009, Lars Leijonborg and his German colleague Annette Schawan signed an MoU concerning "cooperation in materials research and structural biology using neutron and synchrotron radiation," declaring that "the two countries will support each other in the preparation and construction of the planned large facilities, the X-ray laser XFEL in Hamburg, and the spallation neutron source ESS in Lund" (Germany 2009a). A slightly more detailed agreement, signed by Peter Honeth and his German counterpart Frieder Meyer-Krahmer, specified the Swedish commitment to XFEL to 12 million Euro, and noted that Germany is "in principle ready to contribute to the ESS financially with a mix of in-kind and cash contributions in the range of 10–13 % according to the current cost estimates," a commitment that, however, "does not prejudice the final decision about the financing of ESS after the redesign" (Germany 2009b).

An interesting detail concerning this agreement is the increase of the Swedish contribution to the XFEL, between the signing of the XFEL convention in November 2009, where it is set at 12 million Euro (European XFEL 2009), to the figure specified in the Annual Report of the Swedish Research Council for 2013, where it is instead 17.5 million Euro (Swedish Research Council 2014: 32). Lars Leijonborg is unsure about the connection between an increased Swedish commitment to XFEL and the eventual binding agreement by Germany to invest in the ESS, in 2014, but "can go as far as saying that these international negotiations are really complicated, and there might have been a connection" (Interview: Leijonborg).

Germany was particularly deeply involved in making contributions to the Pre-construction Phase of ESS, and although the 20.5 million Euro worth of in-kind contributions that it made ahead of start of construction (see chapter 8) was, of course, a victory for the German Government and the Jülich Research Centre, which administered the contribution, it can also be seen as a very early and very strong German commitment to the ESS in Lund.

The agreements with France were complex and caused some debate in the Swedish Parliament. In April 2011, Swedish public radio reported on a "secret agreement" between the French and Swedish governments that contained Sweden's "single largest investment in nuclear energy research since the 1980s," of SEK 70 million, to French nuclear energy research. According to the news report, it was clearly an offset agreement to gain French support to build the ESS in Lund (Sveriges Radio 2011). Only two weeks earlier, Minister for Enterprise Maud Olofsson and Minister for the Environment Andreas Carlgren (both of the Center Party) had stated very clearly in a debate piece in the major Swedish newspaper *Svenska Dagbladet*, in the wake of the Fukushima nuclear power plant disaster, that "nuclear energy cannot count on any form of [Swedish] government support" (Olofsson and Carlgren 2011), which made member of parliament Kent Persson (of the Left Party) react and direct an interpellation to the Minister for the Environment asking if he "would work to break the existing agreement between Sweden and France," and also asking "if, to the knowledge of the Minister, there are any other secret agreements between Sweden and other countries in the area of nuclear power?" (Persson 2011). Minister for Education Jan Björklund responded to the interpellation on April 26, 2011 (due to the division of labor in the Government, Björklund responded to all interpellations concerning research policy), declaring that the agreement was not secret but "has been handed out to anyone asking for it," and that there are "no secret cooperation agreements between Sweden and other countries, in the area of nuclear energy research." Instead, he claimed, the ambition of the agreement was "to promote and facilitate research collaboration between Sweden and France" including the ESS but also in several other areas. The Minister went on to argue that locating the ESS in Lund "is a great success for Sweden" and that the joint declarations with France and other countries are signs of "an increased interest from other countries in research collaboration with our country" and should neither be changed nor abolished. On Kent Persson's follow-up statement that the agreement went squarely against Swedish policy of turning away from nuclear energy and toward renewable energy sources, Björklund answered that political dealmaking is complicated and that the direct relation between the Swedish investment of SEK 70 million in French nuclear energy research and the French decision to support Lund as the preferred site for the ESS was difficult to fully assess (Swedish Parliament 2011).

The actual agreement is complex in detail. Although its preamble states that the signatories are "taking note of a common understanding that the ESS should be based in Lund and that France has the intention to take part in it," the portal paragraph is broad and vague, stating that the signatories "will encourage, develop and facilitate cooperative activities in fields of common interest where they are pursuing research and development activities in science, technology, and in the field of innovation." The areas that the agreement concerns are Neutron and accelerator sciences, Nuclear and Renewable Energies and Climate change, Nuclear and High Energy Physics, and Space (France 2009). For the first three topics, specific agreements are to be made between the Swedish Research Council and relevant French agencies and laboratories, and cooperation is due to start as early as 2010, "with a successively increased funding from Sweden in the following years coupled to the R&D phase for ESS" (France 2009). Follow-up agreements from September and December 2010 state that "French research teams [...] will participate in the ESS project in Lund" and Swedish researchers will participate in the development of instruments and research activities at a number of French laboratories (France 2010; CEA/CNRS/VR 2010); and while it is clear that these agreements form important parts of the deal that sealed French participation in the ESS, it is also clear that they seek to "increase interaction between French and Swedish researchers" in a number of fields (Swedish Research Council 2011).

Opinions differ, and there are several possible interpretations, on whether these agreements between the French and Swedish governments, and the follow-up agreements between agencies and laboratories, constitute offset agreements with the risk of imbalance in favor of one party. Peter Honeth has repeatedly claimed that the collaborations have been made possible by the ESS but encompass far greater areas and are of benefit to Swedish science in a broader perspective (*Sydsvenskan* 2009n; Interview: Honeth). Lars Leijonborg claims that no agreements with other countries and their research councils or similar have been made unless "they rest on their own scientific merits" (Interview: Leijonborg). There are, nonetheless, signs that some of the collaborations have turned out to be "insignificant" for Swedish scientific communities, and that the interest from other countries has been limited, in spite of significant costs for Sweden, which implies that in the end, much turned out to be direct offset agreements (Interview: Johnsson).

The first multilateral agreement on the ESS in Lund was the Memorandum of Understanding signed by several countries (the "ESS Partner Countries") at the Swedish Embassy in Paris on February 3, 2011. The text of the MoU begins with a few general paragraphs on the importance of the ESS project for European science, a short description of the activities of the ongoing Design Update Phase, and a note that this phase is expected to lead to the start of the Construction Phase (estimated to 2013–2019), which can only start after "a formal decision to proceed" has been taken by the partner countries. The purpose of the MoU is to collect the signatories of the countries expected to participate in the ESS in various ways before and during construction, but the MoU has no legally binding function as it "implies no legal commitment," but only that the ESS Partner Countries "signal their best intentions to pursue these goals" (ESS 2011e). The MoU instructs the ESS organization to produce comprehensive technical documentation, relevant timelines and detailed budgets, a proposal of future organizational form and structure for the ESS, and a "signature-ready agreement on an international partnership for construction and operation of the ESS" (ESS 2011e). The remainder of the MoU contains text on the operations and organization of ESS AB, the responsibilities of the ESS AB board and the Steering Committee, the roles of the advisory committees (see chapter 8), and also a note that the Swedish Government has "given a guarantee to finance the costs of the Design Update Phase, with 30 M€" and that the R&D work of the Design Update Phase will be shared between ESS AB and "internationally qualified R&D institutions predominantly in the ESS Partner Countries." The MoU was to expire on January 1, 2013, but could be extended "by mutual consent of the ESS Partner Countries" (ESS 2011e).

There are nine signatures on the document, of the representatives of Denmark, Estonia, France, Germany, Italy, Lithuania, the Netherlands, Norway, and Sweden – but the Czech Republic, Hungary, Iceland, Latvia, Poland, Spain, and Switzerland are also noted on the final page, although with no signatures. Switzerland signed the MoU in September 2011 (ESS 2011c: 3). The United Kingdom had been "formally adopted as a partner at the seventh meeting of the ESS Steering Committee" on April 15, 2011 (ESS 2011b: 4), but apparently not signed the MoU, since, according to the Swedish Government, there were still ten signatories in October 2011 (Swedish Government 2011a: 78–79). It is unclear if and when additional countries signed on, but the MoU was prolonged at the Steering Committee meeting in Trieste in December 2012, when the TDR and cost report, and other vital documentation, were also approved (chapter 8) (ESS 2012f: 4).

In the summer of 2011, the Swedish Government appointed former ambassador and Volvo executive Lars Anell as new chief negotiator, from September 1, 2011, with the task to "negotiate with other countries about their share and partnership in the project" (Swedish Government 2011c). Simultaneously, the Danish Government appointed bank executive and European Union official Lars Kolte as its chief negotiator, to work together with Anell (ESS 2011c: 5). The challenges facing this negotiating team were many, and the protraction of the process – it would take three years from the appointment of the negotiators to the conclusion of the negotiations, and both chief negotiators would be substituted late in the process – had several interrelated reasons. First, the process of getting the partner countries on board with formally and legally binding agreements was a process deeply intertwined with the Design Update Phase and the buildup of the ESS organization in Lund (chapter 8), and it is quite clear that breakthroughs in the design and project development were necessary to push negotiations forward toward binding commitments. This made negotiations strongly dependent on a complex array of work processes in the ESS organization and its network of collaborations across Europe. Second, the process of making seventeen countries agree is complex in and of itself, with each country having their own mix of incentives and their own election cycles and internal decision-making processes. Third, the Euro crisis would reach its peak in the years 2011–2013, which obviously made all countries more cautious and possibly less generous, causing delays and posing significant risk to the whole project (ESS 2012c: 11; ESS 2013a: 12).

Once the TDR and other documentation had been presented and approved by the Steering Committee in late 2012, it was "time to get serious" in the negotiations (Interview: Börjesson). Jim Yeck, who had arrived as CEO of ESS AB in the early spring of 2013, confirms this picture and emphasizes how the European partners were eager to be informed about the technical and scientific developments, in order to start committing themselves beyond "signal[ing] their best intentions to pursue these goals" (as the 2011 MoU says, see above), and engage for real. The prior organizational culture and strategy of the ESS, summarized by Yeck as "give them as little as you can possibly give them in terms of information" was not conducive to trust-building, and once gears were switched to transparency and clear demonstrations of progress – "everything was on the table" – there was gradually more engagement and commitment (Interview: Yeck).

The officials at the Swedish Ministry of Education, who coordinated the negotiations at European level and worked to turn the general phrases in the 2011 MoU into formally and legally binding commitments, had first-hand experience of the politics of the complex and time-consuming negotiations: "Countries could easily promise that they would be part of it, but finding the money for it was a different issue. [...] The risk was of other countries seeking to delay, in order to pressurize Sweden to give more" (Interview: Johnsson). In-kind contributions during the pre-construction phase can be seen as a rather clear signal of commitment, because they constitute direct investments from partner countries to the ESS. Therefore, the 83 separate contributions from collaborating organizations in the Czech Republic, Denmark, Germany, Italy, the Netherlands, Norway, Spain, and Switzerland, to a total value of 44.6 million Euro (ESS 2015b: 30), can be taken to suggest that these eight countries were more eager to join, and more certain that they would, than others. Some of the commitments had already been made in 2011, for example by the Czech Republic, Denmark, and Germany (*Sydsvenskan* 2011a, 2011c). Denmark is, of course, no surprise, acting as co-host and already a shareholder in ESS AB, and the Czech contribution was comparably minor (1.4 million Euro), but as previously noted, the 20.5 million Euro worth of investments made in kind by Germany in 2011–2014 accounted for some 10% of the amount that it would eventually agree to pay to ESS construction, which shows Germany was already a reliable and committed ESS partner country in 2011. However, this type of commitment, although signaling a strong intention to be part of the ESS and pay a share of its construction costs, says nothing about the size of the eventual final share, which was still subject to negotiation between ministries.

Lars Leijonborg, who had resigned as Minister for Education on June 18, 2009, made a comeback of sorts in September 2013, when the Government appointed him chief negotiator for the ESS. Lars Anell had grown tired of the intensive negotiating work and the frequent traveling, and "had hoped that it would have been possible to resolve faster," himself requesting to leave in the summer of 2013 (*Sydsvenskan* 2013d). The Danish chief negotiator Lars Kolte had passed away in May the same year, after a short period of illness, and was replaced in August by Bo Smith, veteran official in the Danish Government (ESS 2013h). With two new chief negotiators, the final sprint of the intergovernmental negotiations could begin, "this time mainly geared to following up on promises made in 2009" but with the significantly increased costs as key challenge to overcome (Leijonborg 2018: 356).

Accounts differ slightly when it comes to the breakthrough in negotiations, and the point in time when people involved began to feel a certainty that the whole process would be successfully concluded. Governments across Europe "needed to see that their institutes had a stake in the success of ESS because they were going to pay their institutes to do that work," and once they did, "ESS was really beyond the point of no return" (Interview: Argyriou). At some point, in 2013, representatives in the Steering Committee began to "lean forward" rather than "lean backward," which is a metaphor for commitment and eagerness replacing hesitation and cautiousness (Interviews: Yeck, Landelius, Börjesson). Politically, however, the deal was not yet sealed when 2013 came to a close, which had been the original intention. Peter Honeth still had hopes for a successful conclusion of the negotiations within the year when interviewed in September 2013 (*Sydsvenskan* 2013c), but in December, Leijonborg was frank in his realization that "we will clearly not be done by the end of the year" (*Sydsvenskan* 2013e). In February 2014, *Sydsvenskan* reported that the negotiations remained difficult. Leijonborg was "looking for new solutions," saying that "since time is running out, it could be a way forward to speak to additional countries," and that while partner countries "have pledged to pay a share, [...] no exact amounts have been set" (*Sydsvenskan* 2014a). Both France and Germany had previously made clear that they would join the ESS, but the exact amounts were apparently a major stumbling block. France had declared publicly, in the fall of 2013, that they were prepared to sign on, but Leijonborg said the amounts they had pledged were "too low" and that binding commitments from France and several other countries could have been secured already, albeit "on a level that we don't want to accept" (*Sydsvenskan* 2013e). "We want them to pay more, and they want to pay less," he concluded in March 2014 (*Sydsvenskan* 2014b).

The uncertainty was apparently strong in the ESS organization, and when the ESS AB board was preparing the budget for 2014, it made a thorough assessment of risks in order to be prepared for all outcomes. Groundbreaking was planned for 2014, but the owners of the company, the Swedish and Danish states, were unable to give the go-ahead since funding had not been secured. Contingency planning in the board entailed "slowing down and postponing, and delaying," as well as the option of dismantling the project entirely, should more funding not come in (Interview: Landelius). For the Swedish Government, which had initially planned to fund the pre-construction phase with SEK 300 million but by 2014 had paid out close to a billion in shareholders' contributions to the ESS, the risk was already substantial, and continued to grow when financial commitments from other countries were delayed. In March 2014, the Swedish and Danish governments made contact with the European Investment Bank (EIB) to find out if a loan could cover the temporary cashflow. "It sounds more dramatic than it is," Peter Honeth commented (*Sydsvenskan* 2014e).

The first partner countries to sign Letters of Intent with stipulated financial commitments were the Czech Republic and Spain, who pledged to pay 0.3% and 5% of the construction costs, in February 2014 (Czech Republic 2014; Spain 2014). Several countries would follow throughout the year (see below), but the most significant breakthrough seems to have come with an oral commitment, since there is no document to underpin it: In March 2014, the United Kingdom, which had been hesitant toward the ESS for over a decade and neither declared support for Lund in Brussels in May 2009 nor signed the MoU of February 2011, suddenly declared its intention to fund 10% of the construction costs of the ESS (ESS 2014b: 1). Key people say that the UK commitment was the major breakthrough that made them certain that the project would proceed to construction and eventual realization (Interviews: Lindroos, Yeck; *Sydsvenskan* 2014b).

In May, France was reportedly very close to formalizing their commitment of 8% of the construction costs, but the finalization of the agreement was delayed by the local elections in France which led to a reshuffle in the cabinet and a new research minister (*Sydsvenskan* 2014h). A similar, but worse, situation delayed Germany, which had been in political gridlock due to protracted government negotiations after the 2013 election (*Sydsvenskan* 2013e), but the delays continued after the reappointment of Chancellor Angela Merkel, and speculation has it that the new coalition government was reevaluating all their foreign research collaborations (*Sydsvenskan* 2014k).

On May 15, Leijonborg's appointment expired, and the Swedish Ministry commented that the "negotiations phase is now over" and only "summary remains." The Danish chief negotiator added that "we're not quite done, but close. We still lack the final confirmations before we can say that all countries are in" (*Sydsvenskan* 2014g).

Table 1 shows a chronology of the agreements and letters of intent that were secured during 2014, as part of the final sprint of the international negotiations. Although more than half of these did not arrive until later, the Swedish Government decided on July 4, 2014, upon the confirmation from the Federal Republic of Germany that it would contribute with 202.5 million Euro (or approximately 11%) of the construction costs of the ESS (Germany 2014), that it was time to announce that a final funding agreement had been made and that "construction could start" (Swedish Government 2014c). The press release from the Swedish Government contained a table where thirteen countries' shares were detailed (see table 2 below), with only 2.5% missing (or subject to "ongoing discussions" with, among others, the Netherlands, Latvia, Lithuania, and the European Commission) (Swedish Government 2014c).

Table 1: Funding pledges 2014–2015

Date	*Country*	*Contribution*	*Cash/in-kind*
13 Feb 2014	Czech Republic	5.8 M€	mainly in-kind
20 Feb 2014	Spain	5%	combination of cash and in-kind
23 Apr 2014	Estonia	0.25%	up to 70% in-kind
13 Jun 2014	Norway	2.5% (46 M€)	max 2/5 in-kind
4 Jul 2014	Germany	202.5 M€	some in-kind
6 Jul 2014	Denmark	12.5%	
23 Jul 2014	Hungary	1.6 M€	70% in-kind
25 Jul 2014	Poland	2% (max. 36.86 M€)	
27 Jul 2014	United Kingdom	10%	combination of cash and in-kind
27 Jul 2014	Italy	104 M€	in-kind contributions "higher than average"
29 Sep 2014	Switzerland	32.4 MCHF	combination of cash and in-kind
3 Dec 2014	France	147.44 M€	in-kind contributions "above the average level"

(Poland 2014; Norway 2014; Czech Republic 2014; Denmark 2014; Estonia 2014; France 2014; Italy 2014; United Kingdom 2014; Spain 2014; Switzerland 2014; Germany 2014; Hungary 2014.)

Table 2: Funding solution for ESS construction announced by the Swedish Government in July 2014

Country	*Share (%)*
Czech Republic	0.3
Denmark	12.5
Estonia	0.25
France	8.0
Germany	11.0
Hungary	1.5
Italy	6.0
Norway	2.5
Poland	2.0
Spain	5.0
Sweden	35.0
Switzerland	3.5
UK	10.0
"Ongoing discussions"	2.5

(Swedish Government 2014c.)

In August 2014, the ESS announced that "almost 98% of the financing of the ESS is now secure with continued discussions ongoing with partner countries the Netherlands, Latvia, and Lithuania." Discussions were also conducted with additional new partner countries, including Belgium, Finland, Greece, Israel, Portugal, Romania, Russia, Singapore, and Turkey. "The goal is to secure the planned contributions from the current member states and seek new members once construction has started" (ESS 2014c: 2).

The final fixing of the agreements, making the pledges of the respective member countries binding, was the coming into force of the ERIC statutes in August 2015, where 1.506 of the estimated 1.843 billion Euro were accounted for, in a table detailing the contributions of 11 of the partner countries, shown in table 3, where it is also clear that 336.4 million Euro were still needed before the magic number would be reached.

To date, the UK and Spain have joined by amendments to the ERIC statutes (June 10, 2016 and April 26, 2018), contributing 184.3 million Euro and 55.29 million Euro, respectively (ESS 2018b: 37), and bringing down the deficit to 96.81 million Euro.

Table 3: Distribution of construction costs among 11 of the 16 partner countries, as detailed in the August 2015 ERIC statutes (million Euro)

Country	*Sum*
Czech Republic	5.52
Denmark	230.0
Germany	202.5
Estonia	4.61
France	147.0
Italy	110.6
Hungary	17.6
Norway	46.07
Poland	33.2
Sweden	645.0
Switzerland	64.5
Total	1,506.6

(ESS 2015b: 7–8.)

10. Marketing the ESS (2007–2011)

Madison Avenue on Stora Algatan

Chapter 8 chronicled the organizational transition of the ESS in Lund from a campaign organization to a project organization. As campaign, ESS Scandinavia was astoundingly successful, and while a considerable share of the credit for the success should go to the political skills of chief negotiator Allan Larsson, minister Lars Leijonborg, and state secretary Peter Honeth, and the resources of the diplomatic missions in Europe's capitals, the ESS Scandinavia organization conducted an evidently successful campaign under the leadership of Colin Carlile. Composed for a campaign purpose, with substantial scientific competence to enable the crucial mobilization of networks and support across European scientific communities, its core competences lay in communications, public relations, and marketing, as shown by its vast output of brochures, pamphlets, exhibitions, and media appearances. A fair share of these were *pseudo events* and acts of *window dressing* (Alvesson 2013) that drew more public attention to the ESS than the work on renewing the technical and scientific cases and cost estimates, and going ahead with environmental planning and similar things, covered in chapter 8.

ESS Scandinavia was originally run by scientists, and supported by local and regional politicians, which meant that it kept a relatively balanced position in their public marketing, mainly citing the scientific opportunities outlined in the 1997 and 2002 ESS design reports. In the first years, regional growth effects of a future ESS in Lund were only occasionally highlighted by ESS Scandinavia representatives, and opportunist speculations of a future dynamic knowledge-based economy around the ESS were made mainly by those involved in promoting the Øresund Region as a dynamic growth region. The 2002 publication by geography professor Gunnar Törnqvist was discussed briefly in chapter 6, and while it was used by ESS Scandinavia as marketing material in 2002 and on, it was clear that the book was part academic publication and part visionary manifest, and not intended as material for a marketing campaign. ESS Scandinavia remained very much focused on promoting the ESS as a much-needed project for Europe as a whole, that Scandinavian scientists should take an active part in preparing and building, out of need to keep up with the global developments in the concerned fields, and a political opportunity to bring a major European collaborative research facility to Sweden (chapter 6). The brochure produced as part of the ESS Scandinavia campaign ahead of the May 2002 Bonn meeting, to complement the Expression of Interest, positioned the planned ESS facility as crucial for future research in advanced materials science and technology, stating that it would "undoubtedly strongly enhance European science and also contribute in decisive ways to tomorrow's leading technologies" (ESS Scandinavia 2002b: 2). The brochure put forward a fourfold argumentation for an ESS in Lund, namely "an excellent site" with perfect physical conditions; "an excellent scientific environment" with universities and knowledge-intensive industry next door; "an excellent infrastructure with the Øresund Bridge and Copenhagen Airport"; and "an excellent place to live, work, and visit," with the Scandinavian quality of life as a special highlight (ESS Scandinavia 2002b: 4–5). A salesman's message, to be sure, but balanced and earnest in comparison with what was to come.

The investigation by Allan Larsson in 2004–2005, and the report by the Institute of Growth Policy Studies (ITPS) appended to it, were analyzed in great detail in chapter 6, and undoubtedly set a new standard for the ESS Scandinavia campaign, in two regards: First, by switching focus almost entirely to the growth effects and the promotion of the ESS as a generic innovation-boosting project that mainly had its *raison d'être* in enterprise policy (which also showed in Larsson's unrealistic advocacy of a PPP solution for its funding); and second, by the standard it set for the campaign and its PR work: The ITPS report was built on data and analyses that are best described as unscientific, which Larsson repeated unashamedly in his main report.

Wilhelm Agrell (2012) did a formidable job, in the midst of the most intensive marketing phase of the ESS, to analyze the content and conduct of the PR work as seen through the production and pamphlets, brochures, and documents of various sorts, in 2008–2009. He identified two modes of marketing, with an *offensive strategy* "designed to explain and visualize the ESS as an effort with a positive impact on science, the local community and the region as well as on society as a whole," in order to "to find simple convincing arguments to influence the decision makers and interest groups that would materially decide the outcome," and a *defensive strategy* "to counter, or as it turned out largely preempt, criticism concerning waste of research resources, negative environmental impact and unforeseen risk in a complex and untried facility" which was made by presenting problems as "unlikely, limited and manageable" (Agrell 2012: 432). The offensive strategy, unsurprisingly, dominated most public communication. He also identified a seemingly contradictory but apparently very workable strategy to present the future contributions of the ESS to science, innovation, and society as simultaneously completely certain, in that these contributions would be made and would be significant, not to say spectacular, yet mysteriously uncertain, because they were unknown and produced in a scientific context.

This "remarkable imbalance" between uncertainty as something positive and the complete absence of similar uncertainty in the promised future usefulness of the ESS and the science it will (help to) produce, has been analyzed at some length in a previous publication (Hallonsten 2016a: 187ff), and is typical for the marketing of science projects today, with promises of future scientific and technological achievements, and their contributions to innovation and economic development, as powerful tools in the hands of politicians or campaigners who need to mobilize support for a project in order to have it funded (van Lente 2000; Brown and Michael 2003; Borup et al. 2006). Promises and expectations have inherent power, and lend great strength to campaigns, because any opposition to any part of the campaign – no matter how well-balanced the opposition, and how exaggerated or unscientific the target of the criticism – is almost mechanically disregarded as *anti-progress*. This framework for understanding expectations and promises is useful when analyzing the offensive and defensive strategies of the ESS marketing campaign, exactly because the active use of imagery, slogans, and appeals to the public imagination was creatively mixed with logical lapses, relativizing and de-emphasizing. This was a workable combination exactly because the campaign concerned something that did not yet exist and could be filled with all kinds of *expectations*, and so the campaign built almost entirely on *promises* of fulfillment of these expectations, with the main message that all kinds of amazing things would, with certainty, come out of the ESS, and any negative impacts would be either negligible or possible to handle with good planning.

The offensive marketing strategy had four main themes. The first was the presentation of the ESS as a source of knowledge and innovation with direct practical implications for contemporary key challenges concerning energy, climate, environment, consumer products, and health (Agrell 2012: 433). A public document in Swedish, entitled "This is how ESS will improve our everyday life" ("*Så kommer ESS att förbättra vår vardag*"), even lists a number of innovations that will be made possible as a direct contribution from the ESS, among them hydrogen gas technology that will lower greenhouse gas emissions, lithium batteries with improved capacity, and various improvements to cosmetics, cleaners, and paint (ESS Scandinavia 2008d). The 2009–2010 ESS Activity Report similarly portrays the future ESS as a source of good in a broad but palpable way, e.g. claiming that "research based on neutron science helps diverse technologies move forward to improve the fabric of our lives" (ESS 2010a: 4). Two brochures written by scientists and published in 2008 and 2009, reportedly intended to spur interest and mobilize support for the ESS in scientific communities, lend themselves to considerably exaggerated rhetoric, for example by posing the question "Can neutrons help to solve major challenges in health sciences?" and answering it with a huge "YES" (ESS Scandinavia 2008e: 3), and promising, among many other things, "smoother cream cheese," better fuel cells and batteries, and efficient energy use as a result of scientific work at the ESS (ESS Scandinavia 2008e: 22; 2009b: 8). Second, Lund and the Øresund Region including Malmö and Copenhagen, are presented as the perfect host region for ESS, with both advanced knowledge-intensive industry and tradition in the shape of a medieval city center in Lund (ESS Scandinavia 2008c; 2008f). Third, locating the ESS in Lund is presented as a "win-win-win-win situation" (Agrell 2012: 434); in other words, there are only winners: Basic science gets a top research facility, applied science gets innovation, local and regional society certainly get hi-tech startup companies and dynamic growth, and society as a whole gets all the above technological improvements). The one-sided presentation of basic science as a winner is counter-intuitive, given the vast increase in governmental science spending across the board in 2007 and on, since major investments always entail priorities; and a key theme of the governmental science policy offensive in these years, of which the ESS was part, was strategic prioritization (chapter 4). The director of ESS Scandinavia, Colin Carlile, was very keen to deny every risk of budgetary displacement effects, and in fact adopted the strategy of denying that this would indeed be an issue (Interview: Carlile), and while history would in one sense prove Carlile right (see chapter 11), it was a peculiar strategy as part of the marketing campaign, both arrogant and ignorant of the history of science policy priorities in Sweden. It most certainly antagonized a lot of people both in Lund and in Stockholm, with doubtful gains for the future ESS organization in terms of public image, collaboration opportunities in Lund, and harmonization with the Swedish political system and its institutions of science policy.

The fourth theme of the offensive strategy, in Agrell's (2012) analysis, reflected a fundamental problem or classical dilemma in science communication: The physics of elementary particles (like neutrons) and their use in diffraction experiments (see chapter 3) are scientifically and technically very advanced and therefore difficult to explain to the lay public, and it is easy to revert to oversimplifications and misplaced analogies that only lead to confusion in the long run. Most obviously misleading is the likening of the ESS to a giant or large microscope, which was ubiquitously used in 2008 and on; in ESS Activity Reports (ESS 2010a: 4; 2011d: 6; 2013a: 6), in Region Skåne's reports (Region Skåne 2009a: 23; 2012a: 25), by members of the Swedish Parliament (Holm et al. 2005, 2006; Swedish Parliament 2007a, 2013; Wetterstrand et al. 2008), in Government bills (Swedish Government 2012b: 99; 2013b: 97; 2013c: 6; 2014b: 147), not to mention in dozens of news articles in *Sydsvenskan* and other outlets, in 2009–2015. Although microscopes are probably the most well-known scientific instruments for visualizing objects that are too small for the naked human eye to see, the analogy is fundamentally erroneous and can certainly create confusion, let alone appear patronizing to the interested layperson. Microscopes magnify objects, with the use of visible light and lenses, and allow scientists to see the details of objects that are too small to see with the naked eye. Neutron spallation is a completely different thing, as explained in chapter 3, and reveals not only structure but also things like magnetism. Neither light nor lenses are used at neutron facilities – the whole point of using techniques like spallation is that there are structures and dynamics of matter that can never be seen at all (in any real sense of the word) since they happen at a level of resolution far smaller than anything that could ever be seen by the human eye, no matter how sharp lenses a microscope could be equipped with. Calling the ESS a "giant microscope" is an image as inadequate as calling an airplane a "giant bicycle" – while both are means of transportation, the similarities end there.

Adding to the compelling analysis by Agrell (2012), a fifth theme of the offensive strategy can also be discerned in the marketing campaign, namely the habit of presenting the ESS as a *done deal*, as having been finalized, or at least as if binding decisions, including a funding solution, are already in place. The extensive use of computer-rendered imagery of the future ESS facility, with the cities of Lund and Malmö and the Øresund Bridge in the background, is one example, and several such images succeeded each other in marketing pamphlets and brochures (the first, used in the 2002 ESS Scandinavia Expression of Interest, was a handmade drawing). One of these images was printed on the billboard of local newspaper *Kvällsposten* on May 29, 2009, which announced the results of the meeting in Brussels the night before, and contained an image legend that pointed out the facility itself but also a number of smaller buildings where "spin-off companies" were to be located (image reprinted in Hallonsten 2012: 10), as if the ESS organization, or the newspaper that reprinted the image, could know that there would be spin-offs from the ESS significant enough to form companies in numbers great enough to demand designated buildings close to the facility.

An ESS exhibition was created in the spring of 2008 at the ESS Scandinavia offices at Stora Algatan in Lund, aimed at convincing the local population of the advantages of the ESS (*Sydsvenskan* 2008c). It certainly contributed to the oversimplification and exaggeration of the positive effects of an ESS in Lund, most of all in the area of scientific results. The exhibition was focused around the *Science for Society* slogan, and visitors were invited to "look at different examples of benefits of materials research in everyday life" (ESS Scandinavia 2009a: 20). But perhaps the most extreme exponent of the offensive marketing strategy was the 10-minute film from 2010, issued on DVD and distributed freely, where the actor Sir Patrick Stewart, most famous for his role as Captain Jean-Luc Picard in *Star Trek: The Next Generation* and its sequels, promoted the future ESS. Stewart was accurately presented in the short film as Chancellor of the University of Huddersfield, but it was quite clear that the intention of the movie was to make use of his acting skills and his iconic look, which would give the film and the presentation a part-science-fiction aura. Promises and expectations of *scientific and technological progress* are utilized most effectively in the film, in combination with imaginative imagery and visual effects, to present the future ESS as a unique and absolutely vital tool for furthering humanity's understanding of our world:

> Our eyes are our windows on the universe. Throughout our evolution, they have enabled us to observe and make sense of the world around us. And as our understanding has grown, so the questions that we ask about our world have become ever more complex. To help us answer these questions, we have built telescopes to

> study the vastness of outer space, and microscopes to study the smallest objects in the finest detail. But imagine what we could learn, if we were able to build a microscope so powerful that it could look deep inside the very materials from which our world is made. So powerful that it wouldn't just reveal how atoms and molecules are arranged, but how they move and interact with each other, and how they bond together to form metals, alloys, chemicals, plastics, and even the biological substances upon which life itself depends. Such a microscope could unlock the secrets of materials at the atomic level, enabling us to modify and tailor their properties and performance to satisfy the demands of our increasingly technological society. The benefits for medicine, energy, production, transport, electronics, manufacturing, and the environment would be immeasurable. It might be a surprise to learn that such microscopes do already exist, and that they are being used every day to help us understand the structure and properties of materials in more details than ever before. But unlike conventional microscopes, they don't use light and magnifying lenses. Instead, they illuminate the fine structure of the atomic world using the extraordinary quantum physical properties of a subatomic particle, the neutron. [...] However, it is inevitable that whether using the most powerful telescopes or the most powerful microscopes, the more we see, the more we realize there is to see. And now, as we see the atomic world at the very limits of our existing neutron capabilities, we know that there is very much more to be discovered beyond the present neutron horizon. The exciting news is that Europe is about to embark on a voyage of exploration and discovery well beyond the limits of any existing neutron facility. Europe is poised to build one of her most ambitious scientific projects: The European Spallation Source. (University of Huddersfield 2010.)

Contrary to what Agrell (2012) called the offensive strategy, which was summed up rather neatly by the tirade of exaggerations in this quote from the film with Sir Patrick Stewart, the "defensive strategy" of the marketing campaign was geared to presenting the future ESS as a sustainable research facility and to toning down all possible negative aspects of it, especially concerning environmental impact. The strategy was largely preemptive and sought to present all kinds of possible problems – some of which were technically complex and difficult to assess the scope and nature of, like radiation hazards, and some very straightforward, like the use of premium farm land – as either exaggerated, and thus highly unlikely or very limited, or manageable with the kind of prudent and responsible planning by the ESS organization that was also outlined as part of the communication work. As the 2009–2010 ESS Activity Report notes, "[e]nvironmental concerns must be met with respect, and young, curious minds must be met with enthusiasm and encouragement" (ESS 2010a: 36).

Agrell (2012: 435) argues that the key message of the defensive strategy was to present the ESS "not only as a low risk, but also as a virtually no-risk facility." To some extent, the ESS Scandinavia campaign could rely on the fact that Sweden has a well-known, and in international perspective meticulous, procedure for permits and environmental trial. Echoing Allan Larsson's response to the criticism that his 2005 report did not deal (sufficiently) with environmental aspects (chapter 6), the ESS campaign referred to the environmental trial that the project would eventually face: "Before the decision can be taken, the proposal will undergo extensive security and environment trials. A precondition for the construction of the ESS in Lund is, in addition to an international agreement between the other interesting countries, that the proposal lives up to Swedish and European demands of environment and security" (ESS Scandinavia 2008b: 2). The work to present the ESS as an essentially "green" facility, particularly in terms of climate impact, was an important part of the marketing campaign. A brochure from 2008 claimed that the ESS would be "the world's first climate-neutral research facility" (ESS Scandinavia 2008b: 8), which was followed up and substantiated in the 2012 energy report (ESS 2012e; chapter 8). A typical pseudo event was staged in June 2009, in the midst of the hype around the site decision in favor of Lund, when not much existed either in the ESS organization or out on the designated ESS site at Brunnshög. In the presence of local media, Colin Carlile and his associates planted a tree "almost exactly in the middle of the future ESS site" to symbolize the ambition of the ESS to become a "climate neutral facility" (*Skånska Dagbladet* 2009). The relatively realistic promises that all the electric power supply for the ESS would come from renewable sources, and the plans to recycle excess heat into the district heating system of Lund (an idea that had been drafted back in 2001–2002, see chapter 6) were joined by abstract plans to recycle rainwater ecologically, and a ubiquity of green spaces on computer-rendered illustrations of the future ESS facility, where grass "covers practically all the available surfaces, including some of the roofs" (Agrell 2012: 435). The 2009–2010 ESS Activity Report claims that a "strong commitment to sustainability issues was an integral part of the bid to host the ESS in Lund," which it argues makes the ESS "unique among big science projects" and "a beacon for all such future projects around the world" (ESS 2010a: 34). Interestingly, Agrell (2012: 436) notes conflicting goals in the PR campaign: in order not to destroy the image of an "open, green and friendly facility," shielding, security, fencing, and vehicle barriers are remarkably absent from the public PR material.

The central argument made to counter anticipated typical points of criticism concerning environmental impact was that the ESS would contribute significantly to scientific progress and innovation, and thus make decisive contributions to meeting global sustainability challenges, and hence be a net gain for the environment (e.g. ESS Scandinavia 2008b: 11). But the criticism also concerned specific aspects of direct environmental impact, which the campaign generally met or preempted by presenting such impacts as insignificant and temporary. One of the most controversial issues around the ESS plans which had been raised in the Swedish Parliament (chapter 6) was land use. The situation was essentially similar to that of any construction project on premium farm land, and any denial of the fact that the ESS would use some of Sweden's best agricultural land would be pointless. The campaign strategy was to relativize. One key marketing brochure stated that the ESS would only use 0.7 square kilometers of land, which is comparable to the size of a golf course, and which represented as little as 0.2% of the cultivated farmland on the coastal plains around Lund; moreover, the farmland would not be permanently devastated, only temporarily borrowed; after the 40-year period of ESS steady-state operations, the facility was to be dismantled and the land returned to its current use (ESS Scandinavia 2008b: 8). The latter argument in particular is, of course, laughable, both considering historical evidence of how large scientific facilities and institutes renew and expand with a seemingly built-in imperative (e.g. Hallonsten and Heinze 2012, 2016) and considering the vast plans for exploitation of the whole Brunnshög area around the ESS for the Science Village Scandinavia (see below), which were not, of course, the responsibility of the ESS organization but which were drafted in collaboration with the ESS and used indirectly by the ESS in other marketing efforts (see below). Other highly contentious issues were the risks of radiation hazards and the planned use of mercury in the target station, both "far more sensitive than the overall safety issues" (Agrell 2012: 435) due to the historically strong environmental and anti-nuclear movements in Lund. Considerable efforts were therefore made by the ESS organization to explain to whoever would be interested that the ESS was not a nuclear facility, and that it would bring about a reduction in the use of nuclear reactors overall, globally, because it would replace some reactor-based neutron facilities across Europe. The radioactive emission of the ESS itself would be insignificant, which is true especially in comparison with nuclear reactors. The brochures stated that it would in fact be below the natural background exposure, and that the facility's radiation barriers would be so effective that personnel would not even need protective clothing, "not even when working close to the walls of the target station." A comparison with the SNS at Oak Ridge, Tennessee, is used to illustrate this: A major disaster at the SNS would produce a total radioactive contamination below a one-time dose of 10 mSv, which equals the dose of an ordinary medical x-ray scan (ESS Scandinavia 2008b: 4). The tentative plan to place one cubic meter of liquid mercury, or 13 tons, in a building outside Lund in southern Sweden was of course way more problematic from a PR perspective, but eventually resolved by the decision in 2011 to use the significantly less environmentally harmful element tungsten instead of mercury in the target station (chapter 8).

Given the history of environmentalism in Sweden, and in Lund, once an anti-nuclear movement stronghold, in particular, the lack of deeper environmentalist resistance towards the ESS Scandinavia campaign, and the Government campaign from 2007 and on, is in a sense mysterious. In Agrell's (2012: 437) analysis, two key reasons are that the ESS never became a national concern but remained a local issue, and that it likewise never transformed into party politics, which kept it away from the mainstream political debate. Based on Agrell (2012) and Stenborg and Klintman (2012), it is too simple to conclude that the defensive strategy of preempting criticism was successful; instead, it seems the relative silence was due to a sheer lack of interest, especially after 2005–2006, when also most of the noise made in the opinion letters to *Sydsvenskan* (chapter 6) largely died out. In May 2009, no fewer than 160 residents and companies neighboring the designated ESS site were invited to an information meeting, but only 28 people came. In June the same year, a public meeting with a Q&A session was organized and broadly advertised, but only 5 people turned up (*Sydsvenskan* 2009k). Perhaps the act of pretending as if the May 28, 2009 site decision was really the final decision on building the ESS in Lund – which is done, *inter alia*, in the 2009–2010 ESS Activity Report (ESS 2010a) – was the most effective part of the defensive marketing strategy.

But the exaggerations of the marketing campaign, both the offensive and defensive strategy, deserve some further analysis and contextualization. Christian Vettier, who had come to Lund back in 2007 (see chapter 6), argues that the local marketing campaign – which he took active part in – was a means to harmonize the scientific ambitions of the proposed ESS facility with the campaign phase that ESS Scandinavia was in at the time, making a case for how the ESS "could fit in the local picture, the local environment," so that "local people" could think "okay, yes, we could have this here" (Interview: Vettier). Nonetheless, it seems to have created false impressions that created oversimplified views among decisionmakers. Dimitri Argyriou, who succeeded Vettier as science director, had to use some of his time and energy to convince politicians and regional development administrators that the ESS was not "going to solve all the societal problems that we have," because regardless of the message conveyed in information material, people in these positions "extrapolate" and "superimpose their ambitions and their visions on top of that immediately" (Interview: Argyriou). Apparently, the system surrounding and embedding the ESS campaign in these years had a need for more or less exaggerated promises, and it is important to bear this in mind when analyzing the content of the marketing campaign: Part of the blame for unscientific and exaggerated marketing efforts should probably fall on the recipients of the message. Moreover, it can also be argued that the marketing efforts contributed strongly to bringing the ESS to Lund. What is clear is that the whole ESS organization, including its director Colin Carlile, who used very colorful language in public and media appearances, had a deliberate strategy to use the PR work as an integrated part of its operations; as summarized in the *Conceptual Design Report (CDR)* of February 2012:

> The third project is ongoing and continuous and will remain an important element throughout the whole lifetime of the ESS facility. Basically it is an information activity maintaining effective dialogue with our major stakeholders be they scientists, politicians, funders, laypeople, et cetera, and be they local, regional, national or international in origin. The activities involve both the traditional media and the rapidly evolving e-media. In real terms this encompasses press releases, web pages, talks, visits, exhibitions, branding, brochures, annual reports and one-to-one interactions. We have built up a strong communications culture and this is to be constantly updated, appropriate, attractive and relevant. It is very important that the project has many supporters, but it is perhaps even more important that it has few enemies. Public acceptance is crucial to success and we put particular effort into this. (ESS 2012a: 9.)

Of course, none of the promises made were completely baseless. The potential of neutron scattering experiments to make significant or even decisive contributions to progress in fields of great importance for technological progress in a number of areas is great. The potential of a neutron source with unprecedented performance, meeting the demands of a European neutron scattering community already in a world-leading position, is of course formidable. The 2013 TDR lists, in less sales language and with more nuance, a number of areas where it is reasonable that the ESS will enable experiments that will be crucial for scientific and technical development, including novel materials for solar cells, batteries, fuel cells, and data storage, better ways of understanding the molecular mechanisms behind diseases and plant metabolism, delivery systems for new drugs, and in medical implants, among many other things (ESS 2013b: 11). The same report also discusses various types of scientific studies where the special properties of the neutrons produced at the future ESS will be of great use, such as magnetism, low temperature studies, and not least real-world samples and real-world conditions (ESS 2013b: 13–14). Chapter 2 of the TDR (ESS 2013b: 9–147) is recommended reading for anyone interested in the real prospects of the ESS to make those spectacular contributions to scientific and technological development that the marketing campaign envisioned or, in some cases, even promised.

"Growth, Innovation, Availability, Attractiveness"

In the introductory chapter to this book, it was emphasized that the ESS Scandinavia campaign was launched and gained momentum during a time when the discourse of (regional) innovation systems, learning regions, networks and clusters, and dynamic economic effects of colocation of universities, research institutes, and knowledge-intensive industry was perhaps most intensive. The usefulness of the theories that back this up, from economic geography, innovation studies, and entrepreneurship studies, is not questioned here. It is their use in (political) campaigning that can rightly be criticized, since in Sweden this remained very shallow and contributed greatly to the creation of exaggerated images and false impressions of rather spectacular socio-economic effects of the location of knowledge-intensive and high-tech firms and institutes in specific regions. The 2002 book by geography professor Gunnar Törnqvist was discussed in chapter 6, and while Törnqvist is careful not to over-sell the ESS, he also launches the narrative that the ESS can be the center of several "research areas and industrial cluster formations" that can co-produce a range of dynamic effects with the ESS (Törnqvist 2002: 41). It is surprising to note that Allan Larsson did not cite the Törnqvist book in his 2005 report, where he otherwise made a great effort to present the future ESS as a growth engine for the Øresund Region and Southern Sweden. Of course, Larsson procured his own study, from a team of investigators at the Institute for Growth Policy Studies, and its main conclusion, surrounded in the study by many words of caution that Larsson ignored when repeating the conclusion in his main report, was that the Swedish GDP would grow by SEK 4.3 billion, or 0.17%, and that no fewer than 6,000 new jobs would be created, all as a result of the building the ESS in Lund (Larsson 2005a: 116–118).

In 2008–09, Allan Larsson was back in the game of marketing the ESS, apparently not over-burdened by the negotiations on the European stage. On September 25, 2009, in an interview on the Swedish national radio news, he claimed that "from a Swedish perspective" the ESS is a "very good deal," claiming that "for every million we put in we get eight, nine million back" (Sveriges Radio 2009). Larsson repeats the claims in his memoirs (Larsson 2019: 100). The origins of this 8:1 or 9:1 ratio between revenue and investment are unclear; to be sure, the first report of the TITA project (see below) had been published earlier the same year, with its extreme extrapolations and misrepresentations of statistics, but its two estimates of the growth of the gross regional product of the region of Skåne up to 2040 are off the mark of 8:1 or 9:1. Regardless, it is worth noting that investments with a return of eight or nine times the money put in that is almost *guaranteed* (this is at least the impression Larsson gives in the interview) are very good investments, and it is, thus, logical to ask why other countries are allowed to take part and contribute; shouldn't Sweden take the whole investment itself and reap all the benefits? Of course, the calculation and analysis builds on the expectation that foreign investment will come to Sweden and stay in the Swedish economy because procurement of goods and services (especially on the conventional or low-tech side, like transport, catering, cleaning and maintenance, and so on) tends to stay in the close vicinity of facilities. This is, of course, the logical explanation behind the thorny politics around their location and the advanced procedures put in place to secure that benefits are equally distributed among member countries (e.g. fair return policies and in-kind contributions, see chapter 3), but most of the effects are secondary and tertiary, indirect, effects that are expected but impossible to quantify (see e.g. Hallonsten 2016a: 193ff). In other words, "eight, nine million back" per million invested is probably, by most standards, a dramatic exaggeration, and Larsson provides no sources.

A year later, on October 21, 2010, Allan Larsson made another contribution to the local and regional marketing campaign around the ESS, reporting on a conference where a planned visitor's center at the entrance to the ESS and MAX IV site was discussed, under the name "Lundiana," and with the prospects of attracting between 300,000 and 500,000 visitors every year. This figure would make Lundiana one of Sweden's twenty most visited museums – perhaps not a completely unrealistic notion, but highly speculative nonetheless. Larsson makes the comparison with the *Universeum* natural science and technology museum in Gothenburg (*Sydsvenskan* 2010), which has some 500,000 visitors every year but which is located next to

Liseberg amusement park, which is generally regarded as one of the top tourist attractions in Sweden, and therefore probably draws some visitors to Universeum. Moreover, to put the plans in perspective, nobody has heard of Lundiana in the eight years since the October 2010 conference.

While the local newspaper *Sydsvenskan* mainly repeated the exaggerated findings and conclusions of the reports produced by Region Skåne (see below), and debate letters from politicians and industrialists to the paper were largely unchallenged (Kinhult et al. 2008; Müchler and Klarskov 2008), one journalist reacted to the dishonest campaigning by Allan Larsson, and was allowed space to do so in the pages of *Sydsvenskan*, noting that the results and positive regional effects that Larsson promised "can never be guaranteed" and that it was counterproductive, in the long run, to "try to sell the ESS and basic research with the help of obtrusively unscientific guarantees of results" (Samuelsson 2009).

The discourse had, however, been established and there was little or nothing that could penetrate the strong façade that it made up. Therefore, when, in 2008, Region Skåne decided to make a broad investigative project to help prepare for a smooth and dynamic establishment of the ESS (and MAX IV) in Lund, it became most of all a project that fed into the marketing campaign, and very little work was done that could help Region Skåne, the municipalities of Skåne, and all the other actors in the region to prepare and develop the "absorptive capacity" that is absolutely crucial to enable the ESS to contribute to regional growth at a level even remotely approaching what Allan Larsson and others claimed it would.

The first report was the result of a prestudy funded by the European Regional Development Fund (through the Interreg IV program), Region Skåne, the City of Lund, the City of Malmö, the City of Helsingborg, the County Administrative Board in Skåne, and Lund University. It was written by the consultancy firm PricewaterhouseCoopers (PwC), after a bidding process, and published in early 2009. While the political steering group for the project paints the ESS (and MAX IV) as "an opportunity to accelerate growth in the region" and "springboards for future growth" in the preface to the report, they also note that for this to happen, "a large number of players in the region have to work actively and strategically" (Region Skåne 2009a: 3). But even on the first pages of the report, the language becomes more vague and the message less of an appeal for purposeful work, and more of a communication of a "vision" which is summarized in the slogan "Society for Science – Science for Society" and which can be realized with "a creative imagination, a totally committed regional management and an interactive community that melds global opportunities with regional entrepreneurship" (Region Skåne 2009a: 5).

The vision itself is for what the Øresund Region will be like in 2030. Then, the region will be a "world leader through its international materials research institute," with the ESS and MAX IV the "multidimensional springboards that enable the Öresund Region to step into the Europe of the future with all flags flying." The report claims that the region will not only be scientifically excellent and globally attractive to talent, but also "distinguished by a uniquely visionary, innovative and practical cooperation between various sectors in society." The scientific activities in the region, with centers of gravity at the ESS and MAX IV, will by this time have produced Nobel Prize-winning research, and the regional innovation system attracted "international risk capital" (Region Skåne 2009a: 7).

The remainder of the 150-page report is a product typical of a consultancy project, with significant attention paid to layout, imagery, figures and diagrams, and finding the right slogans to sell the report, and little effort on conducting proper underlying analyses and quality control in a deeper meaning, including work to relate the findings and conclusions to the state-of-the-art of social sciences and economics, as well as simpler things like language proofreading. By procuring the report and publishing it in the form it did, Region Skåne made itself an extension of the ESS Scandinavia marketing campaign, simply allowing PwC to use ESS Scandinavia's advertising material, and not least the dubious ITPS study, and repeat the embellished descriptions of what neutrons would do for science and society (above). The report completely lacked even the most rudimentary descriptions of what the ESS would actually do, how, and under what legal, organizational, financial, and political circumstances, which arguably would be the first step in any serious analysis of its impact on science and society (Hallonsten 2016a: 195).

The report uses a myriad of statistics of historical growth patterns, employment numbers, enterprise and industry classifications, commuting, hotel market, and many other things to paint a picture of Skåne and the Øresund Region as very promising and in many ways ideal for the location of the ESS and MAX IV, thus painstakingly avoiding any qualitative and critical discussion about the real challenges pertaining to the legal and organizational dynamics of the research facilities themselves, and the institutions of surrounding society. While pointing to inspirational examples around the world where neutron and synchrotron radiation facilities have been colocated, such as Oxford, Zürich and Grenoble, the report fails completely in drawing conclusions from these cases and comparing them to Lund and the Øresund Region in terms of the preconditions for embedding major research facilities in the regional innovation systems (and other institutional structures) and benefiting from them.

The most famous conclusion of the PwC report, and perhaps the most problematic, was that the gross regional product (GRP) of the region (n.b. this refers only to Skåne, i.e. not the Danish part of the Øresund Region) would be significantly higher in 2040 than in 2010 due to the location of the ESS in Lund. The calculations are based on the 2005 ITPS report and a number of other sources, and use formulae for multiplier and dynamic effects of the location of the ESS in Lund to estimate that the GRP of Skåne will be SEK 16 billion higher in 2040 than in 2010. The estimation is itself not unrealistic, since it implies that the ESS will bring a 0.08 percentage points higher regional growth rate to the region, but the report then launches another figure that consists of the *sum* of the annual increase in GRP, in real terms, over thirty years – in other words, adding up the total annual GRP growth of Skåne and arriving at SEK 214 billion, which it calls the "accumulated higher GRP with the ESS" (Region Skåne 2009a: 73). This figure is then used in conclusions and summaries of the report as the key estimate for the economic effect of the ESS on the region, in two instances noting explicitly that this is an "accumulated effect" (Region Skåne 2009a: 12, 76) but also, once, using vaguer language to claim that "with the ESS, the GRP will be SEK 214 billion higher in 2040" (Region Skåne 2009a: 44).

In itself, the number is harmless, not least because everyone with five minutes available for a proper reading of pages 73–74 in the report will note the faults by which it has been calculated, and hence its inability to say anything meaningful about the future economic impact of the ESS. The problem, of course, lies with the fact that a great many people with influence and widespread communication channels seemed not to read those pages, but only the summarizing paragraphs in the report, quoted above, or the 20-page summary pamphlet which said that "the ESS could result in [...] An accumulative higher GRP of SEK 214 billion by 2040" (Region Skåne 2009b: 12). Representatives of Region Skåne used the figure *SEK 214 billion* in public presentations, comparing it to annual measures of, among other things, the tax base in Skåne and the sum of the planned infrastructure investment in Skåne in the coming decades, with a grossly misleading result. *Sydsvenskan* also repeated the claim uncritically in 2009, stating that "the Gross Regional Product of Skåne is expected to increase by SEK

214 billion, or perhaps as much as SEK 302 billion, by 2040" (*Sydsvenskan* 2009b). Two days after the site decision in Brussels on May 28, 2009, the main op-ed of *Sydsvenskan* lauded the ESS project and painted a bright picture of what it would mean for the region, claiming, among other things, that the "effect of the ESS on the Gross Regional Product" will "in total correspond to SEK 214 billion by 2040" (*Sydsvenskan* 2009h).

The report "ESS in Lund – Effects on Regional Development" was the result of a prestudy, and in the fall of 2010, Region Skåne launched a broader project to investigate and prepare itself and the region for ESS and MAX IV. The project, called TITA, had a total budget of SEK 47 million and was funded by the Regional Development Fund, Region Skåne, and 42 other public actors in the region including all the municipalities of Skåne, Lund University, and the ESS (*Sydsvenskan* 2013b). The TITA acronym represents "*Tillväxt, Innovation, Tillgänglighet, Attraktion*," which translates to "Growth, Innovation, Availability, Attractiveness," and the main result of the project as a whole was nine reports that dealt with a range of issues and were produced by several different consultancy firms and investigation teams internal to the Region Skåne organization. In addition, a final report was published that summarized the work of the subprojects. Some of the reports focused on relatively mundane and basic needs of the region in absorbing the inflow of a high-skill labor force in great numbers, most of all pertaining to transport infrastructure and housing, international schools, and different functions to make immigrants feel at home in the region (an efficient "receiving organization"), as well as coordinating the expansion of the region by supplying information about exploitable land to startup companies and other investors. Some focused on rather superficial activities, such as marketing the two research facilities ESS and MAX IV and the region they are located in, to foreign companies, investors, and suppliers; the assumed potential in exploiting the greater area around the ESS and MAX IV and creating a dynamic new urban environment there; the possibility for the investments to be the driver of a new proud self-image in the region, mainly through rallies where visionary images were presented and discussed; and the idea of creating an "open innovation arena" for materials science around the facilities. But among the TITA reports can also be found a detailed study of the wider effects of the location of the ESS (and MAX IV) in the region in terms of competence supply and how the region must prepare to be able to reap the benefits of the facilities; a study based on an in-depth investigation of comparable cases abroad as well as the preexisting conditions in Skåne, and discussing at length the necessary preparatory work to be undertaken by a variety of actors in the region, in collaboration. In addition, a study of the possibilities of the ESS and MAX IV to become true drivers of growth in a longer perspective, focusing on what the region needs to do in order to maximize its chances of reaping such benefits from the facilities rather than assuming that positive effects will come at almost no cost, was among the TITA studies.

It is difficult to judge how much of the work of the TITA project actually had a purpose beyond the mobilization of resources within Region Skåne to acknowledge the transformative character of the location of the ESS and MAX IV to the region. Many of the studies did not take the issues seriously enough to provide any useful knowledge that could be turned into constructive work to harmonize the facilities with the region. The few studies that managed to do this, and thus to contribute in a deeper sense, led to few or no concrete efforts on the part of the region's actors and organizations responsible. In retrospect, therefore, the words of the chair of the Regional Executive Committee, Pia Kinhult, in the foreword to the TITA final report seem almost like an irony: "We have ever so slightly opened the door to the future and glimpsed a host of possibilities that can follow in the wake of these establishments. Now it is simply a question of daring to take the step over the threshold and invest in concrete initiatives!" (Region Skåne 2012: 5). The message – that a lot has been learned but that nothing happens by itself and a lot of concrete work remains – is repeated in the final report, which lists a number of actions that will be taken, namely "Enhance the competitiveness and innovative capacity of the business sector"; "Build a region strong in education"; "Create dynamic research environments"; "Increase accessibility throughout the region"; and "Develop the international attractiveness of the region" (Region Skåne 2012: 8).

The final report is a synthesis of the nine subproject reports, and hence a variegated collection of results, discussions, conclusions, speculations, and visionary slogans. It is not as superficial as the prestudy, but it still lacks the deeper understanding of how exactly research facilities like the ESS and MAX IV can and cannot interact with surrounding society, based on evidence from historical and contemporary studies, and what positive (and, conceivably, negative) effects the facilities might have on the region. Large research facilities like the ESS (and MAX IV) interact with surrounding society in a number of ways, and their contribution to innovation, regional growth, and other positive developments locally, regionally, nationally, and globally is multifaceted and depends on a great many things, including their technical setups; the epistemic and sociological character of the scientific discipline(s) they cater to; their organizations, financial and legal frameworks, and the traditions they are developed within; plus, of course, all the features of the surrounding society and how well it can match the capabilities of the research facility, which range from the comparably simple question of how well-equipped local and national scientists are to compete for access, to the significantly more complex issue of how scientific and technological development occurs in systems of innovation involving public and private sector actors and linkages across disciplinary, organizational, institutional, and national boundaries (e.g. Andersen and Åberg 2017; Autio et al. 2004; Hallonsten 2016a; Hallonsten and Christensson 2017a, 2017b). Very little of the insight provided by studies of the dynamics of knowledge production and dissemination, and technology development and transfer, was put to use in the TITA project – "it was very much about regional development, and very little about science and what you're going to do with it, so it was a lot of hot air" (Interview: Argyriou).

This is not to say that the TITA reports lacked good ideas. Some of them did a great job of investigating the preconditions of the Øresund Region and suggesting efforts, based on studies of the environments of research facilities like the ESS abroad, on how to improve the regional capacity to make use of the opportunities opened by the launch of the ESS and MAX IV. But it was the implementation, or indeed lack of implementation, that became the major problem. One local industrialist explained to *Sydsvenskan* how he was fooled by the nice words of TITA and wonders where the action is: "There has been a lot of talk, and little action. But I have to admit that I was also naïve to believe in the promises" (*Sydsvenskan* 2013b). Pia Kinhult, who was the highest elected official in Region Skåne during the time of the TITA project, comments that there was probably a need for a "positive view of the future," and that the ESS "is a good thing." So instead of action, "which I now realize is what we really needed," there were exaggerations and over-selling. "This is just like any investment. You make an investment and then you need to make follow-up investments," but "nobody did anything. The Government wasn't prepared to do anything, the municipal authorities weren't prepared to do anything, nobody was prepared to do anything. Not even Lund University. Nobody thought: What follow-up investments do we need?" (Interview: Kinhult).

Sure, there were plans to build up structures and embed the ESS and MAX IV in a fertile environment. The plan to establish the Science Village Scandinavia, exploiting the land area between the ESS and MAX IV and building up a new urban area where universities, institutes, and hi-tech firms could agglomerate, began in 2010 and is ongoing at the time of writing. A far broader and more grandiose plan for the regional development around the ESS and MAX IV was the Science Corridor plan which emerged as a result of the Letter of Intent between the ESS and the German federal research laboratory *DESY* (*Deutsches Elektronen-Synchrotron*) in Hamburg that preceded the agreement between the German and Swedish governments of July and September 2009 (chapter 9). The DESY–ESS agreement concerned "collaboration on matters related to large-scale scientific infrastructures" and "synergies between these two centres of excellence only 250 kms apart" and the "foundation of a Science Corridor stretching from Oslo in the north to Hamburg in the south and beyond to Berlin" (DESY/ESS 2009).

It is unclear where the name "Science Corridor" came from – Allan Larsson claims some credit for it in his memoirs (Larsson 2019: 155) – but the corridor itself was launched in March 2009 and codified in a Memorandum of Understanding from November 25, 2009, which in turn built on discussions at a meeting in Gothenburg during the summer, just days before the first bilateral agreement on the ESS between Germany and Sweden was signed. The Science Corridor MoU mentions the four facilities for materials science research under development in Hamburg and Lund – the ESS, MAX IV, PETRA III, and the European XFEL – and other "existing research facilities," plus "15 large research universities along a corridor from Hamburg northwards to Oslo" (Science Corridor 2009: 4). The purpose of the corridor, according to the MoU, was to "promote synergies between research infrastructures and facilities" and "to do things that are easier to do jointly" and thus "take a global leadership through scientific cooperation" (Science Corridor 2009: 4, 7). Meanwhile, the MoU was very vague and filled with expressions like "encourage," "work towards," and "work together to promote," and a likewise vague sketch of an organization consisting of "natural hubs," each very different and "each according to its own conditions" (Science Corridor 2009: 9). No specific funding was included. It is therefore no surprise that the Science Corridor existed only for a brief period of time, and mainly in advertising and marketing material for the ESS. The ESS Scandinavia Activity Report 2008–2009, itself largely an advertising brochure, mentions "growing media attention" around the Science Corridor, claiming that it "binds together all of these universities and laboratories, creating a strong cluster" (ESS Scandinavia 2009a: 20, 27), and the 2009–2010 ESS Activity Report also mentions the alleged "unique competence within materials and life sciences" that will "support the setting up of the Science Corridor" (ESS 2010a: 39). In *Sydsvenskan*, physics professor Lars Montelius of Lund University makes the bold suggestion that "our part of Europe could become the Silicon Valley of the 21st century," and that the Science Corridor "is a first step towards realizing it" (*Sydsvenskan* 2009q).

The idea of the Science Corridor seems not to have been welcomed in Stockholm, according to people at the center of the developments in the Swedish Ministry of Education in these years (Interview: Johnsson), and there is no mention of it in Government bills or other documentation in these years, or in the annual reports of the Swedish Research Council. After 2010, the concept seems to have disappeared completely – not even Region Skåne mentions it, either in the TITA reports or in other available documentation. Allan Larsson, the possible original author of the Science Corridor plans, notes a lack of interest from Stockholm and Uppsala and is otherwise very vague regarding what the corridor was supposed to be, other than mentioning as a source of inspiration the mythical US Route 66 highway which became a symbol for wealth and prosperity in the early to mid-20th century (Interview: Larsson). Pia Kinhult claims that after the "major meeting in Hamburg" in 2009, with great attendance and everyone on board, "someone calls Hamburg from Stockholm and says 'don't mess with Swedish foreign policy'" (Interview: Kinhult).

Perhaps the Science Corridor was simply too grandiose and not backed up by any real action plan, and too much a product of the hype around the ESS at the time, and the rather large space taken up by the marketing campaign for the ESS. To some extent, the Science Corridor has been replaced by the Röntgen-Ångström Cluster, a collaboration between the Swedish Research Council and DESY where funding is made available for projects that made use of the research facilities in Hamburg and Lund. The total budget, starting in 2010, was 3–4 million Euro annually from Sweden, and a similar amount from Germany, and open calls were issued regularly (Swedish Research Council 2015: 39). Though less advertised, and significantly less surrounded by grandiose notions of creating a "Silicon Valley of the 21st century" and the like (see above), the Röntgen-Ångström Cluster seems attuned to the needs and capabilities of the scientists in Northern Europe with an interest and a competence in neutrons (and synchrotron radiation), which is, after all, the community that will be using the ESS in the future and producing the scientific results that will form the basis of the return for investment in the ESS.

11. Epilogue and conclusions

Groundbreaking and organizational transition

There are two major events in the story of the ESS campaign that mark the final transition from a campaign to a real ESS facility in Lund, and that therefore conclude the chronology of this book. The first, and most physically palpable, is the start of construction of the facility in September 2014, with a groundbreaking ceremony. The second is the transition, in October 2015, of the ESS organization from a Swedish and Danish limited liability company (ESS AB) to a European Research Infrastructure Consortium (ERIC), which was the moment when the ESS organization (and construction project) finally went into pan-European ownership and eleven of the partner countries entered into a binding joint-funding agreement.

Physical preparations on site had, of course, been ongoing for a while, and in parallel with the major design and project planning effort in part chronicled in chapter 8, there had also been a process of planning the visual appearance of the future ESS facility and site, including an architectural design contest, won by *Henning Larsen Architects A/S*, in collaboration with *COBE ApS*, *SLA A/S*, and *NNE Pharmaplan A/S* (ESS 2013b: 483). In March 2013, the City of Lund approved the detailed plan for the ESS site (*Sydsvenskan* 2013a), and almost a year later, the ESS was granted a site permit by the city, which meant ground work for the construction of roads, connection to water and sewage systems, and similar necessary infrastructure could begin, awaiting only the environmental review (chapter 8) (ESS 2014a: 9). An archaeological excavation of the site was made in the fall of 2013, "a giant thing, cost 35 million" (Interview: Landelius). Test piling began on the site in late November 2013 (ESS 2013i).

The civil construction project for the ESS facility was, of course, a major undertaking in its own right, and certainly a task of significant complexity. After a lengthy procurement process, a contract for the construction project was signed with *Skanska ESS Construction HB*, a subsidiary of the major construction company *Skanska AB*, on February 18, 2014. The construction project concerned the buildings for the accelerator, the target station, and instruments, and was set up as a collaboration between the ESS Conventional Facilities Division and the contractor, with management made up of staff from both ESS and Skanska, working in one single organization (ESS 2014a: 9; Garoby et al. 2018: 114–115). Preparatory ground work began in the late summer of 2014, with the actual start of construction at the end of October, after the board of ESS AB had approved the first part of the agreement with Skanska, in accordance with the conditions of the contract (ESS 2015a: 9).

On September 2, 2014, a groundbreaking ceremony was held on site, with Swedish Minister for Education Jan Björklund and Danish Minister for Science and Higher Education Sofie Carsten Nielsen (ESS 2014e). A month later, on October 9, a Foundation Stone Ceremony was held on site, with speeches by ESS CEO James Yeck, chair of the ESS Steering Committee Lars Börjesson, and chair of ESFRI John Womersley. The same day, an ESS Science Symposium was organized, with 500 hundred scientists as guests (ESS 2014g).

The preparations for organizational transition had been underway for several years. It had always been the intention that ESS AB, formed in the spring of 2010 and co-owned by the governments of Sweden and Denmark since December 2010, was to be abolished and the ESS organization to transfer into another form, preferably an ERIC. There was apparently no deeper discussion of the alternatives, at least not in the ESS AB board (Interview: Landelius), although the 2012 "light review" (chapter 8) had not viewed ERIC as the obvious choice (ESS 2012e: 6). The uncertainty around what ERIC status would bring to the ESS organization, and how it would change the organization's relationships with its stakeholders and surrounding society, necessitated preparatory work over several years, both on governmental level and inside the ESS organization, where a task force, in tight collaboration with the Administrative and Finance Committee (AFC), took charge of the work (ESS 2014a: 7; ESS 2014d: 2). In the spring of 2013, the Swedish Government proposed that parliament should authorize it to decide on Swedish membership in an ERIC to operate the ESS (Swedish Government 2013a: 9–10). In the fall of 2014, five of the ESS partner countries (Sweden, Denmark, Spain, Norway, and Hungary, with Norway in "observer" status) submitted an application for ERIC status to the European Commission (ESS 2015a: 5), and on August 19, 2015, the commission approved the application and established the European Spallation Source as an ERIC (ESS 2015c). The prior existence of another ESS ERIC, namely the *European Social Survey* which had been granted ERIC status in 2013 (Duclos Lindstrøm and Kropp 2017), precluded the use of the ESS acronym in the official name of the organization, which is instead European Spallation Source ERIC. In the fall of 2014, the Swedish Parliament had authorized the Government to approve the transfer of all assets of ESS AB to the European Spallation Source ERIC at the time of its creation (Swedish Government 2014b: 195), and in the spring of 2016, ESS AB was liquidated (Interview: Landelius). The transition itself, from ESS AB to European Spallation Source ERIC, took place on October 1, 2015. Eleven countries (the Czech Republic, Denmark, Estonia, France, Germany, Hungary, Italy, Norway, Poland, Sweden, and Switzerland) are noted as "founding members" in the preamble to the ERIC statutes, with four countries (Belgium, the Netherlands, Spain, and the UK) listed as "founding observers" (ESS 2015b: 4). The European Spallation Source ERIC was set up with Lund as its statutory seat, and with the stated mission to construct the ESS facility as described in the 2013 TDR "to a cost not exceeding EUR 1 843 million in January 2013 prices," and to operate, develop, and decommission the facility, all "on a non-economic basis" (ESS 2015b: 5).

Although the ERIC regulation was produced by the EU bodies to stimulate the creation of cross-border research infrastructures in Europe and simplify the process whereby this can be done (European Commission 2008), it is also clear that the ERIC is an entirely new organizational form encumbered with some uncertainty, not least concerning how its legal status shall be interpreted in the member states, in all different areas of interaction with society and its various institutions. It does not fall within the scope of this book to analyze in full how these uncertainties play out, either in general or specifically for the ESS, because it clearly pertains to events and processes that took place after the time period covered here, but some research has already been done on this topic, which the interested reader should consult for further information (Hilling et al. 2017; Duclos Lindstrøm and Kropp 2017; Reichel et al. 2014; Ryan 2015; Yu et al. 2017; Moskovko et al. 2019; Moskovko 2020). But one topic of discussion is apt, and important, as it pertains both to the near-25-year history of the ESS as covered in this book, and to the topic of European collaborative Big Science in general, namely whether the ERIC regulation has helped, in any way, to remedy the past and current problems of taking collaborative European Big Science projects from idea to reality. Assuming that the ERIC regulation functions efficiently in this regard, and reduces the troubles of launching European collaborations around research infrastructures, it can be concluded that it simply came too late in the case of the ESS – such an alleviation was probably most desperately needed for the ESS project in, say, 1997, 2002, or 2008. But this is also a simplification: Assuming that the ERIC regulation functions as intended, it rids projects of lengthy and thorny intergovernmental negotiations on appropriate legal structures, and, by relying entirely on an organizational framework endorsed by the European Commission and with legal basis in the EU treaties, it also preempts some difficulties associated with setting up international organizations on the basis of treaties, such as the potentially cumbersome process of having the treaty ratified by all the national parliaments of the member countries (European Commission 2008: 32). However, there is little in the history of Big Science in Europe to suggest that these particular issues have been especially problematic – the two major stumbling blocks have always been the *funding* and the *location* of facilities (Hallonsten 2014; Krige 2003; Cramer 2020), and the ERIC regulation makes little or no difference with regard to these. Some provisions of the ERIC regulation pertaining to tax issues and the rights and obligations of ERICs in engaging in "economic activities" and issues of intellectual property might, conceivably, mitigate the process of setting up intergovernmental collaborative organizations in science, but they can equally impede the same process, depending on the ambitions of the governments, institutes, and scientific communities with the initiative, and the political and institutional conditions that set the standards for collaboration and development in the field(s) in question at a given point in time.

It also deserves to be pointed out that the ESS is a very unusual ERIC and that it has very little in common with the other 19 ERICs to date, at least nothing in common except ERIC status. Of the 20 ERICs established (by the end of 2018), the ESS stands out not least in terms of its costs. Two of the others have an estimated capital value of 1 billion Euro, and two are in the range of 400–500 million Euro, but most of them (fourteen) cost at maximum a tenth of the ESS, in many cases significantly less. The ESS also stands out among the ERICs in terms of its scientific complexity and scope – among the 20 existing ERICs, only the ESS is a multidisciplinary user facility that operates a number of diverse and specialized instruments designed, developed, and built through an extensive scheme of in-kind contributions and involvement from research groups all over Europe (Moskovko 2020).

Displacement effects and strategic prioritization

As shown in chapter 9, in the almost six years that passed between the Swedish Government launching its ESS campaign in February 2007 and late 2012, when a new and more accurate total construction cost was approved by the ESS Steering Committee, the Swedish commitment grew from 356.7 million Euro to 645 million Euro, an 80% increase produced by revised cost estimates and the raised bid from 30% to 35%. The final price tag is still not known: The 1.843 billion Euro estimate for the total construction is in 2013 prices. Inflation has since pushed the figure up closer to 2 billion Euro, and there are also signs of real cost increases, that seem likely to fall on the Swedish taxpayers; in October 2016, the Swedish Government announced that the construction of the ESS site had become more expensive than expected, and that these increased costs were to be funded with an additional SEK 980 million (approximately 100 million Euro) from the Swedish Governmental budget (Swedish Government 2016: 180). This would mean a total Swedish commitment of close to 750 million Euro, which is more than double the amount the Government envisaged in February 2007 (chapter 7).

Whether or not these increased costs could have been foreseen is a topic outside the scope of this book, but the reader should be reminded of the referral response to the 2005 Larsson investigation from the Swedish Agency for Public Management (*Statskontoret*), which warned about the common issue of cost overruns in large science projects and large construction projects, and the risk that Sweden would have to cover a major share of these due to its role as host of the project (chapter 6). It is also interesting to note what the literature on "megaproject management" says about projects with long planning horizons, complex processes for decision making, planning, and management, the need to use non-standard technology and designs, and some presence of conflicting interests due to the complexity of the undertaking. Flyvbjerg (2017) has shown that in such projects, cost and time overruns are the rule rather than the exception, and has even called it the "Iron Law of Megaprojects." The tendency to avoid, for political and marketing reasons, accounting for unplanned events and assigning an appropriate contingency budget, and the use of "highly, systematically, and significantly deceptive" cost estimates and time plans in media coverage and decision making around large infrastructure projects (Flyvbjerg et al. 2003: 20) seems, at first sight, to be relevant for the ESS story. To begin with, the increased price tag for the construction costs – from 1.478

billion Euro to 1.843 billion Euro between 2008 and 2012 – is a natural consequence of the design update phase and the proper cost calculations made as part of it (chapter 8); put differently, it is unfair to expect realistic cost estimates from a campaign organization that has only done rudimentary analyses of the site characteristics, the construction project ahead, the scientific scope, and the myriad of technical challenges that unavoidably faces those involved in constructing a scientific facility at the cutting edge. This means that cost overruns – especially the doubling of the Swedish commitment – might not be cost overruns per se but perhaps instead, quite simply, the natural consequence of imprecise and optimistic forecasts (see also Hallonsten 2020b). This, in turn, says less about the ability of the ESS organization and its wider network of partners across Europe to conduct a responsible operation and keep within budgetary limits, and more about the incentives of politicians and campaigners to present future investments as lower than they will reasonably be, although of course there are signs that the ESS organization itself, in 2012/2013, also continued to estimate costs at lower levels than appropriate; Sven Landelius speaks of "wishful thinking" in the cost reports presented to the ESS Steering Committee in December 2012, and there has been some downscaling (e.g. of number of instruments from 22 to 16) as part of the efforts not to exceed the budgetary framework of 1.843 billion Euro in 2013 prices, which is laid down in the ERIC statutes and thus cannot be changed.

Leaving aside the price tag as such, and the size of the Swedish investment, it is appropriate to discuss whether the ESS was, or is, a wise investment for Sweden at all. This is, of course, impossible to answer simply, and perhaps even impossible to answer altogether, given all the different variables and valuations that would have to go into the discussion. A number of arguments can, however, be made that will deliver some conclusions that may not be very precise but, nonetheless, possibly illuminating. The fear of displacement effects has lurked in the background of the ESS campaign almost since its very beginning in the early 2000s, and has many reasons. First, science is known for its guarding of turfs and the "more is never enough" principle that nullifies arguments of a balance of priorities and makes virtually every scientist a strongly biased enemy of all suggestions to cut or scale down further investments in his or her particular field (Greenberg 2007; Stephan 2012). Second, Swedish science has traditionally been pluralist and decentralized, largely governed bottom-up and with strong autonomy for university professors and faculties, which has prevented strategic planning and prioritization from becoming a natural element of national research policymaking; therefore, Swedish science seems to suffer particularly from the negative bias of scientists toward any major investment outside their own field. Third, nonetheless, Swedish science has some bad experience of harsh funding priorities in favor of Big Science projects, especially the reprioritization in favor of the expanded CERN in the 1970s (chapter 4), and direct cutbacks due to budget austerity, and although retrospective analysis shows that the ESS has not infringed (so far) on other areas of science funding in Sweden, which also generally benefited from the major resource increase across the board from 2007 and on (chapter 4), it was obviously not possible to foresee this outcome.

It should, however, also be emphasized that displacement effects are natural consequences of strategic prioritization, and that strategic prioritization was the watchword of the research policy renewal of 2004 and on that brought the major resource increase and also the ESS as a "moonlander project" (chapters 4 and 7). The ESS was presented by the campaign and the politicians that pursued it as a major opportunity for Swedish science, and it can well be argued that taking this chance not only requires substantial investment but also a range of follow-up investments in order to take full advantage of the opportunity and what it brings (see Hallonsten 2013b: 53–54). One problem is that the Swedish Government and its agencies almost appear to have considered the match won with the European decision to locate the ESS in Lund, and thus failed to start the work to make the proper complementary investments, a matter that will be returned to below. But sticking, for the moment, to the issue of displacement effects, the risk of such effects can and should be delineated with respect to time frames.

In the *short* and *medium term*, by which is meant a decade or so, or the time frame for ESS construction, the evidence presented in preceding chapters is enough to conclude that the displacement effects of the Swedish ESS bid are insignificant or non-existent, save for the speculative argument that MAX IV has been given inapt funding and organizational arrangements (see *Sydsvenskan* 2018a; Swedish National Audit Office 2012; Swedish Research Council 2018b) in part due to the fact that the Government prioritized the ESS and left MAX IV to a patchwork funding and governance solution, and the possible ripple effects that the Council's funding commitment to MAX IV has caused in terms of cuts in funding to other research infrastructures (Swedish Research Council 2010a: 27). The Swedish share of the ESS itself was funded, by the Swedish Government, on top of other appropriations, and should the campaign not have been successful, this money would most likely not have been available to other projects.

In the *long term*, however, the issue is different, and the risk of displacement effects is higher. That the funding of the Swedish share of ESS construction was made on top of other appropriations can only be proven for the roughly 350–450 million Euro that was the sum of the commitment in 2007–2008 (chapter 9) and that came from the Government budget, not the eventual increases toward 750 million Euro (above) and not the almost 100 million Euro that Region Skåne was coerced or lured into contributing (chapter 9) and for which we have no evidence concerning a possible encroachment on other important areas of investment. Also, we know nothing of the operating costs of the ESS; Sweden has pledged to pay 10% of these, which should amount to something in the range of 15–20 million Euro per annum using the very conservative estimations of the annex to the ERIC statutes, which make no adjustment upward for cost increases and which also are given in 2013 prices (ESS 2018b: 20–21). More worrying are perhaps the signs that Sweden might have to take on a larger share of the operating costs due to negotiation gridlocks and expectations from other partner countries that the host will be making a greater commitment (e.g. *Sydsvenskan* 2018b, 2019), to say nothing of follow-up investments in Swedish science and the Swedish economy to enable the buildup of a strong user base that can reap the benefits of having the ESS located in Sweden. These are impossible to in any way estimate but may be huge. It should be noted that the unprecedented increases in governmental R&D appropriations in 2007 and on (chapter 4) have leveled off, unsurprisingly, and the chance of a similar increase in the near future, that could cover the investments in competence building and similar growth of capacity to make use of the ESS in Swedish science, that are arguably needed, will therefore at least to some extent have to be made "within existing frameworks," in the true meaning of the expression. Once made, however, the benefits for Swedish science of having the ESS in its backyard will very likely be strong and palpable, which brings us to the *very long term*, where displacement effects are low or non-existent since Sweden and its scientific communities will eventually adapt, because scientific communities are organic and adaptive, and so the ESS will become a natural part of Swedish science and of great benefit for it both in fields prominent today, and completely unforeseen areas.

Related to this is the question of the long-term financing of the ESS and the rather evident tendency of these types of Big Science installations to constantly renew and develop, which is necessary in order to keep up with developments in science and new demands, but also expensive (e.g. Hallonsten and Heinze 2012, 2016). In light of this, it can be concluded that the anticipated limited lifetime of the ESS facility – decommissioning is due to start in 2065, according to the ERIC statutes (ESS 2018b: 21) – is unrealistic. What is most likely is that the ESS will remain in place, in one shape or another, far beyond the year 2065 and that Sweden, as its host, will have to somehow contribute with a large share of its funding. Other countries can conceivably (although it is not very likely) withdraw from the ESS collaboration in the future, while other countries can, of course, join and begin paying their share, but for Sweden, the ESS is a given for the foreseeable – indeed, also the unforeseeable – future, physically and as an annual governmental budget line-item.

An important issue that quite colorfully illustrates the duality of displacement effects and strategic prioritization, is whether the Swedish investment in the ESS was a scientific or a political initiative, and whether and to what extent this matters. In the occasional surfacing of criticism of the Swedish commitment to the ESS – the negative referral responses to Allan Larsson's investigation (chapter 6), and the criticism from the Royal Academy of Science (chapter 7) and the Vice-Chancellor of Stockholm University (chapter 9) – the following criticism was more or less pronounced: the ESS was a political initiative with far too limited anchoring in Swedish science. There is some truth to this argument, although it should be emphasized that ESS Scandinavia had its origins in an initiative by scientists in Sweden and Denmark. But the core of the problem is that the scientific arguments for and against a Swedish ESS bid have never been investigated thoroughly, which goes against Swedish tradition and the expectations of the Swedish scientific community. In contrast, it can be mentioned that the Swedish Research Council undertook a comprehensive evaluation of the scientific case for MAX IV in 2005–2006 with the help of an independent panel of international experts (Swedish Research Council 2006b), and its predecessor MAX II was scrutinized in a similar but arguably even more thorough manner in the early 1990s (Hallonsten and Christensson 2017a: 82). The 1997 government investigation that was undertaken to recommend cuts to Swedish membership fees in costly international scientific collaborations, most notably CERN (chapter 4), focused almost entirely on the relative benefit of different collaborations for Swedish science, and led to a consolidation of the support for the particular priority on basis of thorough examination and the weighing of different alternatives against each other (Edqvist 2009: 135–136). It is also interesting to note that although this investigation was made when European scientists were in the midst of the scientific and technical planning for the ESS, with some Swedish involvement, and when the Swedish Neutron Scattering Society (SNSS) reported to a survey made by the European Neutron Scattering Association (ENSA) that it saw the ESS as a high priority for the future (chapter 5), the investigation did not mention neutrons at all, which is a sign of its relative insignificance for Swedish science, at least at the time.

Regardless of the actual eventual benefit of the ESS to Swedish science, which is hard to properly assess, the ESS project never had the opportunity to gain broad and consolidated support in the Swedish scientific communities. The approach of the thorough, transparent, and recurrent reviews championed and implemented by Jim Yeck in the ESS organization from 2013 and on, which arguably helped build the necessary credibility across Europe that was needed to push the intergovernmental negotiations to a close in 2014, was never used domestically. Perhaps the ESS could have achieved a similar consolidated and strong support in the Swedish scientific communities, if properly evaluated from a scientific point of view in 2004–2005 or 2007–2008. One key question is, of course, if such a review would have come out in favor of the project – Allan Larsson seems to have feared the reverse, since he concluded, in his 2005 report, that it would "be hard to get the necessary support for a Swedish bid to host the ESS purely on research policy grounds" (Larsson 2005a: 29). The Swedish Research Council reasoned similarly in a 2012 report, arguing that the "larger benefit from the ESS in Sweden is expected to be regional growth, business development, and other socio-economic gains," implying that science is not the big winner (Swedish Research Council 2012: 32).

Nonetheless, the ESS was often presented as the perfect scientific megaproject for Sweden, and Lund as the perfect location for it (chapter 10). While there is nothing in particular that would rule out Lund and Sweden as the ESS host, or rule out the suitability of a facility project like the ESS for Lund, there are several things to suggest that other priorities would have been better in a Swedish perspective, and that the ESS would be better off in Jülich in Germany or Harwell in the UK. In comparison with those two, and other potential locations in mainland Europe, the Lund site candidature was always seen as an outsider and very unlikely to end up as the winner, which makes the idea of a perfect fit implausible.

Specifically concerning the match, or mismatch, between the ESS and the capabilities of the Swedish scientific communities, several of the referral responses to Allan Larsson's 2005 investigation pointed out the need for far-reaching investments in the scientific environments in academia and industry in order to make these competitive in the future allocation of experimental time at the ESS, and in order to strengthen their absorptive capacity to be able to benefit in a broader meaning from the science that the ESS will enable. As noted in chapter 3, the ESS will be a user facility which means that its scientific use depends on the efforts of its users, who come from universities and institutes across Europe and the world and have access to the instruments at the ESS in stiff competition (the level of industrial use is small, as also noted in chapter 3). The strongholds of neutron use and neutron instrumentation development in Europe, and thus the scientific environments that will likely be most competitive in the allocation of experimental time at the future ESS, are in France, Germany, and the UK, and their involvement in designing and building instruments to the facility through collaborative work and in-kind contributions (chapter 8) further strengthens their position. However, the likely future use of some kind of *fair return* mechanism to balance the relative use of the facility by a country's scientists with the country's financial contribution to operations, could also mean that as much as 10% of the time will be awarded to Swedish scientists almost regardless of their competitiveness. The question then is whether Sweden today has a neutron user community that can make use of such a share.

It is difficult to make a comprehensive assessment of the Swedish scientific demand for neutrons, but a combination of several data sources and some approximation can yield at least a rough estimate. A statistical study by the Niels Bohr Institute in Copenhagen, from 2018, concludes on the basis of a detailed but largely qualitative evaluation of bibliometric data that the Swedish neutron user community has grown significantly in numbers in the past ten years, from roughly 100 in 2007 to way over 200 in 2017 (Lefmann and Buhl Naver 2018). Most of the increase is, however, due to recruitments to the ESS organization, and the 92 neutron users that were employed by the ESS in 2017 (and recurrently do experiments at other facilities elsewhere) do not constitute a strengthening of the Swedish user base and will only be such if they are recruited to other scientific environments in Sweden and continue their work there. Other data, from the ILL in Grenoble, note an increase in the number of Swedish users from 23 in 2008 to 61 in 2016, which is an increase of a factor of 2.65, compared to the growth of the total number of annual users of 4% and a largely unchanged number of users from France, Germany, and the UK in the period (802 in 2008, 794 in 2016) (ILL 2009, 2017). The increase in ILL users from Sweden is, however, certainly in part due also to the vast recruitment of scientists to the ESS organization in this period. Other relevant facilities in Europe, such as ISIS in the UK and the BER II facility in Berlin, do not publish user statistics with national affiliation, but the federal US Department of Energy does. The two neutron scattering facilities within the system of National Labs are the Spallation Neutron Source (SNS) and the High Flux Isotope Reactor (HFIR), both at Oak Ridge National Laboratory, and in the fiscal year 2016 (i.e. October 2015–September 2016) they received a total of 12 users with organizational affiliation in Sweden, of whom 5 were ESS employees (Department of Energy 2019).

The only realistic estimation possible to make on the basis of the incomplete data available is that there are somewhere between 100 and 200 active neutron users in Sweden (not counting ESS employees, for reasons stated above). For comparison, it can be noted that the number of synchrotron radiation users is at least the double: MAX-lab, the predecessor of MAX IV, had 466 users in 2014 with organizational affiliation in Sweden (of which ten were MAX-lab employees) (Hallonsten and Christensson 2017a: 62), and the US synchrotron radiation facilities had a total of 135 users with organizational affiliation in Sweden in 2016 (Department of Energy 2019). Estimations yield that some 5,000 scientific users will visit the ESS every year once the facility is in operation of its full setup of instruments (ESS 2013b: 99), and whether or not a fair return mechanism will be in place to guarantee access for some 500 Swedish scientists to match the envisaged 10% contribution to the operating costs from Sweden, the expectation that 500 Swedish scientists will be capable of making good use of the ESS and its instruments in just a few years' time is an unlikely scenario given the above calculations, unless significant efforts are made to mobilize a greater user community in the coming years. If not, the ESS will be used predominantly by foreigners, and the taxpayers of Sweden, Skåne, and Lund will have to hope for the socio-economic impacts to give them their deserved bang for the buck.

The actors and institutions of science and politics

The overall framing of the story told in this book, and indeed reflected in its title, is that Lund was an unlikely candidate to host the ESS and that the success of the campaign was therefore both impressive (from a layman's point of view) and puzzling (from the perspective of a historian and sociologist of science). Several of the central actors in the story have, in their own accounts as conveyed in their memoirs, presented the unremitting dedication and talent of themselves and other key persons as the key reason for why the ESS ended up in Lund in spite of all the odds (Larsson 2019; Leijonborg 2018; Thomasson and Carlile 2017). Although such personal memories are simplified and strongly biased, and therefore disqualified as sources of relevant and apt historical conclusions concerning the overarching research question of the study presented in this book – *How could it happen?* – there is nonetheless some significance to the claim that a handful of powerful individuals managed to defy logic and common wisdom and bring the ESS to Lund. Sociological rational choice theory can help us evaluate the reach of this statement. Theories of bounded or restricted rationality postulate that individuals are resourceful, evaluating, and maximizing, but also restricted in their actions and both aware and unaware of these restrictions (Lindenberg 1985; Simon 1957). In other words, they act rationally based on their ambitions and competences, but are limited by the contexts of their action and not least by the constraints set by rules, norms, habits, and expectations – everything we call "institutions" – which in some cases also empower individuals and remove barriers for their action. Especially important, in the context of this book, is that the very different agendas, behavioral patterns, spheres of influence, and skills of the various key actors depended significantly on their institutional context; where they came from, what their experiences were, and what their personal networks looked like. The campaign to bring the ESS to Lund clearly made use of a range of different actors with different skill sets, agendas, and spheres of influence. But the story also entails several examples of interplays between institutional contexts and their exponents in (political) decision making – or lack of decision making, i.e. delays, obstructions, and deliberate inaction – far beyond individuals or far beyond what the historian and sociologist is able to ascribe to individual action.

As a European research facility project, the ESS stood on the shoulders of giants. The lab organizations involved in the drafting of a technical design concept and scientific case in 1993–97 marshalled the top scientists and engineers in neutron scattering instrumentation and use, and developed a scientifically and technically solid and cutting-edge ESS that few, if any, competitors in Japan or the United States could have beaten. But the champions of the ESS, represented in the ESS R&D council, were politically naïve and unable to foresee that the 2002 Bonn meeting would not in itself be enough to bring the project to the relevant political level and toward a political decision. Science policymakers in the countries whose standpoints (and expense accounts) mattered most – France, Germany, and the UK – had other plans. No matter the scientific and technical leaps of performance and opportunity that the ESS would bring for Europe as a whole and for the scientific communities of the eventual host country, it would not trump the political self-interest of these countries, expressed in their desire to keep domestic facilities and labs running, without too much competition. Around the turn of the millennium, and going into at least 2002–2003 when the ESS project almost died completely, the story revolved around the interplay of three institutional logics: the scientific and technical planning work undertaken by world-leading lab organizations and scientific communities; the national political and bureaucratic governance systems of three large European countries and their agendas for domestic institutes and research facilities; and, related to both, the overarching science policy context of Europe, where the balance between national self-interest and common good seems to have been tilting strongly in favor of the former, despite attempts to formulate common policy goals within the EU and implement these in programs for the benefit of the continent as a whole. The ESS took almost a quarter of a century from idea (1990) to groundbreaking (2014). The main reason for the delay is obviously the inability of European countries to agree politically, whereas scientists seem to have their house in order. Through the cycles of the ESS project, from design work in the 1990s, through political collapse and despair in the early 2000s, and to the unlikely but impressive completion of international agreements and groundbreaking at the site in Lund in the 2010s, this inability dominated but was also finally rescinded. Key individuals, with personal ambitions and power bases that enabled them, but who were also clearly in just the right place at the right time, was what eventually made the difference.

Because when the ESS project all but collapsed politically in 2002–2003, a window of opportunity certainly opened for Sweden and other smaller players, but two things nonetheless were required for this window to remain open; first of all, that the ESS project remained alive on European stage, and second, of course, that a national campaign of resilient and devoted individuals continued working. Here, the purposeful and maximizing individuals enter our story. Peter Tindemans, chairman of the ESS R&D Council from 1999 to 2003, can possibly be credited with having opened the window of opportunity by keeping the ESS alive through the downscaled ESS Initiative, and certainly had personal devotion and networks and experience to give him some credibility and authority. Concerning the

ESS Scandinavia campaign, it is clear that its formation in 2000 was the result of a grassroots movement that built on the scientific credibility of one of Sweden's very few internationally competitive neutron users, Lars Börjesson, and his Danish colleagues. The plan to locate the ESS in Lund was, however, not their first priority; as scientists, they seem, rather, to have been focused on organizing their communities to gain a seat at the table in the international collaborations that would bring the ESS forward. It was local and regional politicians and academic leaders that realized the opportunity of having the ESS located in Lund and began pushing this as an alternative – this is where Peter Honeth appears for the first time and in his first key role in this story, but also the representatives of Region Skåne – and here too the contrast between science and politics is clear; obviously, the scientists needed to draw the ESS to the attention of the politicians, and once they did, it was the prospects of bringing a major European project to Lund and Sweden that made their hearts beat faster. Whether it was a neutron scattering facility or the Olympic Games mattered little; politicians cared for completely different things than Lars Börjesson and his peers, and therefore gave the campaign what it previously lacked.

The perseverance of politicians and devoted scientists seems, however, also to differ. A large delegation of local and regional politicians accompanied the small core group of scientists and ESS Scandinavia enthusiasts at the May 2002 Bonn meeting, but only one or two years later, their support had waned and ESS Scandinavia was kept alive by the tenacious work of Karl-Fredrik Berggren, Patrik Carlsson, and Carina Wickberg, with the institutional support of Lund University and the financial support of the Swedish Research Council and the Knut and Alice Wallenberg Foundation. Berggren was certainly a scientist – professor of theoretical physics – but also an experienced research policy and funding operative who had worked within the research council structure to bring large projects (though miniscule in comparison with the ESS) to the attention of bureaucrats and politicians. The combination, at this time found in one person, was apparently what ESS Scandinavia needed to stay alive.

Next, the luck or timeliness of the entrance in this story of the greatest political campaigner of them all, Allan Larsson, should not be underestimated. When Larsson was appointed chair of the board of Lund University, starting January 1, 2004, the campaign was clearly no longer at the top of the agenda of many local and regional decisionmakers (or journalists), and was handled with hesitation or indifference by the Swedish Government. Allan Larsson, with decades of experience of political work in the

Swedish Social Democratic party, including a year and a half as Minister for Finance, and with five years as a top bureaucrat in the European Commission, was a formidable ally for the ESS Scandinavia campaign, and the post as chair of the board of Lund University was clearly not a full-time job. Larsson soon devoted his whole political dedication and clout to the ESS Scandinavia cause, to little surprise for his long-time fellows in the Social Democratic party. To the extent that the Government actively sought to marginalize the ESS and kick the can down the road in anticipation of the result of the 2006 election, this strategy apparently worked in the short term, and here the power of institutions also trumped the abilities of a top political trickster. The appeal to the Government to act decisively to make the ESS reality in Sweden, issued by Larsson in his 2005 report, was ignored, seemingly due to the logics of election cycles and fragile parliamentary support.

These same institutional logics of politics worked strongly in favor of the ESS project only a year later, when after the 2006 elections Sweden had its first majority government in 24 years and the energetic Lars Leijonborg was installed as Minister for Education (and Research), determined to make a mark and write himself into Swedish science policy history. Suddenly, electoral politics and the dynamics of party coalitions had made the conditions for pursuing a "moonlander project" favorable, and with the appointment of Peter Honeth as state secretary (deputy) to Lars Leijonborg, moreover, a long-time champion of ESS Scandinavia was placed at the heart of power. In other words, an interplay of the work of purposeful individuals and institutional conditions suddenly made the ESS a top priority within the Government's research policy offensive.

Specifically concerning Peter Honeth, it is quite clear that he knew the restrictions and opportunities built into the institutional logics of the Swedish university system well enough to be able to push them slightly in the direction he wanted and needed. Together with the succession of University Vice-Chancellors in Lund, he secured stable operations for the ESS Scandinavia project secretariat during some five critical years, and in his next role as state secretary in the Ministry of Education, he was able to push the ESS internally at the ministry, but only to the level or point where the Minister had to make the decision whether this was a project for him to stake his reputation on. Once Leijonborg had made his decision, it became possible to mobilize the institutional logics of the four-party majority government, which manifested eagerness and reform ambitions, in research policy and other areas, and the dynamic of the "inner cabinet" of the four party chairs of whom Leijonborg was one, in favor of the ESS project.

The comparably stale organizational structure and institutional logics of Lund University had its advantages, such as the possibility to recruit internationally reputable neutron scientists to comparably prestigious positions as guest professors. Colin Carlile and Christian Vettier, who both stood down from key leadership positions at the ILL in Grenoble, became formidable *de facto* campaign workers for ESS Scandinavia. Their impressive resumés significantly strengthened the scientific credibility of the campaign, including network ties to virtually all relevant scientific environments in mainland Europe whose support had to be earned for their governments to eventually get behind Lund as the ESS site. In Carlile, ESS Scandinavia also had a skilled campaigner and sidekick for Allan Larsson in the political efforts to drum up support across Europe. Carlile is a particularly interesting person from the perspective of how individual competences and ambitions are anchored in different institutional logics; the plan for recruiting Carlile in 2006 was not to find a new director and spokesperson for ESS Scandinavia, but an experienced neutron scientist to raise the profile of the campaign and Lund University; but what the University and the campaign got was clearly a politician, most of all geared to campaigning. Peter Tindemans, who watched ESS Scandinavia gain ground gradually in the years leading up to the site decision in 2009, calls Carlile "a real asset for ESS in Sweden," "a very energetic person" who "has his own ideas about doing things" (Interview: Tindemans). Why Carlile decided to switch tracks from a career in science and science administration (where the role as director general of the ILL was a kind of crowning achievement) to putting his own reputation on the line in campaigning for ESS Scandinavia, including the unscientific "Madison Avenue" advertisement techniques that he personally promoted in the organization, is not known. His own memoirs (Thomasson and Carlile 2017) and testimony (Interview: Carlile) unfortunately give no clue, but remain on the level of a shallow sales pitch.

Another person with "his own ideas about doing things" is certainly Allan Larsson, who laid a kind of groundwork with his investigation in 2004–2005, both for his own devotion and ambition and for the later Swedish Government ESS campaign. During two critical years, from spring of 2007 and to the site decision in May 2009, Larsson led the negotiations on the basis of vast experience and a ruthlessness that certainly helped the campaign to reach its desired goal but also alienated some (chapter 7). His bold assertion to the ESS Site Review Group (SRG) in 2008, that MAX IV would be built "with 99% certainty" is one of very few concrete available examples of Larsson's audaciousness in political campaigning, and it seems to have worked; the Site Review Group (SRG) of 2008 treated MAX IV as a done deal and highlighted it as one of the features that made Lund stand out somewhat in the competition. The assertion also sorts itself neatly into the category of typical exaggerated promises made as part of the campaign, and that Larsson was only one of many authors of (chapter 10). Larsson's power base in the Social Democratic party was apparently enough to make the Government appoint him to do the investigation (and to give him the title "negotiator" instead of "investigator") but not enough to convince it to take an initiative in favor of the ESS in 2005–2006. His experience of European politics probably served him well in the international negotiations in 2007 and on, although we know little about the details of how and to what extent. This is another interesting example of how and when individuals are limited or enabled by institutions and to what extent they can make use of them for campaign purposes.

Larsson's negotiation strategy, to take "such small steps that nobody can say no, and taking so many steps that only yes remains" (Larsson 2019: 118), was ingenious given the state of the ESS project at the time when the negotiations began, and the overall situation for intergovernmental scientific collaboration in Europe (chapters 3 and 5). Given that the ESS project had imploded politically in 2002–2003, and given that there is no "standard cookbook" for European collaborative Big Science (in Yeck's words, see chapter 8), in order to stop the ESS from becoming reality, European politicians simply had to do nothing. This created a specific dynamic – the window of opportunity for smaller countries was such that ESS Scandinavia and its competitors did not have to fight unequal rivals in the shape of France and Germany, but instead keep the idea of an ESS going, build momentum, and eventually present a site candidate with favorable properties. The ESS Scandinavia marketing campaign did not have to compete with, say, the formidable track record of achievement in neutron instrumentation and neutron scattering of the Jülich Research Centre, but could use whatever selling points it wanted and market Lund and the Øresund Region as a generic dynamic region with no particular strength in neutrons but all kinds of other (real or unreal) advantages. But more importantly, Allan Larsson's negotiation strategy of taking small steps and eventually, unnoticeably, passing the point of no return for commitments to the Lund site candidature, worked because its only real competition was the prevalent idea that the ESS was not going to be built at all, and so governments and science administrators that bought into the ESS Scandinavia marketing campaign and played along in Larsson's negotiations, aligning themselves slowly with ESS Scandinavia, were mainly not choosing between different sites for the ESS, but between the idea that the ESS might actually become reality in some distant future, and the prevalent idea at the time, which most did not like but had accepted as a political reality, that the ESS was not going to be built at all. Choosing to slowly move closer to support for the Lund candidature must have appeared as a positive and favorable alternative, especially since it entailed no commitments of funding, or of anything but (moral) support.

What unifies many of the individual actors in this story, and the organizational and institutional arrangements they contributed to shaping and that worked in favor of ESS Scandinavia, is that they seem to have been in the right place at the right time. The ESS Scandinavia Initiative was born out of a Danish desire to have a major research facility located in Scandinavia, and surfed in more than one way on the wave of cross-border integration and optimism for the future that the Øresund Bridge, inaugurated in 2000, had created. Region Skåne, newly created and embodying much of the same optimism, swiftly became a core group of the ESS Scandinavia fanclub, together with Lund University. Börje Johansson and Karl-Fredrik Berggren had just the right devotion and energy left to make use of their reputed standings in Sweden to bring universities on board and continue the campaign when others fell off. Lars Börjesson undertook his own long march through the institutions of the Swedish Research Council and ESFRI, carrying the ESS Scandinavia cause with him and certainly promoting it whenever he had the chance. Project workers Aleksandar Matic, Patrik Carlsson, and Carina Wickberg played immensely important roles as foot soldiers in the campaign. At European level, Germany's decision not to prioritize the ESS opened a window of opportunity for smaller countries, and the UK failed to take a stand until it was too late and Lund had already been declared the designated ESS site. Allan Larsson, retiring from a long career in public administration and known for his vigor and fervor, came in at the best possible moment, when ESS Scandinavia was running out of steam and the Government was dragging its feet. Only a year after the publication of Larsson's very positive report came an electoral defeat for the tired and passive Social Democratic government, and the first majority government in 24 years took office, with a reform agenda and an ambitious and determined new Minister for Education. Leijonborg could, of course, have ended up as foreign minister, should his party have received just a tiny additional slice of the popular vote and become second-largest among the four, and then Peter Honeth would probably have stayed as director of administration at Lund University. Colin Carlile could have decided to go back to science or moved back to the UK after his tenure as ILL director was finishing, but came to Lund, as did Christian Vettier and Mats Lindroos, among many others. Several of these people defied the prevalent institutional logics of the Swedish and European research policy and funding systems, by mobilizing complementary or competing institutions for their cause. Another way of saying this is that the ESS became a reality in Sweden due not to the stability and predictability of known institutional logics in politics and science – if the established practices of the Swedish political system would have had its way, and if the prevalent way of promoting initiatives and pushing agendas in science would have been followed by the ESS Scandinavia campaign, it is fair to say that not much would probably have happened – but by tapping into other power bases and other institutions and, importantly, by being in the right place at the right time. The sacrifices made by not playing entirely by the book or working through established structures and institutions, such as the lack of any thorough evaluation of the scientific arguments for and against a major Swedish commitment to the ESS and a Swedish hosting of the facility, and the relatively unscientific marketing campaign and the risk that it would alienate parts of the scientific establishment, are perhaps a small price to pay for the overall success of the campaign.

It should be noted, before completely leaving the analytical theme of institutional logics and how individuals are restricted by them, mobilize them for their purposes, and move between them, that the processes that started after the 2009 site decision clearly followed (at least) three different institutional logics. First, the development of a technical design and scientific case, the creation of organizational structures and an international network of collaborators, and the work to establish a proper project organization. This involved science and engineering as well as corporate governance and project management. Second, negotiations on the European stage, where high-level politics and diplomacy obviously had their way. For the most part, we neither know or have the ability to systematically map out the logics that rule in politics, other than in a general and schematic sense, i.e. pointing out that power, persuasion, self-interest, and precedent have central roles. Third, the marketing campaign, which almost seems to have been detached from any physical, political, or scientific realness of the ESS, and instead lived its own life in the spheres of grand promises and expectations of scientific and technological achievements and vast dynamic growth effects. One central conclusion of this book is that the ESS became a reality in Lund, Sweden due to the reciprocal work of several different procedures, some of which are difficult or impossible to map out, but all of which operated in accordance with the rather distinct institutional logics that govern processes in different realms of society. Unsurprisingly, the institutional logics of politics and science were the most prominent, but industrial project management – essentially different from both politics and science, and from "Madison Avenue" marketing techniques (see chapter 8) – also played a crucial role, at least in the later stages. In some cases, science beat politics – the handling of the German request to build several instruments simply on the basis of the need to keep domestic institutes busy, is one example. In other cases, politics trumped science or project management – the location of the Data Management and Software Center (DMSC) in Copenhagen most certainly increased the price tag of the whole ESS and produced some administrative and technical challenges, but was politically necessary.

The role of the marketing campaign (chapter 10) is a little less clear. Perhaps it should be seen first and foremost as a natural consequence of the remarkable upswing in the past few decades of ideas and concepts of the knowledge society/economy, dynamic learning regions, and innovation-based sustainable growth, and of the relative lack of real physical and/or scientific progress on site in Lund at the time when the hype around the ESS – in the media and on the agendas of local and regional policymakers – was greatest. Regardless of the seeming lack of need or rational grounds for a marketing campaign – the resistance to the ESS was minimal, as noted at various points in this book – the fact that it took such a prominent place in the ESS Scandinavia campaign must be taken as proof of an apparent need, for some reason, for its existence.

The procedures and regulatory mechanisms, and the norm systems and behavioral patterns that have been given conceptual firmness in institutional theory, including institutional logics, are, of course, only part of the explanation for why things unfolded the way they did in the story of the ESS Scandinavia campaign. Many of the turns of events in the successful campaign also had a certain element of chance. Or at least, so it would seem. But this is where historical institutionalism can help bring about an understanding of the process that accounts for seemingly haphazard events, and places them in the proper context of long-term, gradual and cumulative change. In a personal reflection over the work of ESS Scandinavia in 2003, Lars Börjesson called the efforts to locate the ESS in Lund the work of "getting a snowball to grow to an avalanche" (Börjesson 2003) and in 2019, with the benefit of hindsight, he was able to reflect that this metaphor is quite fitting for what happened (Interview: Börjesson). One of two principal lessons from the historical-institutionalist view on a historical development like the ESS Scandinavia campaign is that seemingly insignificant events can indeed alter the course of history. Such seemingly insignificant events are, on the other hand, difficult to find in historical research precisely because of their insignificance, and once they are found it is still very risky to elevate them to positions of significance in the story, simply because their emergence in the empirical material (e.g. interviews) is due to a biased view of their significance for informants. This is a curse of historical research, and while there is little or no remedy, it is an important realization that serves as a reminder that no historical work, regardless of how extensive and detailed it is, can cover everything.

Another lesson from this story, better understood with the help of historical institutionalism, is the importance of *resilience* and *adaptability*. To a great degree, it was the resilience of the ESS Scandinavia campaign that eventually led to victory – no other site contender lasted all the way from the preparations during the run up to the 2002 Bonn meeting and to the May 2009 meetings in Prague and Brussels. Although its key champions and enthusiasts changed, the idea of an ESS in Lund remained. Historical institutionalism teaches us that long-term survival and success depends on long-term stability and adaptability, and the ESS Scandinavia campaign certainly displayed both over its decade of existence (until transformed into a company in 2010) and also beyond, if the ESS AB and European Spallation Source ERIC (which, at the time of writing, is concluding the construction project and moving into the operations phase) are viewed as extensions of the campaign or metamorphoses of the campaign organization. From a local and regional campaign, to national and government sponsored, to international negotiation and the design update phase, and finally to construction, ESS Scandinavia and its successors demonstrated remarkable stability and adaptability. If viewed as a continuum from 2000 to 2014/15, in spite of the changes in organizational structure, legal status, funding portfolio, top management, and workforce, the campaign was remarkably adaptive. Through several shifts in focus – from a Scandinavian seat at the table, to Scandinavian site candidature and preparations for the Bonn meeting, to national lobbying campaign and underdog role on the European stage, to government sponsored campaign and European negotiating effort, to project and design work and recruitment of an apt workforce, and finally to construction – it did not change the overall aim, and moreover, it eventually succeeded in achieving this aim. The transitions between these different efforts meant transitions from competition with other site candidates to a fight to keep the ESS project alive at all, from science lobby to high-level politics, from campaigning to real project management, from politics to technology and back, and so on. It meant a back-and-forth movement between regional, national, and international (European) levels, and between science, politics, and diplomacy. The ESS Scandinavia campaign adapted and survived. The achievement in this regard is also all the more impressive given the lack of precedent and lack of established procedures for collaborative Big Science in Europe, and for European Big Science in Sweden.

The Campaign

Returning, once again and finally, to the overarching research question for the historical study reported in this book – *How could a European Big Science facility end up in peripheral Lund, Sweden?* – we might just as well also return to the very simple answer to this question, given in the very first pages, namely: *Because of a successful campaign.* The promise in the introduction was that a longer answer would be provided, and hopefully, the reader is satisfied on that account. What remains is a final discussion that highlights and summarizes what the success amounts to, and how it was achieved.

In a preliminary analysis of the political processes that brought the ESS and MAX IV to Lund, Benner (2012: 160–161) suggested that research policy "takes place in two relatively disconnected streams": one which follows the logic of peer review and bottom-up deliberation and typically happens within research councils, other funders, and universities, and one which rather uses a top-down approach and takes "broad decisions which may not be reducible to expert evaluation." The contrast between the two, in terms of mechanisms for decision making and allocation of power and influence, is evident from the story told in this book. Politics entails a unique combination of persuasion and compromise, which also means that as part of political decision making, immensely complex issues are reduced to simple alternatives – often yes/no votes – and once one of them has won, regardless of the form of competition (debate and vote in parliament or committee, or in internal party deliberations; negotiations between parties of a coalition government or between ministries; general elections; referendums; international negotiations and deals), the decision is made and attention quickly shifts to the next issue. This is essentially the logic by which the ESS ended up in Lund – "broad decisions which may not be reducible to expert evaluation" – although of course the process stretched over several years and took a variety of forms. It deserves to be repeated, however, that there was never any thorough science policy evaluation and prioritization process that promoted the ESS as the most viable or favorable investment for Swedish science policy. Such a process had surely taken place on the European stage, from 1990 and on and with the meeting in Bonn in May 2002 as the end point, when it was envisioned that the scientific promotion process would stop and politics take over. Peter Tindemans declared, at the press conference in Bonn on May 16, 2002 that "the ESS is entirely feasible and necessary" (Carsughi and Clausen 2002), and he did so because a decade of work by the scientific communities of Europe, including a basic design and an agreed opinion that the ESS facility was exactly what they needed, backed him up. But the Bonn meeting failed to bring the ESS facility to a site decision, which shows exactly why the success of the process to bring the ESS to Lund had to be a political campaign, driven by politicians, apparently also outside the ordinary framework of science policymaking in Sweden. What, then, were its nuts and bolts, its key campaign assets and selling points, and the discursive framework that it drew on?

The most publicly visible parts of the campaign were perhaps not those that made the crucial difference in creating a favorable outcome, but the tricks of campaigning work not only on the public but also its elected representatives, whose willingness to make a mark on history certainly makes them susceptible to the promises and expectations of campaigning around a major investment in science, and willing to continue the campaigning efforts with the same tools. In the case of the ESS, there seems to be a challenge relating to the relative obscurity of the science of the project – the use of neutrons for the study of materials is a much tougher sell than the previous use of Big Science during the Cold War, to explore the origins of the universe (cf. Martin 2018) – which shows in the oversimplifications of the microscope analogy and the exaggerations of the links between the ESS and product improvements such as "smoother cream cheese" and solving grand challenges such as climate change (chapter 10). As has been shown in detailed studies of the role and function of neutron scattering facilities in science and innovation systems, and the nature of its contributions to scientific and technological development (Hallonsten 2016a, 2016b), it is difficult, to say the least, to associate a scientific breakthrough or a technological innovation directly with a specific facility, and perhaps even more difficult to predict, when the project is only at the idea stage, what types of science it will enable. Indeed, this is key to the whole idea of building it: The ESS was, even at the early planning stage, conceptualized as a neutron source that would improve the performance of experiments by several orders of magnitude, which would mean entirely new experiments in areas which were partly new, in other words unknown, at the time of design and political campaigning.

Probably in part due to this uncertainty, the bold statements made, and great expectations instilled, by the ESS campaigners (chapters 6 and 10) were not mainly about the ESS as a science facility that would make available a number of cutting-edge instruments for use at the very forefront of materials science and life science, but as a motor of progress that would turn Lund, the Øresund Region, and Sweden into a dynamic region and a knowledge economy hotspot in the global arena, competing with Silicon Valley, Route 128, Tsukuba Science City, Oxbridge, or Grenoble. The campaign was under the heavy influence of theories of regional development and economic geography immensely popular at the time. As long as ESS Scandinavia was a mere hope or dream of a number of enthusiasts in the Swedish and Danish condensed matter physics communities, it remained firmly anchored in the scientific traditions it springs from, and the main focus of the campaign was on science and technology. Once politicians entered the campaign, the message largely shifted to economic growth potential and the role of Big Science in creating a dynamic region. This probably made a huge difference, not least on the European stage, where the shift in focus and the taking over of the ESS Scandinavia campaign by skilled politicians was possibly what pushed the campaign over the edge and enabled the site decision in May 2009, in favor of Lund. Meanwhile, the political campaign had provoked an interest from local and regional politicians, and also set new standards inside the ESS Scandinavia organization. Once the site decision had been made (chapter 7), the tasks that remained were the long and hard work to conduct a design update and expand the ESS organization to prepare for start of construction and eventual operation (chapter 8), and negotiate a funding agreement between the seventeen prospective partner countries (chapter 9). At the beginning, when the ESS Scandinavia campaign organization was still in place and local and regional politicians, decision makers and industrialists, and newspapers were still caught up in the euphoria over the 2009 site decision, the campaign simply continued to produce marketing material and events – most of which were *pseudo events* – probably in order to try to make up for the fact that the ESS was still far from a done deal and materialization out on the fields of Brunnshög. The offices of the ESS on Stora Algatan in the center of Lund virtually turned into an advertising firm. Since the 1920s, the street that is squeezed in between Park Avenue and Fifth Avenue on Manhattan has been metonymous with the American advertising industry (Mayer 1991), and so "Madison Avenue techniques" is a colloquial name for "gimmicky, slick use of the communications media to play on emotions" (Safire 1993: 428), which explains – should the reader not have fallen for the temptation to use Google to find the explanation – the headline of the subsection of chapter 10 that deals with this specific part of the story. The campaign claims in these years – that the ESS would in and of itself create an unprecedented boost to the local and regional economy, and produce scientific results that would revolutionize industries and make significant contributions to solving grand challenges of sustainability and health – fit rather neatly with a dominating discourse in growth and development policy at the time, which centered on concepts like the knowledge-based economy, dynamic regional development, and innovation-based growth.

This discussion, and the analysis of the campaign in other parts of this concluding chapter, including the highlighting of the political skill of the key people in the campaign and the fair amount of opportunity that it entailed, should not be allowed to completely obscure the very genuine qualities that the Scandinavian proposal certainly had, and that obviously played an important role. It should, likewise, not be allowed to obscure the fact that the competition was weak. The unlikelihood of Lund as a candidate site for the ESS was neither greater nor smaller than the unlikelihood of Bilbao or Debrecen, and one key factor in the eventual victory for Lund was that no other formidable site candidate – i.e. in France, Germany or the UK – remained in the race. This is also a key reason for the success of the ESS Scandinavia campaign, and were it not so dull and uninviting, the following addendum could probably be made to the short answer of how the ESS could end up in Lund: *Because of weak competition.*

All this points the way to another discussion, with ramifications far greater than what can be contained in the chronicle and analysis that answer the research question in this book, and that picks up some loose ends in the discussion in this final chapter as well as throughout the book, namely, *what do we mean by success?* In a most straightforward sense, the ESS Scandinavia campaign was clearly a success: The ESS is currently being built outside Lund. But few things in science, politics, and society at large are that straightforward to assess the success or failure of – put simply and bluntly, using a typical campaign slogan, the world is not black and white. Neither are historical facts, although they function exemplarily as starting points for insightful discussions.

The May 28, 2009 site decision was the point in time when success for the ESS Scandinavia campaign was most clearly and strongly expressed, and it was certainly a key date in this whole story, for three interrelated reasons. First, because the decision at this point seems to have been whether the ESS was to be built at all; the larger countries with the most likely site proposals for the ESS had withdrawn, and the three remaining candidate sites were evaluated with a different yardstick than if the big players had remained in the race. Put differently, the distinct qualities of the Lund site became assets in their own right rather than compared with qualities that they could not match. Second, because it was a seemingly clear and unequivocal success for the campaign, celebrated and widely communicated, in media reports and by representatives of ESS Scandinavia and the Swedish Government. In retrospect, it is of course possible to pinpoint May 28, 2009 as the decisive moment for the ESS and for the ESS Scandinavia campaign, although as this book has shown, this is not entirely accurate. The project could have collapsed at many points in time in the years to come, and in one sense, it is a surprise that it did not, given the Euro crisis, the relative unpreparedness of the Swedish Government, and the initial lack of the proper competences in the essentially campaign-oriented ESS Scandinavia organization.

The third, and historically most interesting, reason for crowning May 28, 2009 as the key point in time in this whole story is because the event changed the dynamic of the campaign and the mode of operation of both the ESS organization in Lund, which went from ESS Scandinavia to simply ESS, and for the European communities around the ESS project. The campaign had to switch gears, "get serious" (Interview: Vettier; see chapter 8), and the transition is said to have taken up to two years. Here, it is interesting to ponder what success really means. In one sense, the success of the campaign in 2009 created a situation of almost unsurmountable challenges, which means that the success that can be measured by noting that the ESS facility is currently under construction outside Lund is a success that should probably be attributed to the hard work of 2009–2014, described in chapters 8 and 9, rather than the campaign in the prior years or the site decision in May 2009. This not only makes the success of the site decision for the ESS Scandinavia campaign less obviously clear, it also contradicts the simplified and concise answer to the research question as given in the introduction and above, namely that the ESS ended up in Lund because of a successful campaign. But both can be true at the same time, as not least the discussion informed by the theoretical framework of institutional logics in the previous section shows: Parallel processes follow different logics and thus measure success in different ways. But it also brings up the issue of how realistic the ESS campaign was, and whether it mattered that the campaign was to a significant degree based on a shallow and overly positive image of future benefits, with little or no contact with reality. In one perspective, if the end justified the means, it didn't matter: The campaign fulfilled its purpose: Lund won. Another perspective, which this book can only explore very briefly and speculatively (mainly for methodological reasons), is how successful the ESS will be, how successful Sweden will be in reaping the benefits of hosting the ESS, and whether the unrealistic elements of the campaign played a role in preventing politicians and decision makers from taking the issue of follow-up investments seriously and beginning the complementary work of preparing Sweden for the ESS.

What is more, if we stick to *scientific output* as the measure, the assessment of the success of facilities like the ESS runs into trouble. Not primarily because the ESS has yet to commence scientific operation and produce results that can be evaluated – one of the few certainties here is that it will produce science of some quality as soon as it begins operations – but because high-quality science itself is ambiguous in terms of relevance and success in a broader perspective. There can be no doubt that neutron facilities play important roles in science – if they would not be valuable resources for scientific experimentation, governments would not be investing billions of Euro in them – but the broader utility of neutron facilities for society is unpredictable and difficult to gain an overview of because it depends on their momentous use by scientists seeking access in competition. A theoretically relevant way of conceptualizing this is to point out the *functional differentiation* between facility and user community, which means that in a strict sense, the facilities themselves have no scientific output. The inherently dynamic and changing nature of the scientific use (chapter 3) makes it hard to trace the contributions of a neutron facility like the ESS to science;

while neutron facilities are often absolutely crucial tools in the scientific work they facilitate, they are resources used alongside other resources by scientists affiliated with universities and institutes elsewhere. This means that, leaving the idle-headedness of the claim aside, the idea that the future ESS *will contribute to* Nobel Prize-winning work is highly likely but difficult on the verge of impossible to prove. The laureates will, namely, have their affiliation elsewhere (most likely at a prestigious university in the United States, Germany, France, or the UK) and also have made use of a range of other instruments besides those at the ESS, possibly including other neutron facilities, or synchrotron radiation facilities, and most likely their own lab equipment. The ESS will, of course, be swift to claim that their instruments contributed to a Nobel Prize, but so will several other labs and facilities around the world (cf. Hallonsten 2016b: 491). Importantly, from the perspective of the campaign's most ubiquitously used selling points, unless universities and institutes in the close vicinity of the ESS build up world-leading capacity to use the ESS and its instruments, the Nobel Prizes will pass the Øresund Region and Scandinavia by. It matters little, in this regard, that the ESS is the world's most advanced neutron scattering facility and that it happens to be located outside Lund.

Similarly, in order for the future ESS facility to function in a local and regional innovation system and contribute to innovation and economic growth, other actors in the system need to be prepared and equipped to make use of the opportunities offered by the ESS and the knowledge and technology it develops. This includes *absorptive capacity* as well as the ability to make good use of the potential of the facility to form links between local actors and internationally leading innovation environments, become figurative magnets that attract talent and competence, and act as partner in innovation processes on local and regional level. The potential conflict of goals should be noted: The local, regional, and Swedish national mobilization to benefit from the ESS does not entirely cohere with the ambition to attract talent and competence from abroad, since the risk associated with the latter wish is that knowledge and technology from the ESS is extracted to benefit other countries and their innovation systems and markets, unless talent and competence is made to stay in the area and exploit their results and inventions through research and entrepreneurship that stays and grows in the local and national economies (Hallonsten 2016a: 197).

Any effort, either analytical or political (decision making), to make sense of all this and prepare for an embedment of the ESS in apt innovation systems, and society more generally, must of course begin with a rudimentary understanding of what the ESS is and what it does. Here, it is fair to issue a warning that most of the work to establish this understanding among policymakers and decision makers remains, and that the success of the campaign for a long time inadvertently prevented such learning from taking place, by making its oversimplified promises. Here follow, therefore, some instructions on what the politicians and bureaucrats responsible for the appropriate embedment of the ESS in the science and innovation systems of the Øresund Region and Sweden, and broader society, should begin to learn. It must, first, be understood that the ESS produces neutrons to enable a variety of advanced studies of materials (including biomaterials), and that there is a rudimentary but absolutely crucial *functional differentiation* between facility and users, or between facility and scientific community, and that this has consequences for the production and dissemination of results from the experiments that the facility enables. Moving on to the utilization of the results, and the transfer of technology and knowledge from the ESS to the science system and to wider society, it is necessary to establish an understanding of how these processes are typically structured, including knowledge about how scientific projects are contemporarily organized, how results are published, how business opportunities and other potential practical applications of results are identified and exploited, and how these are transformed into improvements in people's living conditions, including as widely disparate entities as better consumer electronics and means to achieve energy production and transportation of people and goods that is less harmful to the climate.

A basic orientation in the European science landscape, and specifically its use of neutrons, is also necessary, in order to understand the power relations and the dynamics of various user groups and their capacities. For member countries, an investment in the ESS is an investment in new and dramatically better experimental conditions for research with neutrons, and each country obviously also wants a return on investment, with or without scientific fair return. National communities of (potential) neutron users will, therefore, probably be supported in their efforts to build up capacity to make use of the dramatically better opportunities. Some countries are, naturally, better equipped than others to do so – the most advanced neutron user environments in Europe are still in France, Germany, and the UK – whereas others need to mobilize strongly to catch up and become competitive. It is not a far-fetched speculation that the relatively low contributions to the ESS construction costs from neutron stronghold countries like France, Germany, and the UK (none of them pays more than 11%) have given these countries some financial leeway to bolster the support of their scientific communities in their efforts to build up capacity to use the future ESS, and that their *in-kind* contributions, which give their scientific communities the ability to design and build the instruments that they will later compete to use, also helps in strengthening their already highly competitive positions.

From a Swedish perspective, and specifically concerning the ESS, it must be known that simply investing in the infrastructure does not secure a scientific output that benefits Swedish science, innovation, and development. As discussed in a previous section, Sweden has a very small neutron user community, and a great challenge for Swedish research policy in the coming years will be to build up competence of the scale and quality necessary to make good use of its share of the total experimental time at the ESS. If it does not master this, and enables Swedish scientific skill in neutron use to grow to a point where it can realistically be competitive in the allocation of experimental time at the facility, the ESS will be located in Lund but used by everyone else, and the success of the campaign will pale and eventually appear as little more than a *pseudo event*.

Returning, again, to the campaign's rhetoric, it was noted in chapter 10 that the campaign combined *uncertainty* as something genuinely and inherently positive in the realm of the future scientific breakthroughs at the ESS, and *certainty* concerning the broad socio-economic benefit of the facility for the region, Sweden, and the world; the science of the ESS was described as "incomprehensible and yet predictable." Harro van Lente (1993: 10–11) has argued that such campaign rhetoric is significative for the current science policy regime, whose focus on strategic investment and priority has elevated expectations and promises to a new level, and increased their use in the promotion of certain projects. As the need for strategic priority grows, projects will have to be promoted and marketed in order to receive priority over other projects, and they can only be considered *strategic* if there is some kind of expectation that they will have beneficial utility. Which, in turn, means that if the utility lies several years ahead and is shrouded in uncertainty, it must be part of the campaign, otherwise the project will not be viewed as strategically important and thus not worth prioritizing. Other analysts have reasoned similarly, pointing out that expectations are what single out strategic research from the purely basic and applied research (Irvine and Martin 1984: 3–5; Stokes 1997), and there is also a connection to be made to "economization" and "commodification" of research (chapter 2), which seem to promote strong beliefs in (over)simplified measures of productivity, excellence, and relevance of research, which in turn further bolsters the importance of expectations and promises. Put differently: A (perceived) harsh prioritization between areas of policy and public expenditure, and within the area of research policy but between projects, strengthens the role of promises and expectations of (preferably measurable) results and benefits, and the need for proponents of projects to use expectations and promises of future impact in advertisement material. The ESS campaign is a bona fide example of a Big Science project in this current science policy regime. The end justified the means, and the largely unscientific campaign to locate the ESS in Lund made perfect sense in the institutional context where it took place.

Everyone said it wouldn't work.
Then along came someone who didn't know that,
and just did it.

List of abbreviations

AFC	Administration and Finance Committee
AFR	*Atomforskningsrådet*, Swedish Research Council for Atomic Affairs
CCLRC	Council for the Central Laboratory of the Research Councils
CDR	Conceptual Design Report
CEA	*Commissariat à l'énergie atomique*, French Commission for Atomic Energy
CERN	European Organization for Nuclear Research (originally the European Council for Nuclear Research, *Conseil Européen pour la Recherche Nucléaire*)
DANSSK	Danish Neutron Scattering Society
DESY	*Deutsches Elektronen-Synchrotron*, German Electron Synchrotron
DMSC	Data Management and Software Center
DOE	Department of Energy
DTU	*Danmarks Tekniske Universitet*, Technical University of Denmark
EC	European Communities
ECNS	European Conference on Neutron Scattering
ERA	European Research Area
ERIC	European Research Infrastructure Consortium
ESFRI	European Strategy Forum on Research Infrastructures
ESO	European Southern Observatory
ESRF	European Synchrotron Radiation Facility
ESS	European Spallation Source
EU	European Union
EWESS	ESFRI Working Group on ESS
IKON	in kind instrumentation
ILL	*Institute Laue Langevin*
ITPS	*Institutet för Tillväxtpolitiska Studier*, Institute for Growth Policy Studies
J-Parc	Japan Proton Accelerator Research Complex
KAW	Knut and Alice Wallenberg Foundation
KFI	*Kommittén för Forskningens Infrastrukturer*, Committee for Research Infrastructures

LHC	Large Hadron Collider
LNF	*Lundabygdens Naturskyddsförening*, Society for Nature Conservation in the Lund District
LoI	Letter of Intent
MoU	Memorandum of Understanding
MP	Member of Parliament
OECD	Organization for Economic Cooperation and Development
PPP	Public Private Partnership
PSI	Paul Scherrer Institute
RAL	Rutherford Appleton Laboratory
RFI	*Rådet för Forskningens Infrastrukturer*, Council for Research Infrastructures
SAC	Scientific Advisory Committee
SciLifeLab	Science for Life Laboratory
SNS	Spallation Neutron Source
SNSS	Swedish Neutron Scattering Society
SRG	Site Review Group
SSNC	Swedish Society for Nature Conservation, *Naturskyddsföreningen*
STAP	Science and Technical Advisory Panel
STU	*Styrelsen för Teknisk Utveckling*, Board of Technical Development
TAC	Technical Advisory Committee
TDR	Technical Design Report
TITA	*Tillväxt, Innovation, Tillgänglighet, Attraktion*, Growth, Innovation, Availability, Attractiveness
VAT	Value-Added Tax
Vinnova	*Verket för Innovationssystem*, Swedish Agency for Innovation Systems
XFEL	European X-ray Free Electron Laser

Acknowledgements

The author would like to thank Harald Lindström, Aleksandar Matic, Karl-Fredrik Berggren, Pia Kinhult, Mats Lindroos, Peter Honeth, Börje Johansson, Johan Holmberg, Mats Johnsson, Lars Börjesson, Peter Tindemans, Sven Landelius, Carina Wickberg, Jim Yeck, Fredrik Melander, Lasse Melin, Katharina Cramer, Marianne Thormälen, David Lindberg, Marion Söderström, MarieLouise Samuelsson, Lina Rönndahl, Ulf Maunsbach, Axel Hilling, Anders Granberg, Wilhelm Agrell, Mikael Eriksson, Christian Vettier, Gunnar Öquist, Dimitri Argyriou, and two anonymous reviewers who read and commented on previous versions of the manuscript – who all, in their own way, have contributed to the finalization of this book.

The research project that produced this book was funded by the Swedish Research Council (grant no. 421-2012-519). The finalization of this book was made possible by funding from the Office of the Vice-Chancellor of Lund University.

References

Interviews, in alphabetical order (only a selection of relevant affiliations listed)

Argyriou, Dimitri: Member of the ESS Science Advisory Group 2009–2010, Chair of the ESS Science Advisory Committee 2010, ESS Science Director 2011–2015. *Lund November 20, 2018.*

Berggren, Karl-Fredrik: Project Director for ESS Scandinavia 2003–2007, Professor of Theoretical Physics, Linköping University 1981–. *Stockholm August 12, 2015.*

Börjesson, Lars: Chairman of the Swedish Neutron Scattering Association 1995–2002, founding member of the ESS Scandinavia Initiative, secretary general of the Swedish Research Council's Committee/Council for Research Infrastructures 2004–2010, Swedish delegate to ESFRI 2002–2010, chairman of the ESS steering committee 2011–2015; chairman of the ESS council, 2015–2019. *Göteborg September 3, 2015 and Lund February 5, 2019.*

Carlile, Colin: Director of the ESS Scandinavia Project Office at Lund University 2007–2010, CEO of ESS AB 2010–2013, Director of ILL 2001–2006. *Lund April 7, 2010.*

Eliasson, Kerstin: State Secretary at the Ministry of Education, 2004–2006. *Stockholm March 10, 2016.*

Flodström, Anders: Professor of Materials Physics, KTH Royal Institute of Technology Stockholm 1985–. *Stockholm March 10, 2016.*

Holmberg, Johan: Research Administrator, Swedish Research Council. *Stockholm August 19, 2015.*

Honeth, Peter: Director of Administration, Lund University, 1990–2006, State Secretary at the Ministry of Education, 2006–2014. *Lund October 5, 2015.*

Johansson, Börje: Professor of Condensed Matter Theory, Uppsala University, 1994–, founding member of ESS Scandinavia. *Uppsala November 9, 2015.*

Johnsson, Mats: Research Administrator, Swedish Ministry of Education. *Stockholm September 28, 2015.*

Karlsson, Per: Research Administrator, Swedish Research Council. *On telephone September 1, 2015.*

Kinhult, Pia: Chair of the Region Skåne Executive Committee 2010–2014, Strategic Advisor to the ESS 2014–. *Lund November 13, 2018.*

Landelius, Sven: Chairman of ESS AB, 2010–2015. *Lund February 6, 2019.*

Larsson, Allan: Special Negotiator for ESS for the Swedish Government, 2004–2005, Chief Negotiator for ESS for the Swedish Government 2007–2009. *Stockholm October 30, 2015.*

Leijonborg, Lars: Swedish Minister for Education 2006–2007, Swedish Minister for Higher Education and Research 2007–2009, Chief Negotiator for ESS for the Swedish Government 2013–2014. *Stockholm September 14, 2015.*

Lindroos, Mats: Head of the ESS Accelerator Division 2009–. *Lund November 29, 2018.*

Lindström, Harald: Chief financial officer, Region Skåne, 2000–2009, Director of ownership issues and strategic investments, Region Skåne, 2010–2016. *Lund November 13, 2018 and Malmö January 29, 2019.*

Matic, Aleksandar: Professor of Physics, Chalmers University of Technology Gothenburg 2010–, Project manager ESS Scandinavia 2000–2002, Chairman of the ESS Science Advisory Committee 2011–2014. *Göteborg December 10, 2018.*

Möller, Kjell: ESS Programme Director 2011–2012. *Lund February 5, 2019.*

Öquist, Gunnar: Professor of plant physiology, Umeå University 1981–, secretary of the Royal Swedish Academy of Sciences 2003–2010. *Umeå January 11, 2017.*

Tindemans, Peter: Chairman of the OECD Megascience Forum 1992–1999, chairman of the ESS R&D Council 1999–2003, chairman of the ESS Initiative 2003–2007. *Amsterdam December 11, 2015.*

Vettier, Christian: ESS Science Director 2007–2011. *Lund April 7, 2010 and Lund November 20, 2018.*

Weeks, Allen: ESS Head of Communications 2012–2017, ESS Head of in-kind 2015–2017. *Lund September 7, 2015.*

Wickberg, Carina: Project coordinator ESS Scandinavia 2002–2007. *Lund October 5, 2015.*

Yeck, James: CEO of ESS AB 2013–2016. *Lund August 25, 2015.*

Printed and online sources

Action Skåne Environment (2005) Referral response to Allan Larsson's investigation (Ds 2005:40), October 10, 2005. Swedish government registry U2005/5686/F.

Administrative Court in Malmö (2017) Judgment in case no 6219-16. May 19, 2017.

ÅF (2004) Ekonomisk analys gällande elförsörjningen av European Spallation Source byggd i Lund, Öresundsregionen. Rapport nr SR-ESS 040107.

Allians för Sverige (2006) *Alliansens valmanifest 2006.*

Berggren K-F (2004) Letter from Karl-Fredrik Berggren to Erna Möller of the Knut and Alice Wallenberg Foundation. August 31, 2004. From the personal archive of Karl-Fredrik Berggren.

Berggren K-F (2006) Letter from Karl-Fredrik Berggren to Christian Vettier. December 14, 2006. From the personal archive of Karl-Fredrik Berggren.

Bernhardsson B et al. (2004) Motion till riksdagen 2004/05:N278 "ESS-anläggningen till Lund och Sverige". September 9, 2004. www.riksdagen.se.

Bernhardsson B et al. (2005) Motion till riksdagen 2005/06:N420 "Lokalisering av ESS-anläggningen". September 27, 2005. www.riksdagen.se.

Bernhardsson B et al. (2007) Motion till riksdagen 2007/08:Ub504 "ESS och MAX IV". September 26, 2007. www.riksdagen.se.

Bernhardsson B et al. (2008) Motion till riksdagen 2008/09:Ub566 "Stöd till ESS och MAX IV". October 1, 2008. www.riksdagen.se.

Bexell G (2006) Letter from Göran Bexell to Colin Carlile. June 8, 2006. From the personal archive of Karl-Fredrik Berggren.

Bexell G (2007a) Letter from Göran Bexell to Ferenc Mezei. January 22, 2007. From the personal archive of Karl-Fredrik Berggren.

Bexell G (2007b) Letter from Göran Bexell to Christian Vettier. January 22, 2007. From the personal archive of Karl-Fredrik Berggren.
Carlile C (2006) Letter from Colin Carlile to Göran Bexell. July 8, 2006. From the personal archive of Karl-Fredrik Berggren.
CEA/CNRS/VR (2010) Cooperation Agreement for the construction of neutron instrumentation. December 13, 2010. Swedish Research Council official registry 813-2010-7489.
CERN (1965) Annual Report 1964. European Organization for Nuclear Research, Geneva.
CERN (1977) Annual Report 1976. European Organization for Nuclear Research, Geneva.
Christophersen H (2004) Letters from Henning Christophersen to Allan Larsson, Anders Fogh Rasmussen, Bendt Bendtsen, Helge Sander, Flemming Hansen, Jørgen Olsen, and Ole Krog. 13 July 2004. From the personal archive of Karl-Fredrik Berggren.
Czech Republic (2014) Letter from Tomáš Hruda, Czech Vice Minister for Education, Youth and Sports, to James Yeck, Director-General of the European Spallation Source. February 13, 2014. Swedish government registry 2014/1398/F.
Denmark (2008) Draft agreement of a Swedish-Danish MoU on establishing ESS. October 13, 2008. Received through personal communication with Mats Johnsson at the Swedish Ministry of Education, 2015.
Denmark (2009) Fælleserklæring (Joint Declaration) om European Spallation Source (ESS), between the Danish and Swedish ministers of education and science. April 3, 2009. Swedish government registry U2009/447/F.
Denmark (2014) Letter from Sofie Carsten Nielsen, Danish Minister for Science and Higher Education to Jan Björklund, Swedish Minister for Education. July 6, 2014. Swedish government registry 2014/1398/F.
Department of Energy (2019) User Statistics Data Archive. Visited January 13, 2019. https://science.energy.gov/user-facilities/user-statistics/data-archive/.
DESY/ESS (2009) Letter of Intent between Deutsches Elektronen-Synchotron – Forschungszentrum der Helmholtz-Gemeinschaft, Germany and The European Spallation Source Scandinavia of Lund University, Sweden concerning collaboration on matters related to large-scale scientific infrastructures. April 14, 2009. Swedish government registry U2009/447/F.
ENSA (1998) *The European Spallation Source. A Statement by the European Neutron Scattering Association (ENSA)*, March 1998.
Eriksson E (2008) *Overview of safety and licensing aspects of the European Spallation Source (ESS), for a site in Lund area.* Studsvik Nuclear AB STUDSVIK/N-08/210.
ESFRI (2003) *Medium to Long-Term Future Scenarios for Neutron-Based Science in Europe.* Working Group on Neutron Facilities, European Strategy Forum on Research Infrastructures, 2003.
ESFRI (2005) *Towards new Research Infrastructures for Europe: The ESFRI "List of Opportunities".* European Strategy Forum on Research Infrastructures, 2005.
ESFRI (2006) *Roadmap 2006.*
ESFRI (2008) *Roadmap 2008.*
ESFRI (2009) Draft Report of the ESFRI Working Group on Site Issues SIG on How to identify, to compare and to choose the best sites for pan-European research infrastructures in proposing the relevant criteria and optimized procedures. European Strategy Forum on Research Infrastructures, 2009.
ESS (2010a) *European Spallation Source Activity Report 2009–2010.* ESS AB.
ESS (2010b) Interim Report for Third Quarter 2010, European Spallation Source ESS AB.
ESS (2010c) Fourth Quarter/Year End Report 2010, European Spallation Source ESS AB.

ESS (2010d) Articles of association, European Spallation Source ESS AB. April 14, 2010. Swedish government registry U2010/2226/F.
ESS (2011a) *European Spallation Source Activity Report 2010–2011*. ESS AB.
ESS (2011b) Interim Report for 2nd Quarter 2011, European Spallation Source ESS AB.
ESS (2011c) Interim Report for 3rd Quarter 2011, European Spallation Source ESS AB.
ESS (2011d) Årsredovisning med hållbarhetsredovisning 2010. ESS AB.
ESS (2011e) A Memorandum of Understanding on participation in the Design Update Phase and the intention to participate in the Construction and Operation of the European Spallation Source (ESS). February 3, 2011. Received through personal communication with Mats Johnsson at the Swedish Ministry of Education, 2015.
ESS (2011f) "ESS Partner Countries in first in-kind instrumentation meeting." Press release from the ESS, November 9, 2011.
ESS (2012a). *Conceptual Design Report*. ESS AB.
ESS (2012b) *European Spallation Source Activity Report 2011–2012*. ESS AB.
ESS (2012c) *European Spallation Source Annual Report 2011 with Sustainability Report*. ESS AB.
ESS (2012d) Interim Report for 1st Quarter 2012, European Spallation Source ESS AB.
ESS (2012e) ESS Project Review May 31st – June 1st 2012.
ESS (2012f) Bokslutskommuniké 2012 för Fjärde Kvartalet, European Spallation Source ESS AB.
ESS (2012g) "James H. Yeck to become new CEO of the ESS project." Press release from the ESS, October 4, 2012.
ESS (2013a) *European Spallation Source Annual Report 2012 with Sustainability Report*. ESS AB.
ESS (2013b) *European Spallation Source Technical Design Report*.
ESS (2013c) Delårsrapport Första Kvartalet 2013, European Spallation Source ESS AB.
ESS (2013d) ESS Energy Design Report. Outcomes from the collaboration between ESS, E.ON and Lunds Energi, 2011–2012.
ESS (2013e) Bokslutskommuniké 2013 för Fjärde Kvartalet, European Spallation Source ESS AB.
ESS (2013f) Call for Expressions of Interest. May 1, 2013. Received through personal communication with James Yeck.
ESS (2013g) "ESS Invites Expressions of Interest for the Construction Phase." Press release from the ESS, May 7, 2013.
ESS (2013h) "Bo Smith will be Denmark's Chief Negotiator for ESS." Press release from the ESS, June 4, 2013.
ESS (2013i) "Test piling on the ESS site." Press release from the ESS, November 21, 2013.
ESS (2014a) *European Spallation Source Annual Report 2013 with Sustainability Report*. ESS AB.
ESS (2014b) Delårsrapport 1 januari – 31 mars 2014, European Spallation Source ESS AB.
ESS (2014c) Delårsrapport 1 januari – 30 juni 2014, European Spallation Source ESS AB.
ESS (2014d) Delårsrapport 1 januari – 30 september 2014, European Spallation Source ESS AB.
ESS (2014e) "The Construction of ESS is Underway." Press release from the ESS, September 2, 2014.
ESS (2014f) Instrument Proposals 2014/2015. September 8, 2014. Received through personal communication with James Yeck.
ESS (2014g) "Foundation Stone Ceremony Marks Scientific Importance of ESS." Press release from the ESS, October 9, 2014.

ESS (2015a) *European Spallation Source Årsredovisning 2014.*
ESS (2015b) Statutes of the European Spallation Source ERIC. Annex to Commission Implementing Decision (EU) 2015/1478 of 19 August 2015 on setting up the European Spallation Source as a European Research Infrastructure Consortium (European Spallation Source ERIC). Consolidated Version.
ESS (2015c) "European Commission Establishes the European Spallation Source as a European Research Infrastructure Consortium (ERIC)." Press release from the ESS, August 25, 2015.
ESS (2018a) *European Spallation Source Activity Report 2017.*
ESS (2018b) Statutes of the European Spallation Source ERIC. Annex to Commission Implementing Decision (EU) 2015/1478 of 19 August 2015 on setting up the European Spallation Source as a European Research Infrastructure Consortium (European Spallation Source ERIC). Consolidated Version. With amendments up to 19 June 2018.
ESS Bilbao (2008) *Responses to the Questionnaire of the ESFRI Working Group on ESS Siting (EWESS).* April 25, 2008.
ESS Council (1997) *ESS. A next generation neutron source for Europe.* Volumes I–III.
ESS Hungary (2008) "Hungary for ESS." European Spallation Source project.
ESS Hungary (2009) "Unfair competition for hosting the European Spallation Source." Statement by the ESS Hungary National Council, June 8, 2009. From the personal archive of Karl-Fredrik Berggren.
ESS Initiative (2004a) Minutes of the first ESS-I meeting on October 11 at ILL in Grenoble, October 11, 2004. From the personal archive of Peter Tindemans.
ESS Initiative (2004b) Articles of Association of the ESS-I. October 11, 2004. From the personal archive of Peter Tindemans.
ESS Initiative (2007) Minutes of the 8th ESS-I meeting, November 13, 2007, Bilbao. From the personal archive of Peter Tindemans.
ESS R&D Council (1997) Minutes of the European Spallation Source 1st Meeting of the ESS R&D Council held in Rise, October 17, 1997. ESS/Ml/97. From the personal archive of Peter Tindemans.
ESS R&D Council (1998a) European Spallation Source ESS R&D Phase Memorandum of Understanding, May 1, 1998. From the personal archive of Peter Tindemans.
ESS R&D Council (1998b) Minutes of the European Spallation Source 2nd Meeting of the ESS R&D Council held in Paris, on May 19, 1998. ESS/Ml/98. From the personal archive of Peter Tindemans.
ESS R&D Council (1998c) Minutes of the European Spallation Source 3rd Meeting of the ESS R&D Council held in Madrid, on October 28, 1998, ESS1M2/98. From the personal archive of Peter Tindemans.
ESS R&D Council (1999) Minutes of the European Spallation Source 5th Meeting of the ESS R&D Council held at IRI, Delft, November 12, 1999. From the personal archive of Peter Tindemans.
ESS R&D Council (2000a) Minutes of the European Spallation Source Special Meeting of the ESS R&D Council, February 1, 2000, FZ Jülich, Germany. From the personal archive of Peter Tindemans.
ESS R&D Council (2000b) Memorandum of Extension of the 1998 ESS R&D Phase Memorandum of Understanding, as agreed by the ESS R&D Council in Berlin on May 24, 2000. From the personal archive of Peter Tindemans.
ESS R&D Council (2001a) Minutes of the Meeting of the ESS R&D Council, Abingdon, June 15, 2001. From the personal archive of Peter Tindemans.

ESS R&D Council (2001b) Guidelines on how to submit an expression of interest to host the European Spallation Source. November 28, 2001. From the personal archive of Peter Tindemans.
ESS R&D Council (2002a). *The ESS Project*, volumes I–IV.
ESS R&D Council (2002b) Minutes of the Meeting of the ESS R&D Council, CNR, Rome, January 21, 2002. From the personal archive of Peter Tindemans.
ESS R&D Council (2002c) Letter from ESS R&D Council chairman Peter Tindemans to the President of the Wissenschaftsrat M Einhäupl, October 25, 2002. From the personal archive of Peter Tindemans.
ESS R&D Council (2002d) Minutes of the Meeting of the ESS R&D Council, Frankfurt, October 26, 2002. From the personal archive of Peter Tindemans.
ESS R&D Council (2003a) Minutes of the Meeting of the ESS R&D Council, Zurich, January 22, 2003. From the personal archive of Peter Tindemans.
ESS R&D Council (2003b) Minutes of the Meeting of the ESS R&D Council, Leipzig, July 2, 2003. From the personal archive of Peter Tindemans.
ESS Scandinavia (2000) Minutes of the inauguration meeting of the ESS-Scandinavia Initiative, October 3, 2000. From the personal archive of Karl-Fredrik Berggren.
ESS Scandinavia (2002a) *Expression of Interest to host the European Spallation Source in Scandinavia.* ESS Scandinavia Consortium.
ESS Scandinavia (2002b) *European Spallation Source Scandinavia.* Brochure. ESS Scandinavia Consortium.
ESS Scandinavia (2002c) ESS-Scandinavia Consortium Agreement March 22, 2002–December 31, 2002. Dated March 22, 2002. From the personal archive of Karl-Fredrik Berggren.
ESS Scandinavia (2002d) Minutes from the meeting of the ESS-Scandinavia "Offentliga Skåne" ("Public Scania") Group, June 25, 2002. From the personal archive of Karl-Fredrik Berggren.
ESS Scandinavia (2002e) Minutes from the meeting of the ESS-Scandinavia "Offentliga Skåne" ("Public Scania") Group, May 30, 2002. From the personal archive of Karl-Fredrik Berggren.
ESS Scandinavia (2002f) Minutes from the ESS Scandinavia Consortium Council meeting September 19, 2002. From the personal archive of Karl-Fredrik Berggren.
ESS Scandinavia (2002g) Minutes from the ESS Scandinavia Consortium Executive Group Meeting September 19, 2002. From the personal archive of Karl-Fredrik Berggren.
ESS Scandinavia (2002h) Vetenskaplig strategi för ESS-Scandinavia. October 17, 2002. From the personal archive of Karl-Fredrik Berggren.
ESS Scandinavia (2002i) Minutes from the meeting of the ESS-Scandinavia "Offentliga Skåne" ("Public Scania") Group, November 12, 2002. From the personal archive of Karl-Fredrik Berggren.
ESS Scandinavia (2002k) Minutes from the ESS Scandinavia Consortium Executive Group Meeting, November 25, 2002. From the personal archive of Karl-Fredrik Berggren.
ESS Scandinavia (2002m) Background material for the meeting of the ESS-Scandinavia Consortium Council, December 4, 2002. From the personal archive of Karl-Fredrik Berggren.
ESS Scandinavia (2002n). ESS-Scandinavia Consortium Agreement January 1, 2003–December 31, 2003. Dated December 4, 2002. From the personal archive of Karl-Fredrik Berggren.
ESS Scandinavia (2002p) Minutes from the ESS Scandinavia Consortium Council meeting, December 14, 2002. From the personal archive of Karl-Fredrik Berggren.

ESS Scandinavia (2003a) Minutes from the ESS Scandinavia Consortium Executive Group Meeting January 7, 2003. From the personal archive of Karl-Fredrik Berggren.
ESS Scandinavia (2003b) Minutes from the ESS Scandinavia Consortium Executive Group Meeting February 3, 2003. From the personal archive of Karl-Fredrik Berggren.
ESS Scandinavia (2003c) Minutes from the meeting of the ESS-Scandinavia Science Group, February 5, 2003. From the personal archive of Karl-Fredrik Berggren.
ESS Scandinavia (2003d) Minutes from the ESS Scandinavia Consortium Executive Group Meeting March 3, 2003. From the personal archive of Karl-Fredrik Berggren.
ESS Scandinavia (2003e) Addendum to the Memorandum of Understanding to the ESS-Scandinavia Consortium Agreement. Dated March 11, 2003. From the personal archive of Karl-Fredrik Berggren.
ESS Scandinavia (2003f) Background material for the meeting of the ESS-Scandinavia Consortium Council, March 18, 2003. From the personal archive of Karl-Fredrik Berggren.
ESS Scandinavia (2003g) Minutes from the ESS Scandinavia Consortium Council meeting March 18, 2003. From the personal archive of Karl-Fredrik Berggren.
ESS Scandinavia (2003h) Minutes from the meeting of the ESS-Scandinavia "Offentliga Skåne" ("Public Scania") Group, March 19, 2003. From the personal archive of Karl-Fredrik Berggren.
ESS Scandinavia (2003i) Minutes from the ESS Scandinavia Consortium Executive Group Meeting April 22, 2003. From the personal archive of Karl-Fredrik Berggren.
ESS Scandinavia (2003k) Background material for the meeting of the ESS Scandinavia Science Group, May 13, 2003. From the personal archive of Karl-Fredrik Berggren.
ESS Scandinavia (2003m) Working material for ESS-Scandinavia. July 8, 2003. From the personal archive of Karl-Fredrik Berggren.
ESS Scandinavia (2003n) Background material for the meeting of the ESS-Scandinavia Consortium Council September 16, 2003. From the personal archive of Karl-Fredrik Berggren.
ESS Scandinavia (2003p) Background material for the meeting of the ESS-Scandinavia Science Group December 8, 2003. Dated December 5, 2003. From the personal archive of Karl-Fredrik Berggren.
ESS Scandinavia (2003q) Background material for the meeting of the ESS-Scandinavia Consortium Council December 8, 2003. From the personal archive of Karl-Fredrik Berggren.
ESS Scandinavia (2004a) Minutes from the ESS Scandinavia Consortium Council meeting, March 29, 2004. From the personal archive of Karl-Fredrik Berggren.
ESS Scandinavia (2004b) Minutes from the ESS Scandinavia Consortium Council meeting, August 30, 2004. From the personal archive of Karl-Fredrik Berggren.
ESS Scandinavia (2005a) Minutes from the ESS Scandinavia Consortium Council meeting, June 14, 2005. From the personal archive of Karl-Fredrik Berggren.
ESS Scandinavia (2005b) Minutes from the ESS Scandinavia Consortium Council meeting September 15, 2005. From the personal archive of Karl-Fredrik Berggren.
ESS Scandinavia (2005c) Arbetsmaterial: ESS-Scandinavia: Kommentarer till frågor kring ESS i remissvaren på rapporten "Svenskt värdskap för ESS". November 29, 2005. From the personal archive of Karl-Fredrik Berggren.
ESS Scandinavia (2005d) Minutes from the ESS Scandinavia Consortium Council meeting, December 6, 2005. From the personal archive of Karl-Fredrik Berggren.
ESS Scandinavia (2006a) Letter from Peter Honeth and Karl-Fredrik Berggren to the regional board, Region Skåne. February 3, 2006. From the personal archive of Karl-Fredrik Berggren.

ESS Scandinavia (2006b) Minutes from the ESS Scandinavia Consortium Council meeting March 6, 2006. From the personal archive of Karl-Fredrik Berggren.
ESS Scandinavia (2006c) Letter from ESS Scandinavia to Region Skåne, asking for financial support for 2007. November 7, 2006. From the personal archive of Karl-Fredrik Berggren.
ESS Scandinavia (2006d) Minutes from the ESS Scandinavia Consortium Executive Group Meeting November 29, 2006. From the personal archive of Karl-Fredrik Berggren.
ESS Scandinavia (2007a) Minutes from the ESS Scandinavia Consortium Council meeting June 25, 2007. From the personal archive of Karl-Fredrik Berggren.
ESS Scandinavia (2007b) Minutes from the ESS Scandinavia Consortium Council meeting September 12, 2007. From the personal archive of Karl-Fredrik Berggren.
ESS Scandinavia (2008a) *The ESS Scandinavia submission to the ESFRI Working Group on ESS siting*. ESS Scandinavia Consortium.
ESS Scandinavia (2008b) *ESS i Lund. ESS, säkerhet och miljö*. Brochure. ESS Scandinavia.
ESS Scandinavia (2008c) *ESS in Lund. The research of the future on your doorstep*. Brochure. ESS Scandinavia.
ESS Scandinavia (2008d) *ESS i Lund. Så kommer ESS att förbättra vår vardag*. Brochure. ESS Scandinavia.
ESS Scandinavia (2008e) *Neutrons for Health*. Brochure.
ESS Scandinavia (2008f) *European Spallation Source Scandinavia 2008*. Brochure.
ESS Scandinavia (2009a) *ESS Scandinavia Activity Report 2008–2009*.
ESS Scandinavia (2009b) *Neutrons for Science*. Brochure.
ESS Site Review Group (2008) *Assessment of the offers to host the European Spallation Source ESS in Lund, Debrecen or Bilbao*. Report from the ESS Site Review Group consisting of Catherine Cesarsky, Norbert Holtkamp, Thom Mason and Peter Tindemans.
Estonia (2008) Memorandum of Understanding between the Government of Sweden and the Ministry of Education and Research of Estonia concerning a Multinational Platform for The European Spallation Source (ESS). July 10, 2008. Swedish government registry U2009/447/F.
Estonia (2014) Letter from Jevgeni Ossinovski, Estonian Minister for Education and Research, to Jan Björklund, Swedish Minister for Education, and Sofie Carsten Nielsen, Danish Minister for Science and Higher Education. April 23, 2014. Swedish government registry 2014/1398/F.
European Commission (2008) Commission Staff Working Document, Accompanying document to the Proposal for a Council Regulation on the Community legal framework for a European Research Infrastructure (ERI). Impact Assessment. COM(2008) 467 final. July 25, 2008. www.europa.eu.
European Commission (2017) Project information: The European Spallation Neutron Source (ESS). Grant agreement ID: 202247. May 29, 2017. https://cordis.europa.eu/project/rcn/88501/factsheet/en.
European Council (2009) Council Regulation No 723/2009 of 25 June 2009 on the Community legal framework for a European Research Infrastructure Consortium (ERIC) [2009] OJ L 206/1, as later amended by Council Regulation (EC) No 1261/2013 of 2 December 2013 [2013] OJ L 326/1. www.europa.eu.
European Union (2012) *Consolidated Version of the Treaty of the Functioning of the European Union*. European Union Official Journal C 326. October 26, 2012. www.europa.eu.
European XFEL (2009) Convention concerning the Construction and Operation of a European X-Ray Free-Electron Laser Facility. November 30, 2009. https://www.xfel.

eu/sites/sites_custom/site_xfel/content/e35152/e49726/e35158/e56713/e60847/xfel_file60848/OTRIS_Convention_AoA_FinalAct_Esigned091130_eng.pdf.

Feidenhans'l, R (2003) *Large Scale Facilities for Synchrotron Radiation and Neutrons – New Possibilities for Denmark.* Risø report MEI-DK-4122.

Flodström A (2009) "Arguments for and financing of MAX IV." Swedish Government official registry U/2009/4845/F.

Folkpartiet (2003) Folkpartiets partiprogram (1997; reviderat 1999, 2001 och 2003)

Folkpartiet (2006) "Folkpartiets valmanifest 2006".

Forskningsstyrelsen (2005) *Fremtidens forskningsinfrastruktur – kortlægning af behov og forslag til strategi.* Report.

France (2009) Declaration of Intention between the Minister for Higher Education and Research of the Government of the Kingdom of Sweden and the Minister for Higher Education and Research of the Government of the French Republic concerning Collaboration in Research and Innovation. October 16, 2009. Swedish government registry U2009/447/F.

France (2010) Declaration of Intention between the Minister for Higher Education and Research of the Kingdom of Sweden and the Minister for Higher Education and Research of the French Republic. September 1, 2010. Swedish government registry U2009/3987/F.

France (2014) Letter of Intent concerning contribution to the European Spallation Source (ESS), from Geneviève Fioraso, French Minister for Higher Education, to Helene Hellmark Knutsson, Swedish Minister for Education. December 3, 2014. Swedish government registry 2014/1398/F.

Germany (2009a) Memorandum of Understanding between the Government of the Kingdom of Sweden and the Federal Ministry of Education and Research of the Federal Republic of Germany: Cooperation in materials research and structural biology using neutron and synchrotron radiation. June 6, 2009. Swedish government registry U2009/447/F.

Germany (2009b) Addition to the Memorandum of Understanding concerning the Cooperation in materials research and structural biology using neutron and synchrotron radiation. September 24, 2009. Swedish government registry U2009/447/F.

Germany (2014) Letter from Dr. Georg Schütte, German State Secretary at the Federal Ministry of Education and Research, to Peter Honeth, Swedish State Secretary of the Ministry of Education and Research, and Uffe Toudal Pedersen, Danish State Secretary of the Ministry of Higher Education and Science. July 4, 2014. Swedish government registry 2014/1398/F.

Granlund M et al. (2008) Motion till riksdagen 2008/09:Ub6 "med anledning av prop. 2008/09:50 Ett lyft för forskning och innovation". November 14, 2008. www.riksdagen.se.

Hallonsten O, Holmberg G and M Benner (2004) *Impacts of Large-Scale Facilities – a Socio-economic analysis.* Research Policy Institute Discussion paper 1/2004.

Helsingborgs Dagblad (2001a) "Storsatsning på forskning i NV-Skåne?" December 6, 2001.

Helsingborgs Dagblad (2001b) "Forskning för miljarder till F10?" December 12, 2001.

Helsingborgs Dagblad (2002) "Hård kamp om ny forskningsbas." January 23, 2002.

Holm U et al. (2005) Motion till riksdagen 2005/06:N470 "ESS i Lund". October 5, 2005. www.riksdagen.se.

Holm U et al. (2006) Motion till riksdagen 2006/07:N258 "Etablering av en ESS-anläggning i Lund". October 29, 2006. www.riksdagen.se.

Hungary (2014) Letter from János Lázár, Minister in the Prime Minister's Office of Hungary, to Jan Björklund, Swedish Minister for Education. July 23, 2014. Swedish government registry 2014/1398/F.

ILL (2006) *ILL Annual Report 2005*. Institut Laue-Langevin, Grenoble.

ILL (2009) *ILL Annual Report 2008*. Institut Laue-Langevin, Grenoble.

ILL (2017) *ILL Annual Report 2016*. Institut Laue-Langevin, Grenoble.

ILL (2018) *ILL Annual Report 2017*. Institut Laue-Langevin, Grenoble.

Italy (2009) Joint Declaration of Intention for Collaboration in Research and Innovation. The Minister for Higher Education and Research of the Kingdom of Sweden and the Minister for Education, University and Research of the Italian Republic. March 25, 2009. Swedish government registry U2009/3987/F.

Italy (2014) Letter from Stefania Giannini, Italian Minister for Education, Universities and Research, to Jan Björklund, Swedish Minister for Education. July 27, 2014. Swedish government registry 2014/1398/F.

Jönsson B (2005) Referral response to Allan Larsson's investigation (Ds 2005:40), October 12, 2005. Swedish government registry U2005/5686/F.

Jönsson C et al. (2006) Motion till riksdagen 2006/07:N262 "European Spallation Source (ESS)-projektet". October 27, 2006. www.riksdagen.se.

Jönsson C et al. (2007) Motion till riksdagen 2007/08:N298 "ESS-projektet". October 2, 2007. www.riksdagen.se.

Jönsson C et al. (2008) Motion till riksdagen 2008/09:Ub457 "ESS och Max-lab". October 2, 2008. www.riksdagen.se.

Karolinska Institute (2005) Referral response to Allan Larsson's investigation (Ds 2005:40), October 4, 2005. Swedish government registry U2005/5686/F.

Kinhult P, Helmfrid M, Reepalu I and P Danielsson (2008) "'ESS ger regional tillväxt.'" Opinion letter, *Sydsvenskan*, October 29, 2008.

Larsson A (2005a) *Svenskt Värdskap för ESS*. Ds 2005:40.

Larsson A (2005b) Bakgrundsinformation om ESS-projektet inför nordiska forskningsministrarnas möte den 5 april 2005. Dated March 29, 2005. Registry of the Nordic Council of Ministers MR-U 04/05.

Latvia (2009) Memorandum of Understanding between the Government of the Kingdom of Sweden and the Government of the Republic of Latvia Concerning a Multinational Platform for The European Spallation Source. January 27, 2009. Swedish government registry U2009/447/F.

Lefmann K and E Buhl Naver (2018) *Nordic and Baltic Neutron Scattering Communities, 2006–2017 – a bibliometric study*. Niels Bohr Institute, University of Copenhagen. https://nnsp.nbi.ku.dk/the-community/Nordic-Baltic-Neutron-users-2017.pdf.

Lindblad L (2003) Motion till riksdagen 2003/04:N371 "European Spallation Space". October 7, 2003. www.riksdagen.se.

Lindblad L (2005) Motion till riksdagen 2005/06:N312 "Forskningsanläggning för avancerad materialanalys". September 30, 2005. www.riksdagen.se.

Lindblad L et al. (2005) Motion till riksdagen 2005/06:N474 "Skåne som en framtidsregion". October 3, 2005. www.riksdagen.se.

Lithuania (2009) Memorandum of Understanding between the Government of Sweden and the Ministry of Education and Science of the Republic of Lithuania concerning a Multinational Platform for the European Spallation Source (ESS). January 22, 2009. Swedish government registry U2009/447/F.

Lund University (2005) Referral response to Allan Larsson's investigation (Ds 2005:40), October 18, 2005. Swedish government registry U2005/5686/F.

Lund University (2006) Viljeförklaring beträffande upplåtelse av mark. June 7, 2006. From the personal archive of Karl-Fredrik Berggren.
Lund University (2007) Inrättande av samt föreskrifter för Sekretariatet for ESS vid Lunds universitet. September 6, 2007. Lund University registry PLAN 2007/73.
Lund University (2009) Avsiktsförklaring avseende etablering av MAX IV. April 27, 2009. Lund University registry LS 2009/431.
Lund University (2012) Rådgivningsuppdrag Styrning och ledning av MAX IV-projektet. July 6, 2012. Lund University registry LS 2012/349.
Lund University (2013) *Årsredovisning Lunds universitet 2012.*
Lund University (2014) *Årsredovisning Lunds universitet 2013.*
Melin L (2016) Begäran om laglighetsprövning av regionfullmäktiges beslut 2016-06-20, ärende 15, om "Medfinansiering European Spallation Source (ESS)". June 21, 2016. Administrative Court in Malmö, case no 6219-16, file appendix 1.
Mezei F (2007) Letter from Ferenc Mezei to Göran Bexell. February 24, 2007. From the personal archive of Karl-Fredrik Berggren.
Müchler S and J Klarskov (2008) "Släpp in Danmark i ett samarbete kring ESS." Opinion letter, *Sydsvenskan,* October 4, 2008.
Müchler S and M Rankka (2014) "Handelskammarnas vd:ar: Därför är ESS en utmärkt affär." Opinion letter, *Sydsvenskan,* April 1, 2014.
Nature (2002a) "Strong case for neutrons." Letter to the editor. July 25, 2002.
Nature (2002b) "Row over neutron source hots up as Germany's advisers cry foul." August 1, 2002.
Nature (2009) "Sweden likely winner for neutron source." June 4, 2009.
Nordic Council of Ministers (2005) Godkänd Referat från Nordiska Ministerrådet (Utdanings- och Forskningsministrene) första möte 2005. København. April 5, 2005. Registry of the Nordic Council of Ministers MR-U 1/05.
Nordic Investment Bank (2003) Letter from Carl Löwenhielm, Executive President of the Nordic Investment Bank and Lars Noren, Vice President of the Nordic Investment Bank, to ESS Scandinavia, May 13, 2003. From the personal archive of Karl-Fredrik Berggren.
Norway (2014) Letter from Torbjørn Røe Isaksen, Norwegian Minister for Education and Research, to Jan Björklund, Swedish Minister for Education, and Sofie Carsten Nielsen, Danish Minister for Science and Higher Education. June 13, 2014. Swedish government registry 2014/1398/F.
Nylander C et al. (2004) Motion 2004/05:N403 "Allians för Skåne". October 4, 2004. www.riksdagen.se.
Olander R (2007) Motion till riksdagen 2007/08:Ub540 "Lokalisering av ESS-anläggningen". October 10, 2007. www.riksdagen.se.
Olander R et al. (2006) Motion till riksdagen 2006/07:N275 "Lokalisering av ESS-anläggningen". October 24, 2006. www.riksdagen.se.
Olofsson M and A Carlgren (2011) "Alliansens energiuppgörelse trygg grund att stå på." Opinion letter, *Svenska Dagbladet,* March 20, 2011.
Olsson M and B af Kleen (2011) "Gyllene affärer för friherren." Opinion letter, *Sydsvenskan,* March 31, 2011.
Pagrotsky L (2005a) Svar på skriftlig fråga 2004/05:1947, "ESS till Skåne". July 1, 2005. www.riksdagen.se.
Pagrotsky L (2005b) Svar på skriftlig fråga 2004/05:2049, "Placering av European Spallation Source". August 19, 2005. www.riksdagen.se.
Pagrotsky L (2006) Svar på skriftlig fråga 2005/06:1985, "den europeiska ESS-anläggningen" och 2005/06:1990, "om forskningsanläggningen ESS". August 22, 2006. www.riksdagen.se.

People's Campaign Against Nuclear Energy and Nuclear Weapons (2005) Referral response to Allan Larsson's investigation (Ds 2005:40), October 12, 2005. Swedish government registry U2005/5686/F.

Persson K (2011) Interpellation 2010/11:307 "Svenskt stöd till kärnkraftsforskning i Frankrike". April 6, 2011. www.riksdagen.se.

Poland (2008) Memorandum of Understanding between the Government of the Kingdom of Sweden and the Government of the Republic of Poland Concerning Polish support to site The European Spallation Source (ESS) in Lund, Sweden and Swedish support to site the Governing Board of the European Institute of innovation and Tedmology (EIT) in Wrodaw, Poland. May 23, 2008. Swedish government registry U2009/447/F.

Poland (2014) Letter from Marck Ratajczak, Polish State Secretary of Education and Research, to Jan Björklund, Swedish Minister for Education. July 25, 2014. Swedish government registry 2014/1398/F.

Region Skåne (2009a) *The ESS in Lund – its effects on regional development.*

Region Skåne (2009b) *The ESS in Lund – its effects on regional development* (Abridged version)

Region Skåne (2009c) Meeting minutes of the Regional Executive Committe of Skåne, 15 December 2009. December 15, 2009. Region Skåne registry 0902713.

Region Skåne (2010a) Markåtkomst för ESS-projektet. Memorandum. February 11, 2010. Region Skåne registry 0902713.

Region Skåne (2010b) Meeting minutes of the Regional Council of Skåne, March 2, 2010. Region Skåne registry 0902713.

Region Skåne (2010c) Köpekontrakt. Appended to the meeting minutes of the Regional Council of Skåne, March 2, 2010. Region Skåne registry 0902713.

Region Skåne (2012) *TITA. Regional mobilisation around ESS and MAX IV.* Final report.

Research Europe (2007) "The long and the short of it." June 6, 2007.

Royal Swedish Academy of Sciences (2005) Referral response to Allan Larsson's investigation (Ds 2005:40), October 19, 2005. Swedish government registry U2005/5686/F.

Royal Swedish Academy of Sciences (2008a) Akademinyheter, no 3, november 2008. Received through personal communication with Olle Edqvist, 2009.

Royal Swedish Academy of Sciences (2008b) Letter from the Royal Academy to the Parliamentary Subcommitte on Education, signed by Bo Sundqvist, Jan S Nilsson, and Gunnar Öquist, November 24, 2008. Received through personal communication with Olle Edqvist, 2009.

Samuelsson M (2009) "Vetenskap säljs med ovetenskap." Opinion piece, *Sydsvenskan*, May 12, 2009.

Science (2009) "European Neutron Source Finally Finds a Home." June 5, 2009.

Science Corridor (2009) The Science Corridor. An Alliance of Hubs. Draft of Memorandum November 23–25 2009 Hamburg. Dated November 25, 2009. DESY Integrated Digital Conferencing (Indico) database: https://indico.desy.de/indico/event/2359/material/2/0.

Skånska Dagbladet (2009) "Ett träd för ESS miljömål." June 26, 2009.

Söderbergh Widding A and A Karlhede (2014) "'ESS – ett hot mot Lund och mot Sveriges forskning'." Opinion letter, *Sydsvenskan*, March 27, 2014.

Spain (2009) Memorandum of Understanding between the Governments of Spain and Sweden: "A Cooperation Towards the European Spallation Source Project." June 10, 2009. Swedish government registry U2009/3987/F.

Spain (2014) Letter from Carmen Vela Olmo, Spanish Secretary of State of Investigation, Development and Innovation, to Peter Honeth, Swedish State Secretary of the

Ministry of Education and Research, and Uffe Toudal Pedersen, Danish State Secretary of the Ministry of Science, Innovation and Higher Education. February 20, 2014. Swedish government registry 2014/1398/F.
Süddeutsche Zeitung (2002) "Wettlauf zur Neutronenquelle." May 16, 2002.
Svenska Dagbladet (2003) "Unik chans för forskningen." January 22, 2003.
Svenska Dagbladet (2016) "Region Skåne ger bort 800 miljoner – utan avtal." August 22, 2016.
Sveriges Radio (2009) "God affär att investera i forskning." September 25, 2009.
Sveriges Radio (2011) "Sverige ger 70 miljoner till kärnkraftsforskning." April 5, 2011.
SWECO FFNS (2002) *En översiktlig studie av förutsättningarna att placera en forskningsanläggning för neutronspridning, European Spallation Source (ESS) till Brunnshögsområdet i Lund.* Malmö March 11, 2002, SWECO FFNS Arkitekter AB, Uppdragsnummer 3830363.
Swedish Agency for Public Management (2005) Referral response to Allan Larsson's investigation (Ds 2005:40), September 12, 2005. Swedish government registry U2005/5686/F.
Swedish Government (1997a) Budgetproposition för 1998. Swedish Governmental Bill 1997/98:1, expenditure area 16, Universities and Research.
Swedish Government (1997b) *Besparingar i stort och smått. Betänkande av storforskningsutredningen.* Swedish public investigation SOU 1997:69.
Swedish Government (2000a) *Om forskningsfinansiering.* Swedish Governmental bill 1999/00:81.
Swedish Government (2000b) *Forskning och förnyelse.* Swedish Governmental bill 2000/01:3.
Swedish Government (2000c) Instruktion för Verket för innovationssystem. Swedish code of statutes 2000:1132.
Swedish Government (2003a) Letter from Claes Ånstrand to Lars Börjesson, June 10, 2003. Swedish government registry 2003/2177/ITFoU.
Swedish Government (2003b) European Spallation Source (ESS). Internal memo. Dated December 8, 2003. Received through personal communication with Mats Johnsson at the Swedish Ministry of Education, 2015.
Swedish Government (2004a) *The Swedish Local Government Act, Kommunallagen på engelska.* Swedish public investigation Ds 2004:31.
Swedish Government (2004b) *Finansiering av starka forskningsmiljöer – en internationell utblick.* Swedish public investigation Ds 2004:21.
Swedish Government (2004c) Regleringsbrev Vetenskapsrådet 2004. Ändring. June 23, 2004. Swedish government registry U2004/2528/F.
Swedish Government (2004d) Uppdrag att undersöka möjligheten att placera European Spallation Source (ESS) i Sverige. July 8, 2004. Swedish government registry Dnr U2004/2555/F.
Swedish Government (2005a) *Forskning för ett bättre liv.* Swedish Governmental bill 2004/05:80.
Swedish Government (2005b) *Svenska miljömål – ett gemensamt uppdrag.* Swedish Governmental bill 2004/05:150.
Swedish Government (2005c) Fortsatt uppdrag att undersöka möjligheten att placera European Spallation Source (ESS) i Sverige. June 30, 2005. Swedish government registry Dnr U2005/5687/F.
Swedish Government (2005d) Regleringsbrev Vetenskapsrådet 2005. Ändring. November 10, 2005. Swedish government registry U2005/5631/BIA.
Swedish Government (2006a) Regeringsförklaring 6 oktober 2006. www.regeringen.se.

Swedish Government (2006b) Budgetproposition för 2007. Swedish Governmental Bill 2006/07:1, expenditure area 16, Universities and Research.
Swedish Government (2007a) Budgetproposition för 2008. Swedish Governmental Bill 2007/08:1, expenditure area 16, Universities and Research.
Swedish Government (2007b) "Regeringen vill bygga stor europeisk forskningsanläggning – partikelacceleratorn ESS blir störst i världen." Press release. February 26, 2007. www.regeringen.se.
Swedish Government (2007c) Memorandum: Locating the European Spallation Source (ESS) in Sweden. Ministry of Education and Research. February 26, 2007. Received through personal communication with Mats Johnsson at the Swedish Ministry of Education, 2015.
Swedish Government (2007d) Uppdrag att som chefsförhandlare erbjuda andra länder medverkan i att konstruera European Spallation Source (ESS) i Sverige. March 15, 2007. Swedish government registry U2005/5686/F.
Swedish Government (2007e) Uppdrag till Lunds universitet att bilda ett sekretariat för planering av European Spallation Source m.m. June 20, 2007. Swedish government registry U2007/4588/F.
Swedish Government (2007f) Förordnande av föreståndare för Sekretariatet för ESS vid Lunds universitet. July 5, 2007. Swedish government registry U2007/4830/F.
Swedish Government (2007g) Förordnande av ledamöter i en rådgivande grupp vid sekretariatet för planering av European Spallation Source vid Lunds universitet. October 4, 2007. Swedish government registry U2007/5786/F.
Swedish Government (2008a) 2008 års ekonomiska vårproposition. Swedish Governmental Bill 2007/08:100.
Swedish Government (2008b) *Forskningsfinansiering – kvalitet och relevans*. Swedish public investigation SOU 2008:30.
Swedish Government (2008c) *Ett lyft för forskning och innovation*. Swedish Governmental Bill 2008/09:50.
Swedish Government (2008d) Budgetproposition för 2009. Swedish Governmental Bill 2008/09:1, expenditure area 16, Universities and Research.
Swedish Government (2008e) Uppdrag om framtagande av beslutsunderlag för finansiering av MAX IV. December 4, 2008. Swedish government registry U2008/7688/F.
Swedish Government (2009a) Budgetproposition för 2010. Swedish Governmental Bill 2009/10:1, expenditure area 16, Universities and Research.
Swedish Government (2009b) Regerinsgbeslut nr 11:12. November 17, 2009. Swedish government registry U2009/7134/F.
Swedish Government (2010a) *En akademi i tiden – ökad frihet för universitet och högskolor*. Swedish Governmental bill 2009/10:149.
Swedish Government (2010b) Budgetproposition för 2011. Swedish Governmental Bill 2010/11:1, expenditure area 16, Universities and Research.
Swedish Government (2010c) Regleringsbrev Lunds universitet 2009. Ändring. January 21, 2010. Swedish government registry U2010/290/UH.
Swedish Government (2010d) Guidelines for European Spallation Source ESS AB. April 15, 2010. Swedish government registry U2010/2226/F.
Swedish Government (2010e) Regleringsbrev Vetenskapsrådet. Ändring. December 9, 2010. Swedish government registry U2010/7507/6847/F.
Swedish Government (2010f) Regleringsbrev Lunds universitet 2010. Ändring. December 16, 2010. Swedish government registry U2010/185.

Swedish Government (2011a) Budgetproposition för 2012. Swedish Governmental Bill 2011/12:1, expenditure area 16, Universities and Research.
Swedish Government (2011b) Regleringsbrev Vetenskapsrådet. Ändring. June 30, 2011. Swedish government registry U2011/3453/F.
Swedish Government (2011c) Uppdrag att som chefsförhandlare erbjuda andra länder deltagande i konstruktion och drift av European Spallation Source (ESS). July 7, 2011. Swedish government registry U2011/4120/F.
Swedish Government (2012a) *Den nationella innovationsstrategin.*
Swedish Government (2012b) Budgetproposition för 2013. Swedish Governmental Bill 2012/13:1, expenditure area 16, Universities and Research.
Swedish Government (2012c) Regleringsbrev Vetenskapsrådet 2012. Ändring. March 29, 2012. Swedish government registry U2012/7485/F.
Swedish Government (2012d) Regleringsbrev Vetenskapsrådet 2012. Ändring. November 29, 2012. Swedish government registry U2012/6597/F.
Swedish Government (2013a) Vårändringsbudget för 2013. Swedish Governmental Bill 2012/13:99.
Swedish Government (2013b) Budgetproposition för 2014. Swedish Governmental Bill 2013/14:1, expenditure area 16, Universities and Research.
Swedish Government (2013c) Avtal med Eric-konsortium för den Europeiska spallationskällan. Swedish Governmental Bill 2012/13:190.
Swedish Government (2013d) Regleringsbrev Vetenskapsrådet 2013. Ändring. February 14, 2013. Swedish government registry U2013/895/F.
Swedish Government (2014a) Kommunal medfinansiering av viss forskningsinfrastruktur. Swedish Governmental Bill 2013/14:136.
Swedish Government (2014b) Budgetproposition för 2015. Swedish Governmental Bill 2014/15:1, expenditure area 16, Universities and Research.
Swedish Government (2014c) "Finanseringen klar för European Spallation Source." Press release, July 4, 2014. www.regeringen.se.
Swedish Government (2014d) Regleringsbrev Vetenskapsrådet 2014. Ändring. December 18, 2014. Swedish government registry U2014/7485/F.
Swedish Government (2014e) Regleringsbrev Lunds universitet 2014. Ändring. December 18, 2014. Regeringsbeslut U2014/7486/UH.
Swedish Government (2016) Budgetproposition för 2017. Swedish Governmental Bill 2017/18:1, expenditure area 16, Universities and Research.
Swedish Government (2017) Budgetproposition för 2018. Swedish Governmental Bill 2017/18:1, annex 1.
Swedish National Audit Office (2012) Lunds universitets årsredovisning samt löpande granskning 2011. Audit report, May 23, 2012. Swedish National Audit Office registry 32-2011-0687.
Swedish National Audit Office (2013) *Svensk rymdverksamhet – en strategisk tillgång?* Report RIR 2013:1.
Swedish Natural Sciences Research Council (1997a) *International evaluation of Swedish National Facilities.* April 1997.
Swedish Natural Sciences Research Council (1997b) *International review of Studsvik Neutron Research Laboratory.* December 1997.
Swedish Parliament (2007a) Riksdagens protokoll 2006/07:52. January 23, 2007. www.riksdagen.se.
Swedish Parliament (2007b) Riksdagens protokoll 2006/07:83. March 28, 2007. www.riksdagen.se.

Swedish Parliament (2009) Lag (2009:47) om vissa kommunala befogenheter. February 5, 2009.
Swedish Parliament (2011) Riksdagens protokoll 2010/11:91. April 26, 2011. www.riksdagen.se.
Swedish Parliament (2013) Riksdagens protokoll 2013/14:27. November 14, 2013. www.riksdagen.se.
Swedish Research Council (2002) *Swedish National Facilities*. Swedish Research Council report 2002:6.
Swedish Research Council (2004) *International Evaluation of Swedish Condensed Matter Physics*, 2004.
Swedish Research Council (2005) Annual Report (Årsredovisning) 2004.
Swedish Research Council (2006a) *An international evaluation of the MAX IV technical concept 2005*, Swedish Research Council report 5:2006.
Swedish Research Council (2006b) *Scientific evaluation of the MAX IV proposal 2006*, Swedish Research Council report 20:2006.
Swedish Research Council (2006c) Annual Report (Årsredovisning) 2005.
Swedish Research Council (2009) Annual Report (Årsredovisning) 2008.
Swedish Research Council (2010a) Annual Report (Årsredovisning) 2009.
Swedish Research Council (2010b) *Report from the review of the MAX laboratory*. Swedish Research Council report 5:2010.
Swedish Research Council (2011) VR:s organisation for att administrera forskningsavtalen med Frankrike. March 9, 2011. Swedish Research Council registry 813-2010-7489.
Swedish Research Council (2012) Swedish Research Council's Guide to Infrastructure 2012. Swedish Research Council report 3:2012.
Swedish Research Council (2013) *MAX IV Project Proces Review*. Report.
Swedish Research Council (2014) Annual Report (Årsredovisning) 2013.
Swedish Research Council (2015) Vetenskapsrådets guide till infrastrukturen 2014. Swedish Research Council report.
Swedish Research Council (2018a) ESS. Svenskt deltagande, nyttjande och kompetensförsörjning. Report.
Swedish Research Council (2018b) Beslut om särskilda villkor för bidrag till MAX IV-laboratoriet. December 19, 2018. Swedish Research Council registry 4.3-2018-7152.
Switzerland (2014) Letter from Mauro Dell'Ambrogio, Swiss State Secretary for Education, Research and Innovation to the Swedish Ministry of Education. September 29, 2014. Swedish government registry 2014/1398/F.
Sydsvenskan (2002a) "Jordberga föreslås hysa forskarjätte." February 9, 2002.
Sydsvenskan (2002b) "Lund favorit till nytt center." February 12, 2002.
Sydsvenskan (2002c) "Nordöstra Lund kan få ny detaljplan." February 26, 2002.
Sydsvenskan (2002d) "Ett litet steg närmare för ESS." March 23, 2002.
Sydsvenskan (2002e) "Kritiska röster väcks mot ESS." May 15, 2002.
Sydsvenskan (2002f) "Lund gör framsteg om ESS-center." May 17, 2002.
Sydsvenskan (2002g) "Bönder vill stoppa storprojekt." June 8, 2002.
Sydsvenskan (2002h) "Miljöpartiet kräver ESS-folkomröstning." August 13, 2002.
Sydsvenskan (2002i) "Partier tveksamma till folkomröstning om ESS i Lund." August 14, 2002.
Sydsvenskan (2002k) "Arbetsgrupp mot ESS-projektet." August 18, 2002.
Sydsvenskan (2002m) "ESS-anläggning kan bli föremål för omröstning." September 7, 2002.

Sydsvenskan (2002n) "Lundaborna är positiva till ESS." September 13, 2002.
Sydsvenskan (2002p) "ESS – riskfyllt skrytbygge eller drömprojekt?" October 4, 2002.
Sydsvenskan (2002q) "Lobbyföretag ska sälja in ESS." October 31, 2002.
Sydsvenskan (2002r) "Beskrivning över hur ESS ska byggas har tagits fram." November 17, 2002.
Sydsvenskan (2002s) "Storbråk om ESS-seminarium." November 29, 2002.
Sydsvenskan (2002t) "Motigt för ESS när nejsägare går ihop." November 30, 2002.
Sydsvenskan (2003a) "Lunds kommun håller fast vid ESS-satsning." February 1, 2003.
Sydsvenskan (2003b) "ESS ingen het fråga för regeringen." March 19, 2003.
Sydsvenskan (2003c) "ESS-anläggning får deadline." May 21, 2003.
Sydsvenskan (2003d) "ESS-konsortiet måste banta." July 26, 2003.
Sydsvenskan (2003e) "Danskt ja viktigt for ESS." September 25, 2003.
Sydsvenskan (2003f) "ESS-anläggning kan bli utan danskt bidrag." October 10, 2003.
Sydsvenskan (2003g) "Pagrotsky får brev om ESS." November 26, 2003.
Sydsvenskan (2004a) "Skatteintäkt från ESS nytt argument." January 24, 2004.
Sydsvenskan (2004b) "Vänstern kräver miljöbeskrivning." February 18, 2004.
Sydsvenskan (2004c) "Universitetet får hemlighålla avtal." April 8, 2004.
Sydsvenskan (2005a) "Lund fortsätter i ESS-konsortiet." January 12, 2005.
Sydsvenskan (2005b) "ESS – himmel för forskarna men helvete för miljön." April 19, 2005.
Sydsvenskan (2005c) "Frän kritik mot Lunds samarbete med ESS." May 26, 2005.
Sydsvenskan (2005d) "Inget om miljön när Larsson förordar ESS." June 23, 2005.
Sydsvenskan (2005e) "'Allan Larsson lurar allmänheten.'" July 23, 2005.
Sydsvenskan (2005f) "Tungt nej till ESS för landets anseende." October 14, 2005.
Sydsvenskan (2005g) "Lokala remissvar sågar ESS." October 18, 2005.
Sydsvenskan (2006a) "ESS-projektet får 1,2 miljoner kronor." August 8, 2006.
Sydsvenskan (2006b) "'Min bedömning är att Lund ligger före sina rivaler.'" August 24, 2006.
Sydsvenskan (2007a) "Det gäller att hinna först till 11 miljarder." February 27, 2007.
Sydsvenskan (2007b) "Säg nej till ESS-anläggningen." March 1, 2007.
Sydsvenskan (2007c) "Miljöpartiet mot ESS." March 11, 2007.
Sydsvenskan (2007d) "Universitetet får fortsätta med ESS." June 26, 2007.
Sydsvenskan (2007e) "ESS kan bli av tidigare än beräknat." October 16, 2007.
Sydsvenskan (2007f) "ESS kritiserades hårt under miljödiskussion." October 21, 2007.
Sydsvenskan (2008a) "Konkurrenterna om ESS går samman." February 4, 2008.
Sydsvenskan (2008b) "Danska staten vill betala för ESS." April 7, 2008.
Sydsvenskan (2008c) "ESS går på charmoffensiv." May 9, 2008.
Sydsvenskan (2008d) "ESS Scandinavia vill skapa nätets efterföljare." May 31, 2008.
Sydsvenskan (2008e) "Lund ett steg närmare ESS." September 20, 2008.
Sydsvenskan (2008f) "120 miljoner kan satsas på ESS." September 22, 2008.
Sydsvenskan (2008g) "Satsa till Max." October 5, 2008.
Sydsvenskan (2008h) "Region Skåne stöttar Max IV." November 21, 2008.
Sydsvenskan (2008i) "Knivskarp kamp om ESS." December 9, 2008.
Sydsvenskan (2009a) "Max IV kan ge ESS draghjälp." January 9, 2009.
Sydsvenskan (2009b) "Mer än ett forskningscentrum." January 14, 2009.
Sydsvenskan (2009c) "'ESS ökar utsläpp av koldioxid'" January 24, 2009.
Sydsvenskan (2009d) "ESS-förhandlaren tror på beslut i maj." February 9, 2009.
Sydsvenskan (2009e) "Drömmen lever om ett ESS-besked." May 5, 2009.
Sydsvenskan (2009f) "Krav på full finansiering kan försena ESS-beslut." May 27, 2009.

Sydsvenskan (2009g) "När orden inte räckte för Leijonborg." May 29, 2009.
Sydsvenskan (2009h) "Ess we can." Op-ed. May 30, 2009.
Sydsvenskan (2009i) "Wallenberg ska säkra Max IV." June 1, 2009.
Sydsvenskan (2009k) "Få ville tycka till om ESS." June 2, 2009.
Sydsvenskan (2009m) "Naturskyddsföreningen vill lägga ESS i Dalby." June 30, 2009.
Sydsvenskan (2009n) "ESS-röster kostar miljoner." November 15, 2009.
Sydsvenskan (2009p) "Alla länder får inte bidra med material." November 17, 2009.
Sydsvenskan (2009q) "Nu skapas en korridor av kunskap." December 7, 2009.
Sydsvenskan (2009r) "Ungern godtar ESS i Lund." December 18, 2009.
Sydsvenskan (2010) "ESS ska locka fler turister än Domkyrkan." October 21, 2010.
Sydsvenskan (2011a) "Tyskland ger 200 miljoner till ESS." March 23, 2011.
Sydsvenskan (2011b) "Testanläggning för ESS byggs i Uppsala." June 14, 2011.
Sydsvenskan (2011c) "Tjeckiska miljoner till ESS." July 7, 2011.
Sydsvenskan (2011d) "Inget kvicksilver i ESS." October 19, 2011.
Sydsvenskan (2012a) "Naturskyddare vill slippa jättelabbet." July 10, 2012.
Sydsvenskan (2012b) "ESS måste lämna in riskanalys." July 27, 2012.
Sydsvenskan (2012c) "Frågetecknen hopar sig." July 28, 2012.
Sydsvenskan (2013a) "Grönt ljus för miljöprövning för ESS." March 17, 2013.
Sydsvenskan (2013b) "Kamp om miljarderna igång." May 18, 2013.
Sydsvenskan (2013c) "ESS saknar pengar." September 20, 2013.
Sydsvenskan (2013d) "Lars Anell: 'Jag har lämnat över det här nu'." September 21, 2013.
Sydsvenskan (2013e) "Få länder har skrivit på notan." December 29, 2013.
Sydsvenskan (2014a) "Kritiskt med pengar för ESS." February 7, 2014.
Sydsvenskan (2014b) "Överraskande ESS-bidrag från Storbritannien." March 12, 2014.
Sydsvenskan (2014c) "'ESS kanske inte borde ligga i Sverige alls.'" March 27, 2014.
Sydsvenskan (2014d) "Björklund: 'Hade ESS legat i Stockholm så hade de aldrig skrivit artikeln'." March 27, 2014.
Sydsvenskan (2014e) "Sverige vill låna för att betala ESS." March 29, 2014.
Sydsvenskan (2014f) "ESS vädjar om snabbt bygglov." April 24, 2014.
Sydsvenskan (2014g) "Förhandlingsdelen är avslutad – nu återstår detaljer innan ESS-finansieringen är på plats." May 20, 2014.
Sydsvenskan (2014h) "Försvunnet ESS-mejl sinkade miljardavtal." May 27, 2014.
Sydsvenskan (2014i) "Miljödomstolen ger klartecken." June 13, 2014.
Sydsvenskan (2014k) "Leijonborg: 'Fler länder är på väg in'." July 5, 2014.
Sydsvenskan (2014m) "ESS får godkänt av Strålsäkerhetsmyndigheten." July 17, 2014.
Sydsvenskan (2014n) "Första spadtaget för ESS." September 2, 2014.
Sydsvenskan (2016a) "Skåne betalar miljard för ESS – kan vara olagligt." December 9, 2016.
Sydsvenskan (2016b) "Det hemliga maktspelet om ESS: Skåne skänkte miljard – fick nytt universitet." December 11, 2016.
Sydsvenskan (2018a) "Så försenades MAX IV." November 10, 2018.
Sydsvenskan (2018b) "Sveriges nota för driften av ESS ökar." December 12, 2018.
Sydsvenskan (2019) "Miljarder på spel i driftskostnaden för ESS." March 27, 2019.
United Kingdom (2014) Letter from Greg Clark, UK Minister for Universities, Science and Cities, to the Swedish Ministry of Education. July 27, 2014. Swedish government registry 2014/1398/F.
University of Huddersfield (2010) "Chancellor Sir Patrick Stewart – European Spallation Source (ESS) research." Youtube video. www.youtube.com/watch?v=KG3Upzc3NGY.
Uppsala University (2005) Referral response to Allan Larsson's investigation (Ds 2005:40), October 11, 2005. Swedish government registry U2005/5686/F.

Valentin F, Larsen MT and N Heineke (2005) *Neutrons and innovations. What benefits will Denmark obtain for its science, technology and competitiveness by co-hosting an advanced large-scale research facility near Lund?* Working Paper 2005-2 from Research Centre on Biotech Business.
Wahlgren M et al. (2005) Motion 2004/05:Ub9 "med anledning av prop. 2004/05:80 Forskning för ett bättre liv". April 6, 2005. www.riksdagen.se.
Wennergren B and M Olsson (2005) Referral response to Allan Larsson's investigation (Ds 2005:40), October 13, 2005. Swedish government registry U2005/5686/F.
Wetterstrand M et al. (2008) Motion till riksdagen 2008/09:Ub8 med anledning av prop. 2008/09:50 "Ett lyft för forskning och innovation". November 14, 2008. www.riksdagen.se.
Wiberg E (2013) "'MAX IV och ESS är nationella angelägenheter – ska inte påverka universitetets övriga anslag'." Opinion letter, *Sydsvenskan*, December 12, 2013.
Wissenschaftsrat (2002) "Stellungnahme zu neun Großgeräten der naturwissenschaftlichen Grundlagenforschung und zur Weiterentwicklung der Investitionsplanung von Großgeräten."

Literature

Agrell W (2012) "Framing prospects and risk in the public promotion of ESS-Scandinavia." *Science and Public Policy* 39: 429–438.
Allison G and P Zelikow (1999/1971) *Essence of Decision: Explaining the Cuban Missile Crisis*. Pearson.
Alvesson M (2013) *The Triumph of Emptiness: Consumption, Higher Education, and Work Organization*. Oxford University Press.
Andersen PH and S Åberg (2017) "Big-science organizations as lead users: A case study of CERN." *Competition & Change* 21(5): 345–363.
Andreasen LE (1995) *Europe's Next Step: Organisational Innovation, Competition and Employment*. Routledge.
Atkinson H (1997) "Commentary on the History of ILL and ESRF." In Krige J and L Guzzetti (eds.) *History of European Scientific and Technological Cooperation*. Office for Official Publications of the European Communities.
Autio E, Hameri A-P and O Vuola (2004) "A framework of industrial knowledge spillovers in big-science centers." *Research Policy* 33: 107–126.
Bacon G E (ed.) (1986) *Fifty Years of Neutron Diffraction. The Advent of Neutron Scattering*. Adam Hilger.
Benner M (2001) *Kontrovers och konsensus. Vetenskap och politik i svenskt 1990-tal*. Nya Doxa.
Benner M (2008) *Kunskapsnation i kris? Politik, pengar och makt i svensk forskning*. Nya Doxa.
Benner M (2012) "Big Science in a small country: constraints and possibilities of research policy." In Hallonsten O (ed.) *In Pursuit of a Promise: Perspectives on the Political Process to Establish the European Spallation Source (ESS) in Lund, Sweden*. Arkiv Academic Press.
Benner M and U Sandström (2000) "Inertia and change in Scandinavian public-sector research systems: the case of biotechnology." *Science and Public Policy* 27(6): 443–454.
Berggren K-F and A Matic (2012) "Science at the ESS: a brief outline." In Hallonsten O (ed.) *In Pursuit of a Promise: Perspectives on the Political Process to Establish the European Spallation Source (ESS) in Lund, Sweden*. Arkiv Academic Press.

Berggren K-F and O Hallonsten (2012) "Timeline of major events." In Hallonsten O (ed.) *In Pursuit of a Promise: Perspectives on the Political Process to Establish the European Spallation Source (ESS) in Lund, Sweden*. Arkiv Academic Press.
Bergh A (2015) *Den kapitalistiska välfärdsstaten*. Studentlitteratur.
Berman EP (2014) "Not Just Neoliberalism: Economization in US Science and Technology Policy Science." *Technology, & Human Values* 39(3): 397–431.
Bernal J D (1939) *The Social Function of Science*. MIT Press.
Bolliger I and A Griffiths (2020) "The introduction of ESFRI and the rise of national Research Infrastructure roadmaps in Europe." In Cramer K and O Hallonsten (eds.) *Big Science and Research Infrastructures in Europe*. Edward Elgar.
Boorstin D (1962) *The Image: A Guide to Pseudo-events in America*. Knopf.
Borge M J B (2016) "Highlights of the ISOLDE facility and the HIE-ISOLDE project." *Nuclear Instruments and Methods in Physics Research Section B: Beam Interactions with Materials and Atoms* 376: 408–412.
Börjesson L (2003) "Hur man får en snöboll att växa till en lavin." In Broberg G, Forkman B and C-M Pålsson (eds.) *Vem styr forskningen?* Lund Studies in the History of Science and Ideas.
Borrás S (2003) *The Innovation Policy of the European Union: From Government to Governance*. Edward Elgar.
Borup M, Brown N, Konrad K and H van Lente (2006) "The Sociology of Expectations in Science and Technology." *Technology Analysis & Strategic Management* 18: 285–298.
Bourdieu P (1977) *Outline Of A Theory Of Practice*. Cambridge University Press.
Brown N and M Michael (2003) "A Sociology of Expectations: Retrospecting Prospects and Prospecting Retrospects." *Technology Analysis & Strategic Management* 15: 3–18.
Bush V (1945) *Science, the Endless Frontier. A Report to the President on a Program for Postwar Scientific Research*. US National Science Foundation.
Capshew JH and KA Rader (1992) "Big Science: Price to the Present." *Osiris* 2nd series 7: 3–25.
Carsughi F and KN Clausen (2002) "Meeting report: The ESS Project is Ready for a Decision." *Neutron News* 13(4): 3–4.
Castells M (1996) *The Rise of the Network Society. The Information Age: Economy, Society and Culture Vol. I*. Blackwell.
Chou M H (2014) "The evolution of the European Research Area as an idea in European integration." In Chou M H and Å Gornitzka (eds.) *Building the Knowledge Economy in Europe. New Constellations in European Research and Higher Education*. Edward Elgar.
Coleman J (1982) *The Asymmetric Society*. Syracuse University Press.
Cramer K (2017) "Lightening Europe: Establishing the European Synchrotron Radiation Facility (ESRF)." *History and Technology* 33(4): 396–427.
Cramer K (2020) *The Other Europe. History and Politics of Collaborative Large-Scale Research, 1977–2009*. Palgrave Macmillan.
Cramer K, Hallonsten O, Bolliger I and A Griffiths (2020) "Big Science and Research Infrastructures in Europe: History and current trends." In Cramer K and O Hallonsten (eds.) *Big Science and Research Infrastructures in Europe*. Edward Elgar.
Crease RP (2001) "Anxious history: The High Flux Beam Reactor and Brookhaven National Laboratory." *Historical Studies in the Physical and Biological Sciences* 32(1): 41–56.
D'Ippolito B and C-C Rüling (2019) "Research collaboration in Large Scale Research Infrastructures: Collaboration types and policy implications." *Research Policy* 48: 1282–1296.

D'Ippolito B and C-C Rüling (2020) "Keeping a neutron source alive: Material, social, and political work at the Institut Laue-Langevin (ILL)." In Cramer K and O Hallonsten (eds.) *Big Science and Research Infrastructures in Europe*. Edward Elgar.

Davies K (2001) *Cracking the Genome: Inside the Race to Unlock Human DNA*. Johns Hopkins University Press.

Delanghe H, Muldur U and L Soete (eds.) (2009) *European Science and Technology Policy: Towards Integration or Fragmentation?* Edward Elgar.

DiMaggio P and WW Powell (1983) "The Iron Cage Revisited: Institutional Isomorphism and Collective Rationality in Organizational Fields." *American Sociological Review* 48: 147–160.

Duclos Lindstrøm M and K Kropp (2017) "Understanding the infrastructure of European Research Infrastructures—The case of the European Social Survey (ESS-ERIC)." *Science and Public Policy* 44(6): 855–864.

Durkheim É (1982/1895) *The Rules of Sociological Method. And selected texts on sociology and its method*. Macmillan.

Edler J, Kuhlmann S and M Behrens (eds.) (2003) *Changing Governance of Research and Technology Policy: The European Research Area*. Edward Elgar.

Edquist C and M McKelvey (1998) "High R&D Intensity Without High Tech Products: A Swedish Paradox." In Nielsen K and B Johnson (eds.) *Institutions and Economic Change: New Perspectives on Markets, Firms and Technology*. Edward Elgar.

Edqvist O (2009) *Gränslös forskning*. Nya Doxa.

Eklund M (2007) *Adoption of the Innovation System Concept in Sweden*. Doctoral thesis, Uppsala University.

Elzinga A (1993) "Universities, research and the transformation of the State in Sweden." In Rothblatt S and B Wittrock (eds.) *The European and American University Since 1800. Historical and Sociological Essays*. Cambridge University Press.

Elzinga A (1997) "The science-society contract in historical transformation: with special reference to 'epistemic drift'." *Social Science Information* 36: 411–445.

Elzinga A (2012) "Features of the current science policy regime: Viewed in historical perspective." *Science and Public Policy* 39(4): 416–428.

Esser H (1993) "The Rationality of Everyday Behavior." *Rationality and Society* 5(1): 7–31.

Faherty W B (2002) *Florida's Space Coast. The Impact of NASA on the Sunshine State*. University Press of Florida.

Fioretos O, Falleti T G and A Sheingate (eds.) (2016) *The Oxford Handbook of Historical Institutionalism*. Oxford University Press.

Fjæstad M (2010) *Visionen om outtömlig energi. Bridreaktorn i Svensk kärnkraftshistoria 1945–80*. Gidlunds Förlag.

Florida R (2002) *The Rise of the Creative Class*. Basic Books.

Flyvbjerg B (2017) "Introduction: The Iron Law of Megaproject Management" In Flyvbjerg B (ed.) *The Oxford Handbook of Megaproject Management*. Oxford University Press.

Flyvbjerg B, Bruzelius N and W Rothengatter (2003) *Megaprojects and Risk*. Cambridge University Press.

Fridlund M (1999) *Den gemensamma utvecklingen: Staten, storföretaget och samarbetet kring den svenska elkrafttekniken*. Doctoral dissertation, KTH Royal Institute of Technology.

Friedland R and R R Alford (1991) "Bringing Society Back In: Symbols, Practices, and Institutional Contradictions." In Powell W W and P DiMaggio (eds.) *The New Institutionalism in Organizational Analysis*. The University of Chicago Press.

Galison P and B Hevly (eds.) (1992) *Big Science – The Growth of Large-Scale Research*. Stanford University Press.

Garoby et al. (2018) "The European Spallation Source Design." *Phys. Scr.* 93 014001.
Gaubert A and A Lebeau (2009) "Reforming European space governance." *Space Policy* 25: 37–44.
Giddens A (1979) *Central problems in Social Theory. Action, Structure and Contradiction in Social Analysis.* Macmillan.
Giddens A (1984) *The Constitution of Society. Outline of the Theory of Structuration.* Polity.
Goldthorpe, J H (1998) "Rational action theory for sociology." *British Journal of Sociology* 49: 167–192.
Graham L R (1992) "Big Science in the Last Years of the Soviet Union." *Osiris* 2nd series 7: 49–71.
Granberg A (2012) "ESS as a creator of conflict and collaboration in the Swedish scientific community." In O Hallonsten (ed.) *In Pursuit of a Promise: Perspectives on the Political Process to Establish the European Spallation Source (ESS) in Lund, Sweden.* Arkiv Academic Press.
Greenberg DS (1999/1967) *The Politics of Pure Science*, 2nd ed. The University of Chicago Press.
Greenberg DS (2001) *Science, Money and Politics: Political Triumph and Ethical Erosion.* The University of Chicago Press.
Greenberg DS (2007) *Science for Sale: The Perils, Rewards, and Delusions of Campus Capitalism.* The University of Chicago Press.
Gribbe J and O Hallonsten (2017) "The emergence and growth of materials science in Swedish universities." *Historical Studies in the Natural Sciences* 47(4): 459–493.
Guellec D and B van Pottelsberghe de la Potterie (2004) "From R&D to Productivity Growth: Do the Institutional Settings and the Source of Funds of R&D matter?" *Oxford Bulletin of Economics and Statistics* 66(3): 353–378.
Guston DH (2000) *Between Politics and Science: Assuring the Integrity and Productivity of Research.* Cambridge University Press.
Guzzetti L (1995) *A Brief History of European Union Research Policy.* European Commission.
Hallonsten O (2009) *Small science on big machines: Politics and practices of synchrotron radiation laboratories.* Doctoral thesis, Lund University, Lund, Sweden.
Hallonsten O (2011) "Growing Big Science in a Small Country: MAX-lab and the Swedish Research Policy System." *Historical Studies in the Natural Sciences* 41(2): 179–215.
Hallonsten O (ed.) (2012) *In Pursuit of a Promise: Perspectives on the Political Process to Establish the European Spallation Source (ESS) in Lund, Sweden.* Arkiv Academic Press.
Hallonsten O (2013a) "Introducing facilitymetrics: A first review and analysis of commonly used measures of scientific leadership among synchrotron radiation facilities worldwide." *Scientometrics* 96(2): 497–513.
Hallonsten O (2013b) "Myths and realities of the ESS project: A systematic scrutiny of readily accepted 'truths'." In Kaiserfeld T and T O'Dell (eds.) *Legitimizing ESS: Big Science as a Collaboration Across Boundaries.* Nordic Academic Press.
Hallonsten O (2014) "The Politics of European Collaboration in Big Science." In Mayer M, Carpes M and R Knoblich (eds.) *The Global Politics of Science and Technology – Vol. 2.* Springer.
Hallonsten O (2015a) "Unpreparedness and risk in Big Science policy: Sweden and the European Spallation Source." *Science and Public Policy* 42(3): 415–426.
Hallonsten O (2015b) "The parasites: Synchrotron radiation at SLAC, 1972–1992." *Historical Studies in the Natural Sciences* 45(2): 217–272.
Hallonsten O (2016a) *Big Science Transformed: Science, Politics and Organization in Europe and the United States.* Palgrave Macmillan.

Hallonsten O (2016b) "Use and productivity of contemporary, multidisciplinary Big Science." *Research Evaluation* 25: 486–495.
Hallonsten O (2018) "Development and transformation of the third sector of R&D in Sweden, 1942–2017." *Science and Public Policy* 45(5): 634–644.
Hallonsten O (2020a) "Research infrastructures in Europe: The hype and the field." *European Review* 28(4): 617–635.
Hallonsten O (2020b) "Is there an 'Iron Law' of Big Science?" In Cramer K and O Hallonsten (eds.) *Big Science and Research Infrastructures in Europe*. Edward Elgar.
Hallonsten O and C Silander (2012) "Commissioning the university of excellence: Swedish research policy and new public research funding programs." *Quality in Higher Education* 18(3): 367–381.
Hallonsten O and O Christensson (2017a) *An ex post impact study of MAX-lab*. Lund University.
Hallonsten O and O Christensson (2017b) "Collaborative technological innovation in an academic, user-oriented Big Science facility." *Industry and Higher Education* 31(6): 399–408.
Hallonsten O and T Heinze (2012) "Institutional persistence through gradual adaptation: analysis of national laboratories in the USA and Germany." *Science and Public Policy* 39: 450–463.
Hallonsten O and T Heinze (2015) "Formation and Expansion of a New Organizational Field in Experimental Science." *Science and Public Policy* 42(6): 841–854.
Hallonsten O and T Heinze (2016) "'Preservation of the laboratory is not a mission'. Gradual organizational renewal in national laboratories in Germany and the United States." In Heinze T and R Münch (eds.) *Innovation in Science and Organizational Renewal. Historical and Sociological Perspectives*. Palgrave Macmillan.
Hanel T and M Hård (2015) "Inventing traditions: interests, parables and nostalgia in the history of nuclear energy." *History and Technology* 31(2): 84–107.
Hazelkorn E (2011) *Rankings and the Reshaping of Higher Education*. Palgrave Macmillan.
Heinecke S (2016) "The Gradual Transformation of the Polish Public Science System." *PLoS ONE* 11(4): e0153260.
Heinze T, Hallonsten O and S Heinecke (2017) "Turning the Ship: The Transformation of DESY, 1993–2009." *Physics in Perspective* 19: 424–451.
Hermann A, Krige J, Mersits U and D Pestre (eds.) (1987) *History of CERN. Volume I: Launching the European Organization for Nuclear Research*. North-Holland.
Hessels L K, van Lente H and Smits R (2009) "In search of relevance: the changing contract between science and society." *Science and Public Policy* 36(5): 387–401.
Hevly B (1992) "Introduction: The Many Faces of Big Science." In Galison P and B Hevly (eds.) *Big Science – The Growth of Large-Scale Research*. Stanford University Press.
Hewlett RG and JM Holl (1989) *Atoms for Peace and War, 1953–1961*. University of California Press.
Hewlett RG and OE Anderson (1962) *A History of the United States Atomic Energy Commission. Volume 1. The New World, 1939–1946*. The Pennsylvania State University Press.
Hilling A, Hallonsten O and K Engholm Elgaard (2017) "Tax policy issues in connection with The European Spallation Source Project and other European Research Infrastructure Consortiums." *Skattepolitisk Oversigt* no 8/2017.
Hiltzik M (2015) *Big Science. Ernest Lawrence and the Invention that Launched Military-Industrial Complex*. Simon & Schuster.
Hobsbawm E and T Ranger (eds.) (1983) *The Invention of Tradition*. Cambridge University Press.

Hoddeson L, Brown L, Riordan M, and M Dresden (eds.) (1997) *The Rise of the Standard Model: Particle Physics in the 1960s and 1970s.* Cambridge University Press.
Hoddeson L, Kolb AW and C Westfall (2008) *Fermilab: Physics, the Frontier & Megascience.* The University of Chicago Press.
Hoerber TC (2009) "The European Space Agency and the European Union: The Next Step on the Road to the Stars." *Journal of Contemporary European Research* 5(3): 405–414.
Holmberg D (2012) *Vem formar den stategiska forskningen? Institutionell dynamik och anslagspolitisk utveckling i KK-stiftelsen.* Doctoral Dissertation, Lund University.
Holmberg D and O Hallonsten (2015) "Policy reform and academic drift: Research mission and institutional legitimacy in the development of the Swedish higher education system 1977–2012." *European Journal of Higher Education* 5(2): 181–196.
Irvine J and BR Martin (1984) *Foresight in Science. Picking the Winners.* Pinter.
Jacobsson S and E Perez Vico (2010) "Towards a systemic framework for capturing and explaining the effects of academic R&D." *Technology Analysis & Strategic Management* 22(7): 765–787.
Johnson A (2004) "The End of Pure Science: Science Policy from Bayh-Dole to the NNI." In Baird D, Nordmann A and J Schummer (eds.) *Discovering the Nanoscale.* IOS Press.
Judt T (2005) *Postwar. A History of Europe Since 1945.* Pimlico.
Kaiserfeld T (2013) "ESS from Neutron Gap to Global Strategy: Plans for an international research facility after the cold war." In Kaiserfeld T and T O'Dell (eds.) *Legitimizing ESS: Big Science as a Collaboration Across Boundaries.* Nordic Academic Press.
Krige J (2003) "The Politics of European Scientific Collaboration." In Krige J and D Pestre (eds.) *Companion to Science in the Twentieth Century.* Routledge.
Krige J (2006) *American Hegemony and the Postwar Reconstruction of Science in Europe.* MIT Press.
Krige J and L Guzzetti (1997) *History of European Scientific and Technological Cooperation.* European Commission.
Krige J and D Pestre (1987) "The how and the why of the birth of CERN." In Hermann A et al. (eds.) *History of CERN. Volume I.* North-Holland.
Larsson A (2019) *I vetenskapens värld.* Gotlandsboken.
Leijonborg L (2018) *Kris och framgång: mitt halvsekel i politiken.* Ekerlids.
Lindenberg S (1985) "An Assessment of the New Political Economy: Its Potential for the Social Sciences and for Sociology in Particular." *Sociological Theory* 3(1): 99–114.
Luhmann N (1995) *Social Systems.* Stanford University Press.
Mahoney J and K Thelen (eds.) (2010) *Explaining Institutional Change: Ambiguity, Agency, and Power.* Cambridge University Press.
Maier-Leibnitz H (1986) "The birth of the Institut Max von Laue-Paul Langevin in Grenoble" In Bacon G E (ed.) *Fifty Years of Neutron Diffraction. The Advent of Neutron Scattering.* Adam Hilger.
March J and H Simon (1958) *Organizations.* Blackwell.
Martin J (2018) "Prestige Asymmetry in American Physics: Aspirations, Applications, and the Purloined Letter Effect." *Science in Context* 30(4): 475–506.
Maskell P, Eskelinen H, Hannibalsson I, Malmberg A and E Vatne (1998) *Competitiveness, Localised Learning and Regional Development: Specialisation and Prosperity in Small Open Economies.* Routledge.
Maskell P and G Törnqvist (1999) *Building a cross-border learning region. Emergence of the North European Øresund region.* Handelshøjskolens forlag.
Mayer M (1991) *Whatever Happened to Madison Avenue: Advertising in the '90s.* Little, Brown.

McCray WP (2006) *Giant Telescopes. Astronomical Ambition and the Promise of Technology.* Harvard University Press.
McCray WP (2010) "'Globalization with hardware': ITER's fusion of technology, policy, and politics." *History and Technology* 26(4): 283–312.
Melander F (2006) *Lokal forskningspolitik. Institutionell dynamik och organisatorisk omvandling vid Lunds universitet 1980–2005.* Doctoral Dissertation, Lund University.
Merton RK (1948) "The self-fulfilling prophecy." *The Antioch Review* 8: 193–210.
Merton RK (1968/1949) *Social Theory and Social Structure.* The Free Press.
Merton RK (1973) *The Sociology of Science.* The University of Chicago Press.
Meusel EJ (1990) "Einrichtungen der Großforschung und Wissenstransfer." In Schuster HJ (ed.) *Handbuch des Wissenschaftstransfer.* Springer.
Meyer JW and B Rowan (1977) "Institutionalized Organizations: Formal Structure as Myth and Ceremony." *The American Journal of Sociology* 83(2): 340–363.
Middlemas K (1995) *Orchestrating Europe: The Informal Politics of the European Union 1973–95.* Fontana Press.
Mintzberg H (1983) *Structure in Fives: Designing Effective Organizations.* Prentice-Hall.
Moskovko M (2020) "Intensified role of the EU? European Research Infrastructure Consortium (ERIC) as a legal framework for contemporary multinational research collaboration." In Cramer K and O Hallonsten (eds.) *Big Science and Research Infrastructures in Europe.* Edward Elgar.
Moskovko M, Astvaldsson A and O Hallonsten (2019) "Who is ERIC? The politics and jurisprudence of a governance tool for collaborative European research infrastructures." *Journal of Contemporary European Research* 15(3): 249–268.
Nielsen H and H Knudsen (2010) "The troublesome life of peaceful atoms in Denmark." *History and Technology* 26(2): 91–118.
North D (1990) *Institutions, Institutional Change, and Economic Performance.* Cambridge University Press.
Nowotny H, Pestre D, Schmidt-Aßmann E, Schulze-Fielitz H and H-H Trute (2005) *The Public Nature of Science under Assault. Politics, Markets, Science and the Law.* Springer.
Olshov A (2013) *Øresundsregionen: Københavns outnyttjade möjlighet.* Gyldendal Business.
Opp K-D (1999) "Contending conceptions of the theory of rational action." *Journal of Theoretical Politics* 11(2): 171–202.
Östling J (2018) *Humboldt and the modern German university. An intellectual history.* Lund University Press.
Papon P (2004) "European Scientific Cooperation and Research Infrastructures: Past Tendencies and Future Prospects." *Minerva* 42(1): 61–76.
Parsons T (1977) *Social Systems and the Evolution of Action Theory.* Free Press.
Persson G (2007) *Min väg, mina val.* Albert Bonniers Förlag.
Pestre D (1997) "Prehistory of the Franco-German Laue-Langevin Institute." In Krige J and L Guzzetti (eds.) *History of European Scientific and Technological Cooperation.* Office for Official Publications of the European Communities.
Pettersson I (2012) *Handslaget. Svensk industriell forskningspolitik 1940–1980.* Doctoral dissertation, KTH Royal Institute of Technology.
Pfeffer J (1982) *Organizations and Organization Theory.* Pitman.
Premfors R (1986) *Svensk forskningspolitik.* Studentlitteratur.
Presthus (1979) *The Organizational Society.* Knopf.
Price DJ dS (1986/1963) *Little Science, Big Science ... and beyond.* Columbia University Press.

Pugh DS (1973) "The measurement of organization structures: does context determine form?" *Organizational Dynamics* 1(4): 19–34.
Radder H (ed.) (2010) *The Commodification of Academic Research. Science and the Modern Universities*. Harvard University Press.
Reichel J, Lind A-S and M Hansson (2014) "ERIC: a New Governance Tool for Biobanking." *European Journal of Human Genetics* 22: 1055–1057.
Reinfeldt F (2015) *Halvvägs*. Albert Bonniers Förlag.
Riordan M, Hoddeson L and A Kolb (2015) *Tunnel Visions. The Rise and Fall of the Superconducting Super Collider*. University of Chicago Press.
Rothstein B and S Steinmo (2002) *Restructuring the Welfare State. Political Institutions and Policy Change*. Palgrave Macmillan.
Rush JJ (2015) "US Neutron Facility Development in the Last Half-Century: A Cautionary Tale." *Physics in Perspective* 17: 135–155.
Ryan L (2015) "Governance of EU Research Policy: Charting Forms of Scientific Democracy in the European Research Area." *Science and Public Policy* 42: 300–314.
Safire W (1993) *Safire's New Political Dictionary: The Definitive Guide to the New Language of Politics*. Random House.
Salter AJ and BR Martin (2001) "The economic benefits of publicly funded basic research: A critical review." *Research Policy* 30: 509–532.
Schmied H (1982) "Results of Attempts to Quantify the Secondary Economic Effects Generated by Big Research Centers." *IEEE Transactions on Engineering Management* 29(4): 154–165.
Schön L (2007) *En modern svensk ekonomisk historia*. 2nd ed. SNS.
Scott WR (2001) *Institutions and Organizations*. 2nd ed. Sage.
Selznick P (1957) *Leadership in Administration. A Sociological Interpretation*. Harper & Row.
Simon H (1957) *Models of Man*. Wiley.
Smith RW (1989) *The Space Telescope: A Study of NASA, Science, Technology, and Politics*. Cambridge University Press.
Steinmo S (2010) *The Evolution of Modern States: Sweden, Japan, and the United States*. Cambridge University Press.
Stenborg E and M Klintman (2012) "Organized local resistance: investigating a local environmental movement's activities against the ESS." In Hallonsten O (ed.) *In Pursuit of a Promise: Perspectives on the Political Process to Establish the European Spallation Source (ESS) in Lund, Sweden*. Arkiv Academic Press.
Stephan P (2012) *How Economics Shapes Science*. Harvard University Press.
Stevens H (2003) "Fundamental physics and its justifications, 1945–1993." *Historical Studies in the Physical and Biological Sciences* 34(1): 151–197.
Stevrin P (1978) *Den Samhällsstyrda Forskningen. En samhällsorganisatorisk studie av den sektoriella forskningspolitikens framväxt och tillämpning i Sverige*. Almkvist & Wiksell.
Stokes D (1997) *Pasteur's Quadrant: Basic Science and Technological Innovation*. Brookings.
Streeck W and K Thelen (eds.) (2005) *Beyond Continuity. Institutional Change in Advanced Political Economies*. Oxford University Press.
Svenning O (2005) *Göran Persson och hans värld*. Norstedts.
Thomasson A and C Carlile (2017) "Science facilities and stakeholder management: how a pan-European research facility ended up in a small Swedish university town." *Physica Scripta* 92: 062501.
Tindemans P (2009) "Postwar research education and innovation policymaking in Europe." In Delanghe H, Muldur U and L Soete (eds.) *European Science and Technology Policy: Towards Integration or Fragmentation?* Edward Elgar.

Törnqvist G (2002) *Science at the Cutting Edge: The Future of the Øresund Region*. Copenhagen Business School Press.
van Lente H (1993) *Promising technology. The Dynamics of Expectations in Technological Developments*. Doctoral Dissertation, University of Twente.
van Lente H (2000) "Forceful futures: From promise to requirement." In Brown N, Rappert B and A Webster (eds.) *Contested Futures*. Ashgate.
Weber M (1920) *Gesammelte Aufsätze zur Religionssoziologie*, Band 1. Mohr.
Weinberg A (1961) "Impact of Large-Scale Science on the United States." *Science* 134(3473): 161–164.
Weinberg A (1967) *Reflections on Big Science*. Pergamon Press.
Westfall C (2003) "Rethinking big science: Modest, mezzo, grand science and the development of the Bevalac, 1971–1993." *Isis* 94: 30–56.
Westfall C (2008) "Surviving the squeeze: National laboratories in the 1970s and 1980s." *Historical Studies in the Natural Sciences* 38(4): 475–478.
Westfall C (2010) "Surviving to Tell the Tale: Argonne's Intense Pulsed Neutron Source from an Ecosystem Perspective." *Historical Studies in the Natural Sciences* 40(3): 350–398.
Westwick PJ (2003) *The National Labs: Science in an American System 1947–1974*. Harvard University Press.
Wetterberg G (2017) *Skånes historia. Del 2, 1376–1720*. Albert Bonniers Förlag.
Wetterberg G (2018) *Skånes historia. Del 3, 1720–2017*. Albert Bonniers Förlag.
Whitley R and J Gläser (eds.) (2007) *The Changing Governance of the Sciences. The Advent of Research Evaluation Systems*. Springer.
Whitley R, Gläser J, and L Engwall (2010) *Reconfiguring Knowledge Production. Changing Authority Relationships in the Sciences and their Consequences for Intellectual Innovation*. Oxford University Press.
Widmalm S (1993) "Big Science in a Small Country: Sweden and CERN II." In Lindqvist S (ed.) *Center on the Periphery: Historical Aspects of 20th Century Swedish Physics*. Watson Publishing International.
Widmalm S (2013) "Innovation and Control: Performative Research Policy in Sweden." In Rider et al. (eds.), *Transformations in Research, Higher Education and the Academic Market*. Springer.
Yu H, Wested J B and T Minssen (2017) "Innovation and intellectual property policies in European Research Infrastructure Consortia—Part I: The Case of the European Spallation Source ERIC." *Journal of Intellectual Property Law & Practice* 12(5): 384–397.
Ziman J (1994) *Prometheus Bound. Science in a Dynamic Steady State*. Cambridge University Press.

Name index

Arkiv Academic Press

Arkiv Academic Press is an imprint of the Swedish publishing house Arkiv förlag. For up-to-date information on distribution and available titles, please visit:

www.arkivacademicpress.com

Published books

Ericka Johnson, *Situating Simulators. The Integration of Simulations in Medical Practice* (paperback 2012 [original edition by Arkiv förlag 2004])

Olof Hallonsten (ed.), *In Pursuit of a Promise. Perspectives on the Political Process to Establish the European Spallation Source (ESS) in Lund, Sweden* (paperback 2012)

Rebecca Selberg, *Femininity at Work. Gender, Labour, and Changing Relations of Power in a Swedish Hospital* (paperback 2012)

Sven E O Hort (birth name Olsson), *Social Policy, Welfare State, and Civil Society in Sweden*. Volume I: *History, Policies, and Institutions 1884–1988* (hardcover & paperback 2014, 3rd enlarged edition [1st edition by Arkiv förlag 1990])

Sven E O Hort (birth name Olsson), *Social Policy, Welfare State, and Civil Society in Sweden*. Volume II: *The Lost World of Social Democracy 1988–2015* (hardcover & paperback 2014, 3rd enlarged edition [1st edition by Arkiv förlag 1990])

Lisa Lindén, *Communicating Care. The Contradictions of HPV Vaccination Campaigns* (paperback 2016)

Gunnar Olofsson & Sven Hort (eds.), *Class, Sex and Revolutions. Göran Therborn – A Critical Appraisal* (paperback 2016)

Kristofer Hansson & Markus Idvall (eds.), *Interpreting the Brain in Society. Cultural Reflections on Neuroscientific Practices* (paperback 2017)

Olof Hallonsten, *The Campaign. How a European Big Science Facility Ended up on the Peripheral Farmlands of Southern Sweden* (paperback 2020)

www.ingramcontent.com/pod-product-compliance
Ingram Content Group UK Ltd.
Pitfield, Milton Keynes, MK11 3LW, UK
UKHW041633190726
13854UKWH00006B/2477